Green Energy and Technology

Climate change, environmental impact and the limited natural resources urge scientific research and novel technical solutions. The monograph series Green Energy and Technology serves as a publishing platform for scientific and technological approaches to "green"—i.e. environmentally friendly and sustainable—technologies. While a focus lies on energy and power supply, it also covers "green" solutions in industrial engineering and engineering design. Green Energy and Technology addresses researchers, advanced students, technical consultants as well as decision makers in industries and politics. Hence, the level of presentation spans from instructional to highly technical.

Indexed in Scopus.

More information about this series at http://www.springer.com/series/8059

Nil Patel · Akash Kumar Bhoi ·
Sanjeevikumar Padmanaban ·
Jens Bo Holm-Nielsen
Editors

Electric Vehicles

Modern Technologies and Trends

Springer

Editors
Nil Patel
Cismac Solution Pvt. Ltd.
Ahmedabad, India

Sanjeevikumar Padmanaban
Aalborg University
Esbjerg, Denmark

Akash Kumar Bhoi
Sikkim Manipal Institute of Technology
Sikkim Manipal University
Rangpo, India

Jens Bo Holm-Nielsen
Aalborg University
Esbjerg, Denmark

ISSN 1865-3529 ISSN 1865-3537 (electronic)
Green Energy and Technology
ISBN 978-981-15-9250-8 ISBN 978-981-15-9251-5 (eBook)
https://doi.org/10.1007/978-981-15-9251-5

This Springer imprint is published by the registered company Springer Nature Singapore Pte Ltd.
The registered company address is: 152 Beach Road, #21-01/04 Gateway East, Singapore 189721, Singapore

Preface

Recently, electrical vehicle (EV), hybrid electrical vehicle (HEV), battery electrical vehicle (BEV), fuel cell electrical vehicle (FCEV), and plug-in hybrid electrical vehicle (PHEV) are proliferating in the transportation sector in today's era. As per present trends, this mode of transport supplants internal combustion engine (ICE) vehicles in near future. Each component of electrical vehicles has a numerous technology that is currently in use or can be conspicuous in the future. The power system is going through huge instabilities with limited EV incursion; however, with appropriate coordination and management, EVs can be converted into prima contributor to the flourishing effectuation of the smart grid concept. There are myriad of possibilities of large amount of environmental advantages, and the EVs can in a widespread way mitigate the greenhouse gas emissions generated by the conventional transportation sector. Nonetheless, there are few hinders for EVs to extenuate before totally supplanting ICE vehicles. This book is emphasized on reviewing all useful and important techniques and data available on EV configurations, conventional electrical machines, advanced electrical machines, battery energy sources, on-board charging techniques, off-board charging techniques, optimization techniques, trends, impacts and possible ways of future technological development to aid in researchers in this domain.

In the first chapter "Design and Modeling of Fuel Cell Hybrid Electric Vehicle for Urban Transportation", the authors describe the architecture and its dynamic modelling of FCHEV. Also, simulation results are validated with theoretical analysis. The advantages of the proposed fuel cell hybrid electric vehicle have been validated with these simulation results. It has been shown that the proposed architecture and modulation methods work satisfactorily in all the modes of operation of the FCHEV. Chandana Sasidharan et al. in the second chapter "Light Electric Vehicles and Their Charging Aspects" mention that traditionally, these vehicles were slowly charged due to the use of lead acid batteries. With the adoption of advanced chemistry, faster charging becomes possible but thermal management continues to be the biggest challenge. On the other hand, battery swapping is an option that can assist in reducing the charging time of the batteries. The proposed chapter would cover details on LEV models and their charging and swapping aspects and possible recommendations. The third chapter "Plug-In Hybrid Electric Vehicles (PHEVs)" demonstrates that PHEVs come with a lot of features that makes it exclusive, such as different modes of operation, charging, independent choice of fuel (electricity or gasoline) etc. The fourth chapter "Power Electronics—EV Battery Charging" elaborates power system layouts of EV battery charging systems, different categories of power electronic converters for such applications and working principles of basic power electronic converters. At last, several new topologies of recently developed power converters with many of the features mentioned above have been presented and explained briefly. In the fifth chapter "An Overview of Solar-Powered Electric Vehicle Charging in Vehicular Adhoc Network", the authors introduce the VANET environment in which the communication between EVs on the road, road side unit (RSU) and a traffic server takes place. Then the overview of solar PV integration with grid to meet the demand in peak hour has been discussed.

V. Dhinakaran et al. in the sixth chapter "Study on Electric Vehicle (EV) and Its Developments Based on Batteries, Drive System and Charging Methodologies in Modern World" depicted the use of numerous batteries, benefits of several drive and charging methodologies towards electric vehicles (EV). In recent time has been made along with its pros and cons to provide a better knowledge for researchers to promote substantial augmentation of EV. Jay Patel and Rajesh Patel in the seventh chapter "An Overview on the Prominence of Phase Change Material Based Battery Cooling and Role of Novel Composite Phase Change Material in Future Battery Thermal Management System" illustrate that novel CPCM not only owns great thermal conductivity but also gives contribution to heat storage. It will give excellent temperature control on the battery pack in contexts of peak battery cell temperature and temperature uniformity in the battery module. The eighth chapter "Battery Electric Vehicles (BEVs)" describes that battery electric vehicles (BEVs) do not produce any sort of unsafe dangers, not at all like fuel-controlled vehicles or ordinary petroleum/diesel vehicles, and in this manner, they are profoundly condition cordial. Likewise, they have a low running expense of around 33% per kilometre when contrasted with vehicles. Considering every one of these variables, battery worked vehicles are probably going to supplant regular ICE vehicles soon.

Lipi Chhaya in the ninth chapter titled "Communication Standards for Interconnections of Smart Grid Infrastructure and Intelligent Electric Transportation System" focuses on vehicular communication technologies which are categorized to discuss the suitability of the specific set of standards for particular application. Moreover, this paper focuses on integration of smart grid technology and plug-in vehicles. Various communication standards can be used for charging of plug-in vehicles in the public charging system or in the home area network of smart grid infrastructure. The tenth chapter "Smart Grid, V2G and Renewable Integration" describes the cost benefits from customer and utility side in terms of operating 2G. Also, optimization of the peak demand and the economic scheduling is achieved through many intelligent algorithms and a modified dynamic programming type of mathematical modelling was also discussed in this chapter. Manish Kumar Thukral in the eleventh chapter titled "Blockchain-Based Smart Contract Design for

Crowdfunding of Electrical Vehicle Charging Station Setup" presents the research work on the blockchain-based approch towards EV charging. The designed smart contract has been coded in solidity language. It has been implemented and tested on REMIX IDE Ethereum platform. Whereas the twelfth chapter "Modeling and Designing of E-bike for Local Use" emphasizes on two-wheeler e-bikes which are really useful in present society for students and old age people for local uses and it is also environment-friendly. Mohd Parvez et al. in the thirteenth chapter "A Novel Energy and Exergy Assessments of Solar Operated Combined Power and Absorption Refrigeration Cogeneration Cycle" present that the particular research is widely based on the concept of cogeneration cycle which has been developed by integrating a solar field consisting of heliostats coupled with the $LiNO_3$-H_2O absorption refrigeration cycle. Energy and exergy analyses have been performed on this conceptual cycle. Estimates for the irreversibility of all the major energy-consuming components of the cycle lead to possible measures for performance enhancement. The fourteenth chapter "The Effectiveness of Smart Grids V2G and Integration of Renewable Energy Sources" deals with the effectiveness of smart grid V2G, its integration with renewable energy systems along with some of the smart grid technologies. B. Hemanth Kumar et al. in the fifteenth chapter "A New Series-Parallel Switched Capacitor Configuration of a DC–DC Converter for Variable Voltage Applications" proposed a new topology of the switched-capacitor (SCC) circuit. The converter has a basic construction that does not require any magnetic elements which is powerful enough to produce 21 specific voltage conversion ratios. The last chapter "A Comprehensive Study on Electrical Vehicle in Charging Infrastructure, Challenges and Future Scope" by Piyush Sharma et al. focuses to explore substances, industry experts and venture delegates as a prepared reference of the remote charging arrangement of EVs, with data on significant attributes and guidelines.

The editors wish to thank all the authors who have contributed to this book and sincerely thank the Springer Nature editorial and production team for the constant support. We send our best wishes to our readers!

Ahmedabad, India — Nil Patel
Rangpo, India — Akash Kumar Bhoi
Esbjerg, Denmark — Sanjeevikumar Padmanaban
Esbjerg, Denmark — Jens Bo Holm-Nielsen

Contents

About the Editors

Nil Patel (S'16, IEEE), received the Bachelors of Engineering (B.E.) degree in Electrical and Electronics Engineering from the Gujarat Technological University, Gujarat, India, in 2014, and the Master of Technology (M.Tech.) degree (Hons.) in Electrical Engineering with specialization in Power Electronics, Machines and Drives from Institute of Technology, Nirma Universty, India, in 2016. During his master studies, he was graduate research project student at Hitachi, Gandhinagar, India, from 2015-2016. He was awarded the Gold Medal for securing highest academic standing for his bachelors studies. He is currently working as Research Engineer in Research and Development Department (Power Electronics and Drives) at Cismac Solution Pvt. Ltd, Ahmedabad, India. From July 2016 to January 2017, he was a Lecturer in Electrical Engineering Department at Institute of Technology, Ganpat University, Gujarat, India. From January 2017 to December 2017, he was an Assistant Professor and Faculty-in charge of Power Electronics and Drives Research Laboratory in Electrical Engineering Department at Institute of Technology, Ganpat University, Gujarat, India. His current research interests include power electronics converters, converter control techniques, advanced electrical machines, motor-drives, renewable energy and transportation electrification. He is regular reviewer for multiple IEEE Transactions, Conferences and IET journals.

Akash Kumar Bhoi has completed his B.Tech. (Biomedical Engineering) from Trident Academy of Technology, BPUT University, Odisha and M.Tech. (Biomedical Instrumentation) from Karunya University, Coimbatore in the year 2009 and 2011 respectively. In 2019 he is awarded Ph.D. in the area of Biomedical Signal Processing from Dept. of Electrical and Electronics Engineering, Sikkim Manipal Institute of Technology, Sikkim Manipal University, India. He is working as Assistant Professor-I in the Department of Electrical and Electronics Engineering at Sikkim Manipal Institute of Technology (SMIT), India since 2012. He is the University Ph.D. Course Coordinator for "Research & Publication Ethics (RPE)". He is member of IEEE, ISEIS and IAENG, associate member of IEI, UACEE and editorial board member reviewer of Indian and international journals. He is also a regular reviewer of journals of repute namely IEEE, Springer, Elsevier, Taylor and Francis, Inderscience etc. His areas of research are Biomedical Signal Processing, Optimization Algorithms, Computational Intelligence, Antenna, Renewable Energy. He has published several papers in national and international journals and conferences. He has 90+ documents registered in Scopus database. He has also served on numerous organizing panels for the international conferences and workshops. He is currently editing several books with Springer Nature and serving as Guest editor for special issues of the journal like Springer Nature and Inderscience.

Sanjeevikumar Padmanaban (Member'12–Senior Member'15, IEEE) received the bachelor's degree in electrical engineering from the University of Madras, Chennai, India, in 2002, the master's degree (Hons.) in electrical engineering from Pondicherry University, Puducherry, India, in 2006, and the PhD degree in electrical engineering from the University of Bologna, Bologna, Italy, in 2012. He was an Associate Professor with VIT University from 2012 to 2013. In 2013, he joined the National Institute of Technology, India, as a Faculty Member. In 2014, he was invited as a Visiting Researcher at the Department of Electrical Engineering, Qatar University, Doha, Qatar, funded by the Qatar

National Research Foundation (Government of Qatar). He continued his research activities with the Dublin Institute of Technology, Dublin, Ireland, in 2014. Further, he served an Associate Professor with the Department of Electrical and Electronics Engineering, University of Johannesburg, Johannesburg, South Africa, from 2016 to 2018. Since 2018, he has been a Faculty Member with the Department of Energy Technology, Aalborg University, Esbjerg, Denmark. He has authored more than 300 scientific papers.

S. Padmanaban was the recipient of the Best Paper cum Most Excellence Research Paper Award from IET-SEISCON'13, IET-CEAT'16, IEEE-EECSI'19, IEEE-CENCON'19 and five best paper awards from ETAEERE'16 sponsored Lecture Notes in Electrical Engineering, Springer book. He is a Fellow of the Institution of Engineers, India, the Institution of Electronics and Telecommunication Engineers, India, and the Institution of Engineering and Technology, UK. He is an Editor/Associate Editor/Editorial Board for refereed journals, in particular the IEEE SYSTEMS JOURNAL, IEEE Transaction on Industry Applications, IEEE ACCESS, *IET Power Electronics, IET Electronics Letters*, and Wiley-*International Transactions on Electrical Energy Systems*, Subject Editorial Board Member – *Energy Sources – Energies Journal*, MDPI, and the Subject Editor for the *IET Renewable Power Generation, IET Generation, Transmission and Distribution*, and *FACTS* journal (Canada).

Jens Bo Holm-Nielsen currently works at the Department of Energy Technology, Aalborg University and Head of the Esbjerg Energy Section. On this research, activities established the Center for Bioenergy and Green Engineering in 2009 and serve as the Head of the research group. He has vast experience in the field of Biorefinery concepts and Biogas production–Anaerobic Digestion. Implementation projects of Bio-energy systems in Denmark with provinces and European states. He served as the technical advisory for many industries in this field. He has executed many large-scale European Union and United Nations projects in research aspects of Bioenergy, biorefinery processes, the full chain of biogas and Green Engineering. He has authored more than 300 scientific papers. He was a member on the invitation with various capacities in the committee for over 500 various international conferences and Organizer of international conferences, workshops and training programmes in Europe, Central Asia and China. Focus areas Renewable Energy - Sustainability - Green jobs for all.

Design and Modeling of Fuel Cell Hybrid Electric Vehicle for Urban Transportation

Mallikarjunareddy Bandi, Naveenkumar Marati, Balraj Vaithilingam, and Kathirvel Karuppazhagi

1 Introduction

With the adamant of severe regulations related to global warming, fuel economy and conventional energy limitations, electric, hybrid and FC vehicles have been drawing attention from vehicle manufacturers and governments [1–3]. Especially, proton exchange membrane (PEM) [4]-based FCHEV has been gaining attention in this decade due to increasing concern about pollution and subsequent issues regarding environment [5, 6]. Even though, development of powertrain is difficult in FC vehicle and keeping on board hydrogen tank is a challenge [7–9].

In the end of nineteenth century, the effects of greenhouse gases are dominant, which are key to the research community to take decision on the emission of such gases with concern of governments [10]. The governments have been taken decision to cut down the thermal plants, usage of ICEs and similar polluting elements in the Kyoto policy [11]. The FC vehicles have zero emission and longer life [12].

The PEMFC is operated within 100 °C [13–15] and full load efficiency of FC is reported to be around 46% [16]. With these parameters, the PEMFC is suitable for the EV or HEV due to its lower temperature and better efficiency.

FCs are particularly attentive in EVs because of their cleanliness, lesser noise and higher relative efficiency compared with photovoltaic (PV) system. However,

M. Bandi · N. Marati (✉) · B. Vaithilingam · K. Karuppazhagi
Valeo India Private Ltd., Chennai, India
e-mail: naveen0724@gmail.com; naveenkumar.marati@valeo.com

M. Bandi
e-mail: mallikarjunareddy.bandi@valeo.com; mrmalleswari67@gmail.com

B. Vaithilingam
e-mail: balraj.vaithilingam@valeo.com

K. Karuppazhagi
e-mail: kathirvel.karuppazhagi@valeo.com

N. Patel et al. (eds.), *Electric Vehicles*, Green Energy and Technology,
https://doi.org/10.1007/978-981-15-9251-5_1

FCHEV has following limitations [17–19]: (1) Lack of storage capability. (2) Sluggish dynamic response compare to battery but pretty similar response in the steady state. (3) Lower tolerance to ripple. (4) Problems with cold start. For aforementioned problems, FCs are integrated with an auxiliary power source (battery) to mitigate these issues. The FC and battery are integrated to DC bus bar through unidirectional converter (UDC) and bidirectional converter (BDC), respectively [20]. The architecture of FC vehicle integrated with battery has been discussed in the further sections of the manuscript. The parts of the manuscript deal with detailed discussion on the architecture of fuel cell hybrid electric vehicle (FCHEV) [21, 22]. This article is mainly focused on design and developed the 1.4 kW FCHEV for urban transportation in the three-wheeler commercial section. The FCHEV can decrease the emission as well as sound pollution in the urbanization. Because, in cosmopolitan cities, there is lot of sound pollution which has been creating too much disturbance to the highly skilled labor for their environment [23, 24]. To avoid aforementioned drawbacks, need to design suitable three-wheeler vehicle based on fuel cell [25].

The organization of the article mainly divided into five sections. Section 1 discussed about the literature of the fuel cell and its associates and drawbacks caused due to sound pollution and emissions in the urban society. Section 2, describes the type of architecture used in FCHEV and design of DC/DC converters used in unidirectional and bidirectional interfaces. Different modes of operation and its model graphs of FCHEV described in Sect. 3. The simulation results of 1.4 kW FCHEV are present in Sect. 4, and finally conclusion and its applications presented in Sect. 5.

2 Architecture of FCHEV

The main purpose of FC integrated vehicle is that it would provide supply to both load as well as charge the battery under normal running condition. The load can get supply from both FC and battery when vehicle is in acceleration mode. In non-hybrid system, cold start, acceleration and deceleration performance is not satisfactory. Figure 1 shows the architecture of the FC hybrid electric vehicle and it consists of mainly seven parts, which are FC, unidirectional DC/DC converter (UDC), battery pack, [22] bidirectional converter (BDC), an inverter, single-phase electric machine and the chassis section of vehicle. In this section, the architecture of FCHEV is described in detail.

2.1 PEMFC Modeling and Design

The PEMFC is the best FC type out of several FC technologies like DMFC, AFC and SOFC, for design of hybrid electric vehicle because it operates under lower temperature range below 100 °C. It is used specially in portable applications and could be used as well in applications like three-wheeler vehicles like e-rickshaws. The

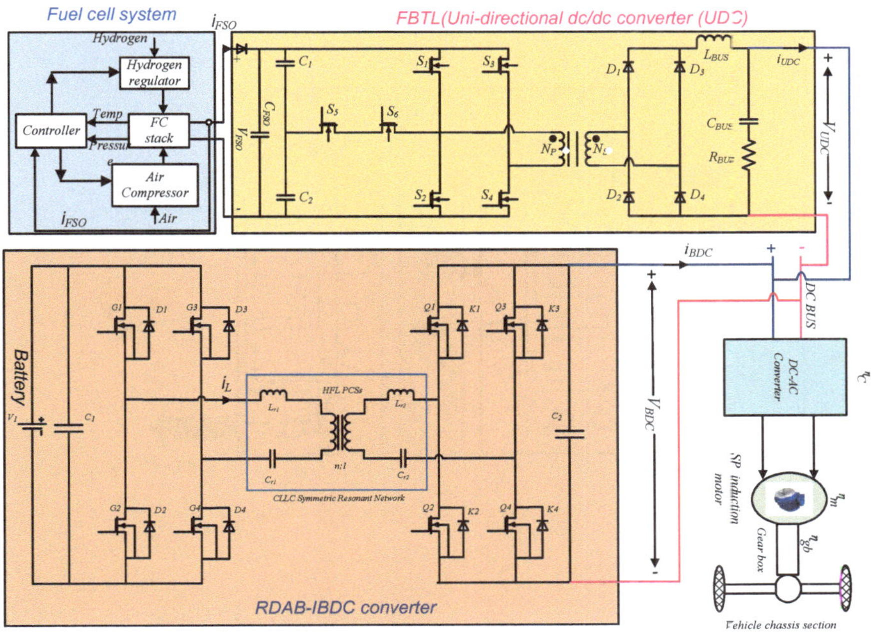

Fig. 1 The proposed architecture of FCHEV

construction of PEMFC is simpler than the other FC type categories and maintenance and repair is easier than SOFC and molten carbonate fuel cell (MCFC). In further, the detailed discussion on the design of PEMFC in the following subsections.

2.1.1 Fuel Cell Modeling

The FC with its characteristics is shown in Fig. 2. In the FC, hydrogen and oxygen are used to develop electricity by using electrolysis process. The PEMFC has seven layers like anode bipolar plates (ABP), anode gas diffusion layer (AGDL), anode carbon catalyst (ACC), membrane electrode assembly (MEA), cathode carbon catalyst (CCC), cathode gas diffusion layer (CGDL) and cathode bipolar plates (CBP) as shown in Fig. 2. The hydrogen (H_2) is taken from reformer and provided to the anode bipolar plates (ABP). The anode bipolar plates can supply it to the anode gas diffusion layer and AGDL can diffuse hydrogen into the anode carbon catalyst (ACC) layer. The anode carbon catalyst layer can break the hydrogen into hydrogen ions (H^+) and electrons (e^-) in ACC layer as shown in Fig. 2. The membrane electrode assembly (MEA) has higher conductivity for ions in comparison to electrons. Due to this, hydrogen electrons flow through the field plates and electrical appliance and hydrogen ions go through the MEA. Both hydrogen ions and electrons merge at the cathode. In the cathode section, hydrogen and oxygen merge at the cathode bipolar

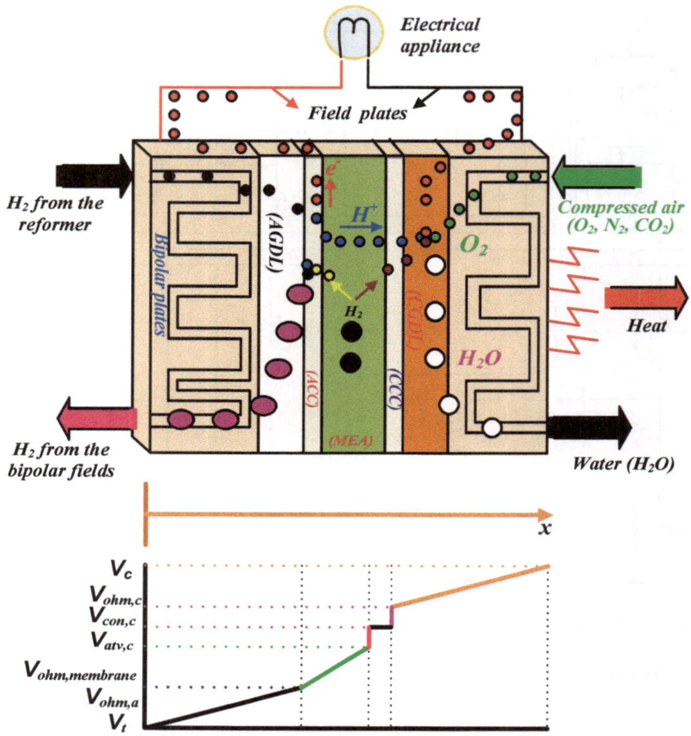

Fig. 2 The PEMFC with its characteristics

plates and form water as a by-product as shown in Fig. 2. The internal characteristics of a single fuel cell are given in Fig. 2, where the Nernst voltage or Nernst equation (V_C) is internal voltage of single FC.

The terminal voltage (V_t) is calculated from the Nernst voltage by subtracting voltage drops due to losses and is shown in Fig. 2. The voltage drops of single FC due to ohmic losses ($V_{ohm,c}$), concentration losses ($V_{con,c}$) and activation losses ($V_{atv,c}$) are subtracted from the Nernst equation to get terminal voltage (V_t).

The FC internal voltage (or) Nernst voltage is calculated by using empirical formula or mechanistic analysis. In order to determine FC output voltage, one needs to decide partial pressure of H_2 and O_2. The mixture of gas containing of 'n' species, diffusion element 'i' through the permeable electrode is described by 1D Stefan–Maxwell concept. The characteristics of single PEMFC have been shown in Fig. 3 with random change in channel molar fraction of H_2O for better understanding of internal operation of FC. The channel molar fraction of H_2O has been changed randomly to know whether corresponding change occurs in effective partial pressure of O_2 and Nernst voltage (V_c). The effective partial pressure of O_2 is inversely proportional to effective molar fraction of H_2O and the detailed explanation for the

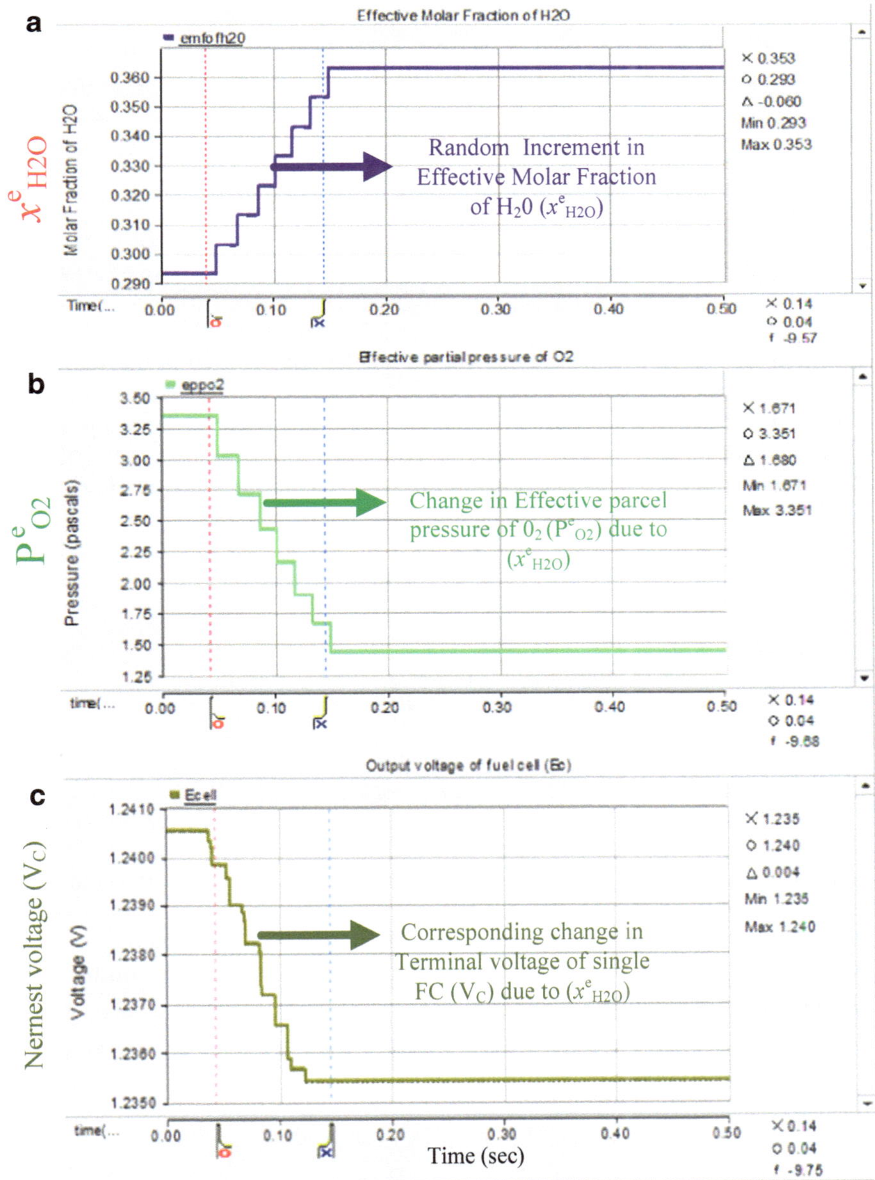

Fig. 3 Characteristics of single PEMFC with random change in channel molar fraction of H_2O **a** Effective channel molar fraction of H_2O ($x^e_{H_2O}$), **b** effective partial pressure of O_2 ($P^e_{O_2}$), **c** Nernst or internal voltage (V_c)

Table 1 Parameters of PEMFC used in EV

Specifications of the PEMFC	
Output power	1.2 kW
Output voltage	80–140 V
Fuel source	Hydrogen
Fuel pressure	0.04 MPa
Operating temperature	0-40 °C
Cooling mode	Forced air cooling
Size	300 mm × 80 mm × 200 mm

same is in later parts of this section. The parameters of FC used in the vehicle are listed in Table 1.

The effective partial pressure of H_2 and O_2 can be calculated by using Stefan–Maxwell equation and Faradays law. The 1D Stefan–Maxwell equation is given in Eq. 1 and gas mixture constant is given in Eq. 2. The molar fraction of '*i*' can be calculated from Stefan–Maxwell equation and given by Eq. 3.

$$\Delta x_i = \frac{RT}{P}\sum_{J=1}^{n}\frac{x_i N_j - x_j N_i}{D_{i,j}} \tag{1}$$

$$M = \frac{RT}{P} \tag{2}$$

$$\Delta x_i = M\sum_{J=1}^{n}\frac{x_i N_j - x_j N_i}{D_{i,j}} \tag{3}$$

where Δx_i, R, T, P, M, N_i, N_j, x_i, x_j, $D_{i,j}$, n, are molar fraction of diffusion component '*i*', gas constant, temperature (Kelvin), pressure (Pascal), gas mixture constant, molar flux of diffusion component '*i*' (mol/(m^2 s)), molar flux of diffusion component '*j*', mole fractions of species '*i*', mole fractions of species '*j*', effective diffusivity of '*i-j*' pair (m^2/s) and number of species.

In the 1D process along '*x*' axis, shown in Fig. 2, diffusion of water can be written in Eq. 4

$$\frac{dx_{H_2O}}{d} = \frac{RT}{P_a}\left(\frac{x_{H_2O}N_{H_2} - x_{H_2}N_{H_2O}}{D_{H_2O,H_2}}\right) \tag{4}$$

where P_a, N_{H_2}, N_{H_2O}, x_{H_2O}, x_{H_2}, D_{H_2O,H_2}, are anode side pressure, molar flux of diffusion component H_2, molar flux of diffusion component H_2O, mole fractions of species H_2O, mole fractions of species H_2, effective diffusivity of H_2O-H_2 pair.

In anode side, the gas stream is a combination of H_2 and $H_2O_{(g)}$. The N_{H_2O} normal to anode surface, which set to zero according to FC internal operation. After substitution of molar flux of water is zero in Eq. 4, the diffusion of water can be rewritten in Eq. 5.

$$\frac{\mathrm{d}x_{H_2O}}{\mathrm{d}x} = \frac{RT}{P_a}\left(\frac{x_{H_2O}N_{H_2}}{D_{H_2O,H_2}}\right) \tag{5}$$

The molar flux of H_2 is determined by Faradays law and given in Eq. 6.

$$N_{H_2} = \frac{I_d}{2 * F} \tag{6}$$

where I_d, F are current density (A/cm^2), and Faraday constant.

By combining Eqs. 5 and 6 and integrating the expression with respect to 'x' from the anode channel (anode bipolar plate) to catalyst surface, the effective molar fraction of H_2O is obtained and given in Eq. 7.

$$x^e_{H_2O} = x^{cha}_{H_2O}\, e^{\left(\frac{RT I_d L_{ac}}{2F P_a D_{H_2O,H_2}}\right)} \tag{7}$$

where $x^{cha}_{H_2O}$, L_{AC} are channel molar fraction of water, and effective length of anode-catalyst in meters.

Since $x^e_{H_2O} + x^e_{H_2} = 1$, effective partial pressure of H_2 is given in Eq. 8.

$$P^e_{H_2} = \frac{P^e_{H_2O}}{x^e_{H_2O}}\left(1 - x^e_{H_2O}\right) \tag{8}$$

where $P^e_{H_2}$, $P^e_{H_2O}$, are effective partial pressure of H_2 (Pascal), effective partial pressure of H_2O (Pascal).

According to FC internal characteristics, the effective partial pressure of H_2 at anode is half of the saturated partial pressure of H_2O. Therefore, effective partial pressure of H_2 is given by Eq. 9.

$$P^e_{H_2} = 0.5\, P^{str}_{H_2O}\left(\frac{1}{x^{cha}_{H_2O}\, e^{\left(\frac{RT I_d L_{ac}}{2F P_a D_{H_2O,H_2}}\right)}} - 1\right) \tag{9}$$

where $p^{Str}_{H_2O}$, are saturated partial pressure of H_2O (Pascal).

In the cathode channel, gases are coming from fuel reformer and compressed air, which are H_2, $H_2O_{(g)}$, O_2 and CO_2. Using Eq. 1, $H_2O_{(g)}$ diffusion in the cathode side can be obtained from Eq. 10 and given by Eq. 11.

$$\frac{dx_{H_2O}}{dx} = \frac{RT}{P_k}\left(\frac{x_{O_2}N_{H_2O} - x_{H_2O}N_{O_2}}{D_{H_2O,O_2}}\right) \tag{10}$$

$$\frac{dx_{H_2O}}{dx} = \frac{RT}{P_k}\left(\frac{-x_{H_2O}N_{O_2}}{D_{H_2O,O_2}}\right) \tag{11}$$

where $P_{K,}$ N_{O_2}, x_{O_2}, D_{H_2O,O_2}, are cathode side pressure (pascal), molar flux of diffusion component O_2, mole fractions of species O_2, effective diffusivity of H_2O-O_2 pair.

The analysis of cathode side is similar to anode side; the effective molar fraction of water can be derived by channel molar fraction of water and can be written in Eq. 12.

$$x^e_{H_2O} = x^{cha}_{H_2O}\, e^{\left(\frac{RTI_dL_{kc}}{2FP_kD_{H_2O,O_2}}\right)} \tag{12}$$

where $L_{kc,}$ is the length in meters between cathode to catalyst.

Through similar meticulous calculations, the effective molar fractions of N_2 and CO_2 are derived and given by Eqs. 13, and 14, respectively.

$$x^e_{N_2} = x^{cha}_{N_2}\, e^{\left(\frac{RTI_dL_{kc}}{2FP_kD_{N_2,O_2}}\right)} \tag{13}$$

$$x^e_{CO_2} = x^{cha}_{CO_2}\, e^{\left(\frac{RTI_dL_{kc}}{2FP_kD_{CO_2,O_2}}\right)} \tag{14}$$

where $x^{cha}_{N_2}$, $x^{cha}_{CO_2}$, $x^e_{N_2}$, $x^e_{CO_2}$, D_{CO_2,O_2}, D_{N_2,O_2} are channel molar fraction of N_2, channel molar fraction of CO_2, effective molar fraction of N_2, effective molar fraction of CO_2, effective diffusivity of CO_2-O_2pair, effective diffusivity of N_2-O_2 pair.

The effective molar fraction of O_2 can be calculated by subtracting effective molar fraction of N_2, CO_2, and H_2O and given in Eq. 15.

$$x^e_{O_2} = 1 - x^e_{N_2} - x^e_{CO_2} - x^e_{H_2O} \tag{15}$$

The equivalent effective partial pressure of O_2 is given in Eq. 16.

$$P^e_{O_2} = \frac{P^e_{H_2O}}{x^e_{H_2O}}x^e_{O_2} = \frac{P^e_{H_2O}}{x^e_{H_2O}}\left(1 - x^e_{H_2O} - x^e_{N_2} - x^e_{CO_2}\right) \tag{16}$$

where $P^e_{O_2}$, is effective partial pressure of O_2.

According to FC internal characteristics, effective partial pressure of O_2 at the cathode equals to the saturated partial pressure of H_2O, and the Eq. 16 can be modified and given by Eq. 17.

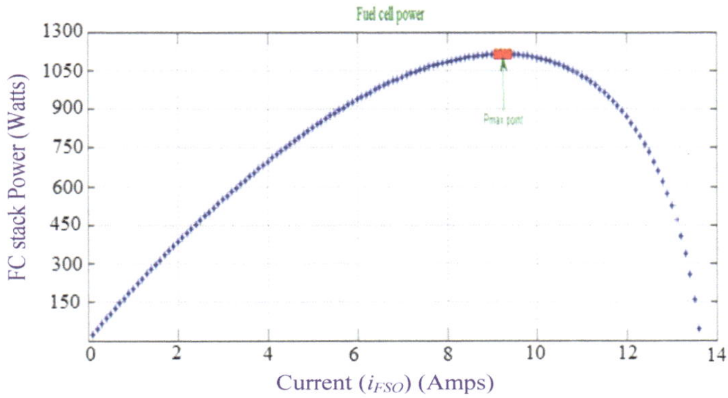

Fig. 4 Current (i_{FSO}) verses power characteristics of FC stack

$$P^{e}_{\mathrm{O}_2} = P^{\mathrm{str}}_{\mathrm{H_2O}}\left(\frac{1 - x^{e}_{\mathrm{N}_2} - x^{e}_{\mathrm{CO}_2}}{x^{e}_{\mathrm{H_2O}}} - 1\right) \tag{17}$$

where $p^{e}_{\mathrm{H}_2}$ and $p^{e}_{\mathrm{O}_2}$, calculated from Eqs. 9 and 17, is used in the Nernst equation to find FC internal voltage or Nernst voltage (V_{c}). The FC stack power verses current (i_{FSO}) is shown in Fig. 4. P_{Max} is reached at 9.5 A and FC stack power falls down further if current is increased. Hence, it is performed to limit current (i_{FSO}) to maintain maximum power level of FC stack power as shown in Fig. 4. In subsequent section, material conservation equations are developed for the calculation of FC internal voltage (V_{c}).

2.1.2 Material Conservation Equations

The internal process of FC is slower than battery. The FC internal voltage or Nernst voltage (V_{c}) is not affected by delays in normal mode operation but FC internal voltage (V_{c}) can affect the performance in the acceleration or deceleration modes as given in Sect. 2.1.3. So there is need to consider effect of material conservation in design of Nernst equation. The changes in partial pressure of H_2 and O_2 are determined by the gas concepts and given by Eqs. 18 and 19, respectively;

$$\begin{aligned}\frac{V_{\mathrm{a}}}{RT}\frac{\mathrm{d}p^{e}_{\mathrm{H_2O}}}{\mathrm{d}t} &= M_{\mathrm{H}_2,\mathrm{in}} - M_{\mathrm{H}_2,\mathrm{out}} - \frac{i_{\mathrm{d}}}{2 * F} \\ &= M_{\mathrm{H}_2,\mathrm{net}} - \frac{i_{\mathrm{d}}}{2 * F}\end{aligned} \tag{18}$$

$$\frac{V_k}{RT}\frac{\mathrm{d}p^e_{O_2}}{\mathrm{d}t} = M_{O_2,\mathrm{in}} - M_{O_2,\mathrm{out}} - \frac{i_\mathrm{d}}{2*F}$$
$$= M_{O_2,\mathrm{net}} - \frac{i_\mathrm{d}}{4*F} \quad (19)$$

where V_a, V_k, $M_{H_2,\mathrm{in}}$, $M_{H_2,\mathrm{Out}}$, $M_{H_2,\mathrm{net}}$, $M_{O_2,\mathrm{in}}$, $M_{O_2,\mathrm{Out}}$, $M_{O_2,\mathrm{net}}$, are anode volume, cathode volume, inlet flow rate of H_2, outlet flow rate of H_2, net flow rate of H_2, inlet flow rate of O_2, outlet flow rate of O_2, net flow rate of O_2.

Fundamentally, the instantaneous change of partial pressure of H_2 and O_2 should be zero in steady-state operation of EV and given by Eq. 20.

$$\frac{\mathrm{d}p^e_{H_2}}{\mathrm{d}t} = \frac{\mathrm{d}p^e_{O_2}}{\mathrm{d}t} = 0 \quad (20)$$

Hence, net flow rates of H_2 and O_2 at normal mode operation are given in Eq. 21.

$$M_{H_2,\mathrm{net}} = 2M_{O_2,\mathrm{net}} = \frac{I_\mathrm{d}}{2*F} \quad (21)$$

Under acceleration or deceleration mode, there are slight delays between the fuel inlet and FC stack current (i_{FSO}). The following equations represent these slight delay effects in FC design and given in Eq. 22.

$$\tau_\mathrm{F}\frac{\mathrm{d}M_{H_2,\mathrm{net}}}{\mathrm{d}t} = \frac{i_\mathrm{d}}{2*F} - M_{H_2,\mathrm{net}}$$
$$\tau_\mathrm{O}\frac{\mathrm{d}M_{O_2,\mathrm{net}}}{\mathrm{d}t} = \frac{i_\mathrm{d}}{4*F} - M_{O_2,\mathrm{net}} \quad (22)$$

where τ_F, τ_O are fuel flow delay, and oxidant flow delay.

The effective partial pressure of H_2 and effective partial pressure of O_2 are used in the FC stack internal voltage or Nernst voltage calculation. The material conservation and delay effects are used in the calculation of FC stack voltage or Nernst voltage. In the next section, FC terminal voltage (V_t) is determined.

2.1.3 Fuel Cell Terminal Voltage

In the PEMFC, both electricity and water are developed by the reverse electrolysis process in the membrane electrode assembly. Hydrogen (H_2) is injected at the anode end of FC and compressed air is injected at cathode end of FC to get electricity. In this process, water is produced in the cathode side as a by-product as shown in Fig. 2, and the standard equation of water in FC is given in Eq. 23.

The complete reaction in a FC can be written as Eq. 23.

$$H_2 + \frac{1}{2}O_2 = H_2O_{(l)} \tag{23}$$

The Nernst equation used to calculate the FC internal voltage or Nernst voltage is given in Eq. 24.

$$V_C = V_{O,C} + \frac{RT}{2 * F}\ln\left[p^e_{H_2} \cdot (p^e_{O_2})^{0.5}\right] \tag{24}$$

where $V_{O,C}$ is a function of temperature (T) and is expressed in Eq. 25.

$$V_{O,C} = V^0_{O,C} - B_E(T - 298) \tag{25}$$

where $V^0_{O,C}$, B_E are the reference standard potential at normal temperature and pressure of 1-atm and 298 K, empirical constant in calculating $V_{O,C}$, respectively. To simplify analysis, a $V_{d,C}$ is considered to be subtracted from right side of Eq. 24 for overall effect of fuel and oxidant delay. In the steady state, $V_{d,C}$ is zero, but it will show influence of fuel and oxidant delays on FC internal voltage (V_c) during acceleration and deceleration modes. It can be written as in Eq. 26.

$$V_{d,C} = F_C\left[i(t) - i(t)e^{\left(-\frac{\tau_a}{t}\right)}\right] \tag{26}$$

The $V_{d,C}$ is a transient term and will reduce to zero when $t = \infty$. Where $V_{d,C}$, F_C, $i(t)$, τ_a, are voltage due to fuel and oxidant delay, empirical constant in calculating $V_{d,C}$, instantaneous current of FC stack, over all flow delay. Equation 26 is subtracted from Eq. 24 to get internal voltage (V_c) and given by Eq. 27.

$$V_C = V_{O,C} + \frac{RT}{2 * F}\ln\left[P^e_{H_2}.(p^e_{O_2})^{0.5}\right] - V_{d,C} \tag{27}$$

where V_C, calculated from Eq. 27, is actually internal voltage of FC. However, FC terminal voltage (V_t) is less than internal voltage, V_C. The activation loss ($V_{atv,c}$), ohmic loss ($V_{ohm,c}$) and concentration loss ($V_{con,c}$) are voltage drops across FC. Hence, the terminal voltage of FC can be written in Eq. 28.

$$V_t = V_C - V_{atv,c} - V_{ohm,c} - V_{con,c} \tag{28}$$

The terminal voltage of FC stack is written as in Eq. 29.

$$V_{FSO} = N_T V_t \tag{29}$$

where N_T, V_{FSO}, are number of cells in FC stack, FC stack voltage, respectively. FC stack voltage varies between 80 and 140 V. A typical FC stack may have 100 fuel cells to develop 80–140 V, each FC having voltage of 0.8–1.4 V as displayed in Fig. 3. The full bridge three-level (FBTL) converter has presented for renewable applications. The FBTL converter has been used as a unidirectional DC/DC converter (UDC) in FC vehicle. In the subsequent subsections of the manuscript, UDC design for FC vehicle is discussed in detailed manner.

2.2 *Unidirectional Converter (UDC)*

The unidirectional DC/DC converter is needed between FC stack and DC bus bar to protect FC stack under deceleration mode. In deceleration mode, reverse inrush current can develop and damage the FC stack if there is no unidirectional converter.

The FC stack terminal voltage (V_{FSO}) varies with load, and deviation range is wide, like 80–140 V, [26–28] due to its internal voltage (V_c) and activation, ohmic and concentration losses. Therefore, FC is unable to supply inverter directly due to its poor regulation. For this reason, UDC is necessary for FC hybrid electric vehicle. The parameters of UDC converter are tabulated in Table 2. Moreover, transformer in UDC provides galvanic isolation between DC bus bar and FC stack to protect the FC stack, this extra to voltage boost. The UDC converter is shown in Fig. 1.

The FBTL converter is utilized as a UDC and following are few of the advantages of converter [29]: (1) It works at both two-level and three-level modes. (2) When FC terminal voltage is high, UDC converter can work at two-level mode. (3) High efficiency can be accomplished over a wider input voltage range.

The output of UDC is connected to DC bus bar, which is followed by an inverter with output of 220 V AC, 50 Hz. The efficiency of H-bridge inverter has been reported as 95% [30]. The input DC voltage for output voltage of a 220 V AC inverter is normally 120 V; hence, V_{UDC} is set to 120 V.

The rating of FC hybrid electric vehicle is taken to be 1.4 kW sufficient to drive a three-wheel e-rickshaw. Thus, unidirectional converter current (i_{UDC}) can be calculated and given in Eq. 30.

Table 2 Parameters of unidirectional converter

Parameters of unidirectional converter	
Input voltage	(80–140) V
Output voltage	120 V
Output power	1.2 kW
N_P:N_S	1:1.2
Switches (S_1, S_2, S_3, S_4)	100 V
Switches (S_5, S_6)	50 V
Power diodes ($D1, D2, D3, D4$)	100 V
Filter components (L_{BUS}, C_{BUS})	320 mH, 900 μF

$$i_{\mathrm{UDC}} = \frac{P_{\mathrm{Out}}}{\eta_{\mathrm{Inv}} \times V_{\mathrm{UDC}}} = \frac{1400}{0.95 \times 120} = 12.27\,\mathrm{A} \tag{30}$$

Here, i_{UDC} selected as 13 A.

In the subsequent section, BDC has been designed for the above FC vehicle.

2.3 Bidirectional Converter (BDC)

The BDC interfaces DC bus bar and battery [31–34]. Different types of BDCs are available; non-isolated BDCs and isolated BDCs. Isolated BDCs are generally utilized if voltage variation between DC bus bar and battery is higher [35]. Transformer in isolated BDC assumes a vital part in galvanic isolation and enhance output voltage [36–38].

The RDAB-IBDC is used as a BDC converter in the FCHEV. The RDAB-IBDC has been derived from GDAB-IBDC. In RDAB-IBDC, both bridges have three-level voltages and soft switching compared to GDAB-IBDC. The RDAB-IBDC has resonance in both sides of isolation transformer due to its resonance network as shown in Fig. 1. The parameters of BDC are listed in Table 3. The RDAB-IBDC is needed between battery and DC bus bar for avoiding damage to battery due to reverse inrush current which can arise in acceleration and deceleration modes of FCHEV. The design of BDC is based on buck and boost mode as follows [39].

In buck mode, output voltage of BDC is battery voltage V_{Bat}, and current is i_{BDC}. The rating of battery is 60 V, so $V_{\mathrm{Bat,}}$ is set to 60 V. Battery instruction manual [22] explains that charging current of battery is around 0.1 C_{Bat}. C_{Bat} is capacity of battery

Table 3 Parameters of bidirectional DC/DC converter

Circuit parameters	Value
Capacity	1.4 kW
Input voltage (V_1)	60 V
Output voltage (V_{out})	120 V
Load resistance (R_{Load})	(50–100) Ω
Switching frequency (f_S)	20 kHz
Capacitor (C_1)	600 MFD, 60 V
Capacitor (C_2)	500 MFD, 120 V
Inductors (L_{r1}, L_{r2})	320 nH
Capacitors (C_{r1}, C_{r2})	400 MFD
Parameters of isolation transformer	
Capacity	3.5 kVA
Frequency (f_S)	20 kHz
Magnetic current (i_μ)	1% rated current

and is taken as 30 Ah, which is addressed in subsequent section. Hence, i_{BDC} can be expressed as Eq. 31.

$$i_{BDC} = 0.1 \times C_{Bat} = 0.1 \times 30 = 3\,A \tag{31}$$

where $i_{BDC}, V_{Bat}, C_{Bat}$, are bidirectional converter current, battery voltage and capacity of battery.

In boost mode, output voltage and current of BDC are V_{BDC} and i_{BDC} respectively. Hypothetically, V_{BDC} should be same as output voltage of UDC (V_{UDC}), yet it is hard to guarantee that they have a similar value in actual circuit. During starting period, battery supplies to DC bus bar at first, and BDC works in constant forward voltage mode [40]. In case, V_{BDC} is larger than V_{UDC}, UDC cannot work and it cannot provide output voltage, V_{UDC}. Hence, V_{BDC} should be lower than $V_{UDC} = 120$ V, so as to promise that inverter output is 220 V, and with conduction voltage drop and parasitic components accounted for, V_{BDC} must be larger than 105 V, due to which V_{BDC} is kept as 110 V [41, 42]. The efficiency of inverter has been reported as 95%. The efficiency of BDC is considered 95% to calculate the BDC current (i_{BDC}). To ensure that the vehicle works under acceleration mode, battery has to supply 1.4 kW, so i_{BDC} can be calculated and given in Eq. 32.

$$i_{BDC} = \frac{P_{Out}}{\eta_{Inv} \times \eta_{BDC} \times V_{min-Bat}} = \frac{1400}{0.95 \times 0.95 \times 60} = 25.14\,A \tag{32}$$

Here, i_{BDC} comes out as 26 A.

Where, η_{Inv}, η_{BDC}, $V_{min\text{-}Bat}$, are efficiency of inverter, efficiency of BDC converter and minimum battery voltage. Either a battery or an ultra-capacitor can be utilized in a vehicle as auxiliary energy source. Ultra-capacitor has better dynamic characteristic and higher power density. However, a battery costs considerably lesser and has higher energy density. In view of the cost and energy density consideration, a battery is utilized here as assisting energy source. In subsequent sections of manuscript, battery rating has been selected for FCHEV.

2.4 Battery Pack Sizing

The selection of C_{Bat} has been done based on power requirement during the cold start and acceleration mode. As per battery handbook [22], battery maximum discharge current for tens of seconds can be controlled to (0.5–0.7) C_{Bat} to keep a considered distance from battery damage point due to deep discharge. Currently, 0.7 C_{Bat} is set as the lowest capacity after discharge. The base voltage of single cell is 12 V; therefore, base voltage of a 5 battery stack is $V_{min\text{-}Bat} = 5 \times 12\text{ V} = 60$ V. Then, $V_{min\text{-}Bat}$ is set at 60 V. Efficiencies of the BDC and inverter are $\eta_{BDC} = 95\%$ and $\eta_{Inv} = 95\%$, respectively. Then C_{Bat} can be calculated and given by Eq. 33.

Table 4 Parameters of battery pack

Specifications of the battery pack	
Category	Lead acid battery
Number of batteries	05
Single battery voltage	12 V
Rated voltage	60 V
Rated capacity	30 Ah
Length	15 cm
Width	13 cm
Height	8 cm
Size of battery package	150 mm × 130 mm × 80 mm

$$C_{\text{Bat}} = \frac{P_{\text{Out}}}{0.7 \times \eta_{\text{Inv}} \times \eta_{\text{BDC}} \times V_{\text{min}-\text{Bat}}} = \frac{1000}{0.7 \times 0.95 \times 0.95 \times 54} = 28.78\ \text{Ah} \quad (33)$$

A 30 Ah battery is considered, and its parameters are kept in Table 4. The FCHEV specifications and parameters are listed in Table 5. The electric motor sizing has been discussed in further subsections of the manuscript.

Table 5 Specifications of PEM fuel cell hybrid electric vehicle

Specifications and parameters for the FC hybrid electric vehicle		
Fuel cell	Power rating (P_{Out})	1.2 kW
	Output voltage (V_{FSO})	80–140 V
Battery	Capacity (C_{Bat})	30 Ah
	Voltage rating (V_{Bat})	60 V
Unidirectional converter (UDC)	Input voltage (V_{FSO})	80–140 V
	Output voltage (V_{UDC})	120 V
	Output current (i_{UDC})	13 A
Bidirectional converter (BDC)	Output voltage in boost mode (V_{BDC})	110 V
	Output current in boost mode (i_{BDC})	26 A
	Output voltage in buck mode (V_{Bat})	60 V
	Output current in buck mode (i_{BDC})	3 A
Inverter	Output voltage	220 V
	Power rating	1.5 kVA
	Output frequency	50 Hz

2.5 Electric Motor Sizing

In this section, the electric motor has been designed for the FC hybrid-based three-wheel e-rickshaws and the normal and worst-case conditions are tabulated in Tables 6 and 7. The mathematical modeling of electric motor has been developed with the data given in Tables 6 and 7 for calculation of the power and torque of electric motor. The mass (vehicle mass and load mass) has been considered while calculating the power and torque demanded of the electric motor in FC hybrid electric vehicle. The normal and worst-case conditions have been considered to analyze the FC hybrid electric vehicle in detail. To determine the torque and power of electric motor, one needs to calculate the rolling resistance force, drag force, acceleration force, ramp force and vehicle mass and load mass. The four forces of FC hybrid electric vehicle are derived and given in Eqs. 34–37.

$$F_{rr} = m \times g \times K_{rr} \tag{34}$$

where F_{rr}, m, g, K_{rr}, are rolling resistance force, vehicle mass and load mass in kg, gravitational constant (9.81 m/s^2) and co-efficient of rolling resistance force.

$$F_{\mathrm{d}} = \frac{1}{2} \times \rho_{\mathrm{air}} \times A_{\mathrm{F}} \times K_{\mathrm{d}} \times N^2 \tag{35}$$

where F_{d}, ρ_{air}, A_{F}, K_{d}, N, are drag force, air density, frontal area of vehicle and co-efficient of drag force, motor speed, respectively.

Table 6 FC hybrid electric vehicle during normal conditions

Mass (kg)	Speed, *N* (km/h)	Acceleration (m/s^2)	Ramp, θ (Degrees)
450	40	0.5	2
450	30	1.7	2
500	30	2.8	3
500	20	2.1	3

Table 7 FC hybrid electric vehicle during worst-case conditions

Mass (kg)	Speed, *N* (km/h)	Acceleration (m/s^2)	Ramp, θ (Degrees)
700	10	0.5	5
700	10	0.5	5
900	5	1	10
900	5	1	10

$$F_{\mathrm{a}} = m \times a \tag{36}$$

where F_{a}, a, are acceleration force, acceleration, respectively.

$$F_r = m \times g \times \sin\left(\theta \times \frac{\pi}{180}\right) \tag{37}$$

where F_r, θ, are ramp force, ramp angle in degrees, respectively.

The torque of electric motor is determined by using Eqs. 34, 35, 36, and 37, and given by 38.

$$T = (F_{rr} + F_{\mathrm{d}} + F_{\mathrm{a}} + F_{\mathrm{r}}) \times r_{\mathrm{w}} \tag{38}$$

where T, r_{w}, are torque of electric motor, radii of wheel, respectively.

The power demanded by the electric motor of FC hybrid electric vehicle is derived and given by Eq. 39.

$$P_{\mathrm{Out}} = \omega \times T \tag{39}$$

The FCHEV has been operated in four modes. The rating of the electric motor is 1.4 kW, which needs to be supplied by the FC and battery to develop torque of 132 N m. The rated speed of the electric motor is 300 rpm during the normal mode. The speed of electric motor is 200 rpm during acceleration mode as shown in Table 8. In the deceleration mode, the motor torque is similar to the starting torque but the motor is operated in the breaking region. The motor power, speed and torque verses modes are given in Fig. 5a–c, respectively.

The electric motor torque is inversely proportional to speed to maintain constant power during the normal mode as shown in Fig. 6a, i.e., power required by electric motor during acceleration mode is supplied by both FC and battery. The battery rating should be decided by peak demand and FC should be designed for steady-state requirement of FCHEV. Power verses speed of electric motor has been shown in Fig. 6b. From above discussion, motor rating has been selected as 1.4 kW as it is maximum power demanded during acceleration mode. In the subsequent sections, modes of operation of FC vehicle have been discussed.

Table 8 Characteristics of electric drive of FCHEV during four modes

Mode	Motor speed, N (rpm)	Motor torque, T (Nm)	Power, P_{Out} (kW)
Cold start	100	132	0.6
Normal	300	47	1.2
Acceleration	200	70	1.4
Deceleration	−100 (breaking mode)	132	−0.6 (stores in battery pack)

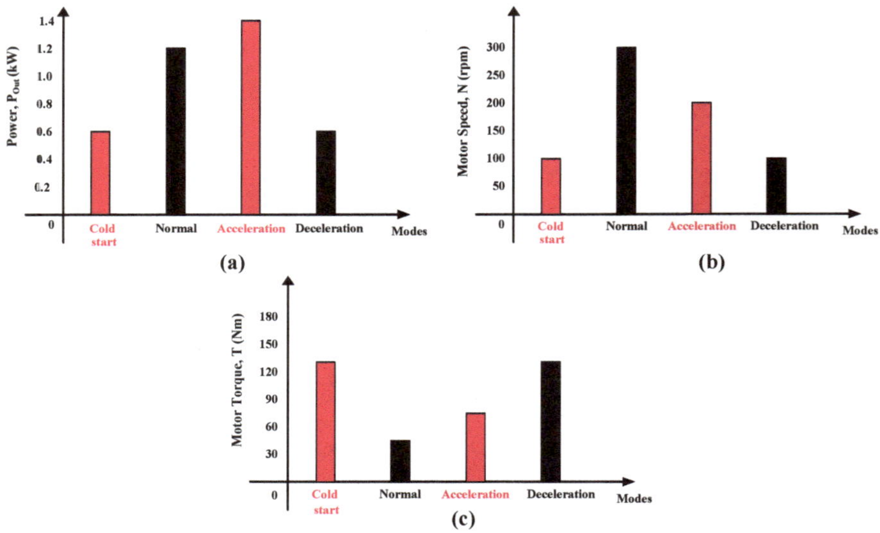

Fig. 5 Characteristics of electric motor of FC hybrid electric vehicle during four modes **a** Motor power verse modes, **b** Motor Speed verses modes, **c** motor torque verses modes

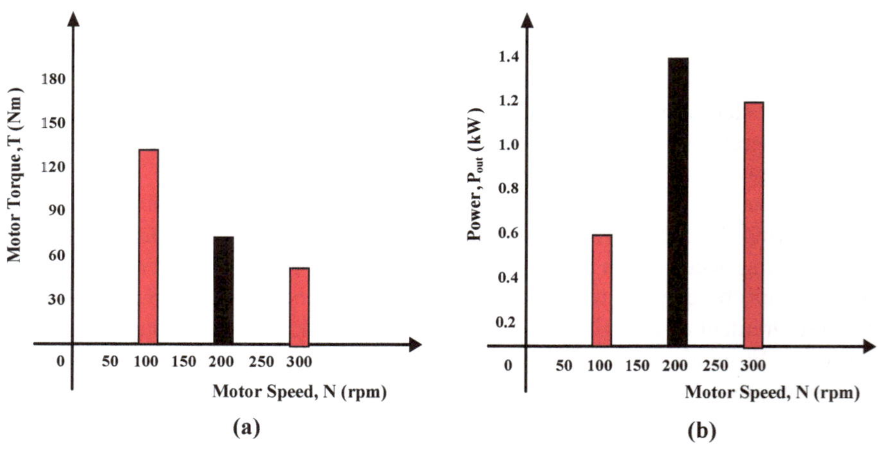

Fig. 6 Characteristics of electric motor of FCHEV **a** Motor torque verses speed, **b** power verse motor speed

3 Different Modes of Operation of FC Vehicle

The FCHEV is operated in four modes like cold start, normal, acceleration and deceleration modes as shown in Fig. 7. Detailed operation of four modes is analyzed

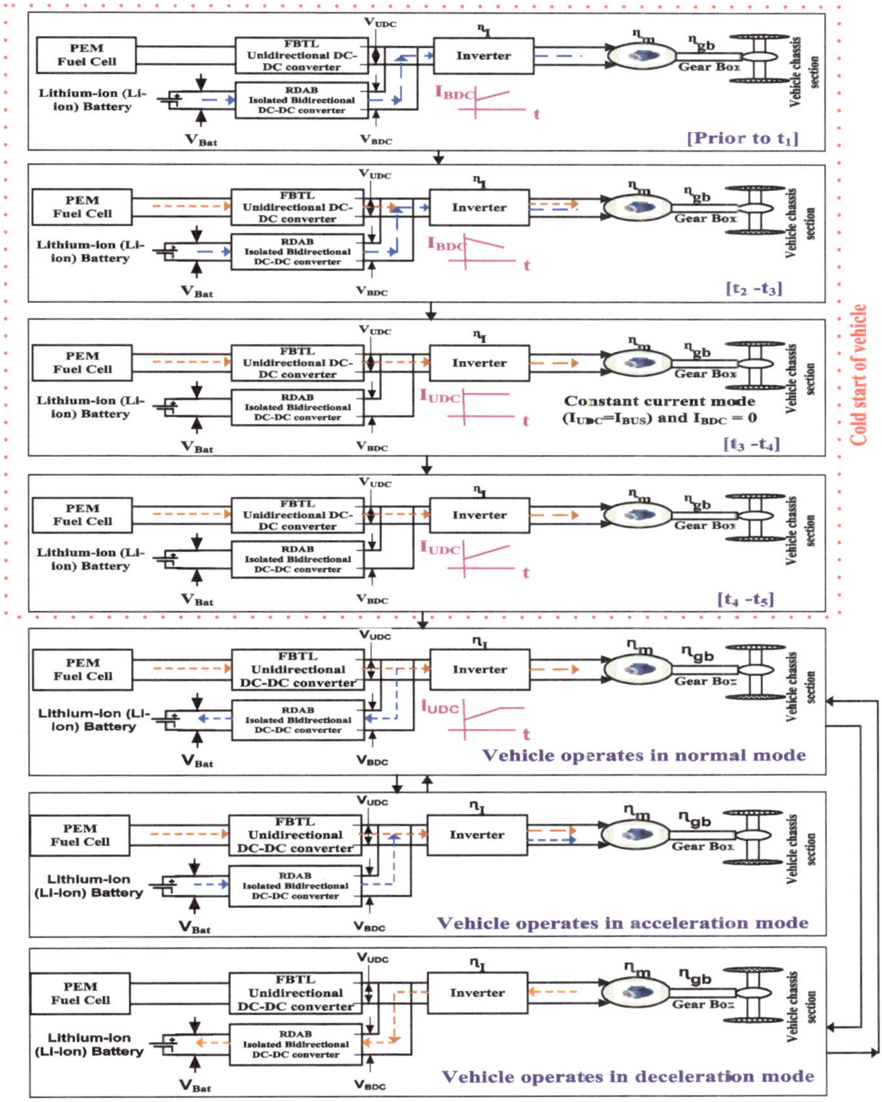

Fig. 7 Modes of operation of FC hybrid electric vehicle

in this section. The unidirectional and bidirectional currents and DC bus-bar voltage of cold start and normal mode are shown in Fig. 8.

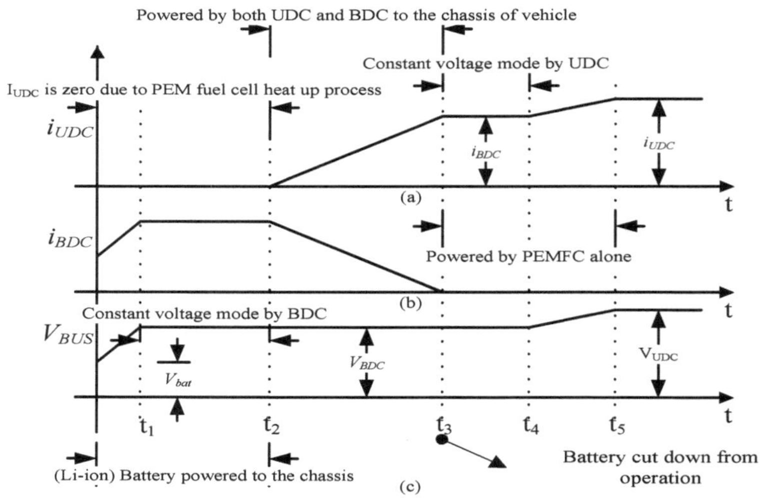

Fig. 8 Unidirectional and bidirectional currents and DC bus-bar voltage of FC hybrid electric vehicle in cold start and normal mode

3.1 Cold Start and Normal Mode

i. **Period 0 [Prior to t_1]**: When the vehicle gets start, the FC needs certain time, to activate the gas diffusion, process and break down of hydrogen into ions (H^+) and electrons (H^-). Due to the catalyst of anode and cathode, and internal fan requires certain time to collect atmospheric air and convert it into compressed air to the FC; hence, the FC cannot power vehicle instantly, and UDC does not work. Battery supplies vehicle through BDC converter, which works in boost mode.

ii. **Period 1 [t_1, t_2]**: At t_1, DC bus-bar voltage (V_{BUS}) reached output voltage of BDC converter (V_{BDC}). The BDC completes its soft start stage and works in constant voltage mode to regulate constant DC voltage across an inverter. FC still functions in self-start phase and cannot supply vehicle

iii. **Period 2 [t_2, t_3]**: At t_2, FC finishes the heat up process and is ready to supply to the vehicle. The UDC converter also has a soft start; therefore, its output current (i_{UDC}) increases slowly, and BDC current (i_{BDC}) decreases slowly. FC and battery both supply to vehicle.

iv. **Period 3 [t_3, t_4]**: At t_3, BDC converter (i_{BDC}) current is reached to zero and i_{UDC} reached rated current of BDC converter (i_{BDC}) value at which provides constant current to the vehicle. The duration between t_3 and t_4 has a constant current mode.

v. **Period 4 [t_4, t_5]**: At t_4, i_{UDC} starts increasing linear to reach the rated current of UDC converter, which needs to operate vehicle without battery support and vehicle runs under normal mode. At this situation, battery power disconnected from the vehicle. Therefore, V_{BUS} and i_{UDC} increases. At t_5, DC bus-bar voltage V_{BUS} reaches V_{UDC}. The i_{UDC} operates under constant current mode. FC powers the vehicle.

3.2 *Acceleration and Deceleration Modes of FC Vehicle*

Figure 9 shows unidirectional and bidirectional currents and DC bus-bar voltage of FC hybrid electric vehicle in acceleration and deceleration modes.

i. **Period 0 [Prior to t_1]**: Prior to t_1, FC alone supplies to the vehicle, UDC functions under constant voltage mode and provides constant voltage to an inverter section, and $V_{BUS} = V_{UDC}$.
ii. **Period 1 [t_1, t_2]**: At t_1, the torque of vehicle is suddenly increases, FC power (P_{Out}) is limited and cannot supply more than the rated, and i_{UDC} increases with some slope; therefore, FC cannot provide energy immediately and lacking power is provided by input capacitor C_{FSO} of UDC and DC bus capacitor C_{BUS} and V_{BUS} voltage drops.
iii. **Period 2 [t_2, t_3]**: At t_2, V_{BUS} drops to V_{SIP}, and the BDC starts to operate in boost mode. FC and battery are providing power to the vehicle.
iv. **Period 3 [t_3, t_4]**: At t_3, V_{BUS} reached V_{BDC}, and BDC works under constant voltage mode. FC and battery provide power to vehicle simultaneously. During this region, the i_{UDC} keeps on increasing, and i_{BDC} keeps on decreasing accordingly.

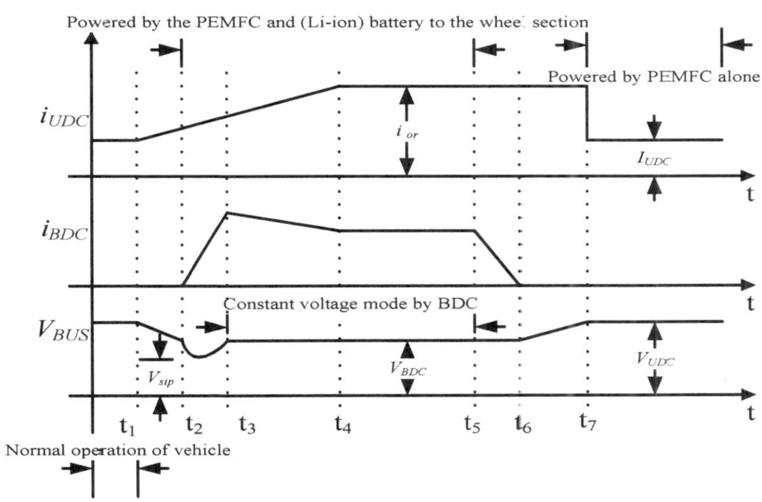

Fig. 9 Unidirectional and bidirectional currents and DC bus-bar voltage of FC hybrid electric vehicle in acceleration and deceleration mode

v. **Period 4 [t_4, t_5]**: At t_4, i_{UDC} increases to over rated current (i_{Or}), UDC works under limited current mode, $V_{\text{BUS}} = V_{\text{BDC}}$, and $i_{\text{UDC}} = i_{\text{Or}}$.
vi. **Period 5 [t_5, t_6]**: At t_5, the load of vehicle steps down to be lower than rated power, and i_{BDC} reduces.
vii. **Period 6 [t_6, t_7]**: i_{BDC} is reduced to zero at t_6, battery cut down from the vehicle operation, and only FC supply to the vehicle. V_{BUS} increases and reaches V_{UDC}. The UDC works in constant voltage mode, and i_{UDC} decreases to rated current of UDC converter. Acceleration and deceleration modes are completed.

4 Simulations and Results Discussion

Simulations of FC hybrid electric vehicle have been carried out in PSCAD for the four different case studies. The input voltage of UDC converter is varied from 80 to 140 V. The DC bus-bar voltage (V_{BUS}) is 120 V. The four cases considered for vehicle are cold start, normal, acceleration and deceleration modes. The four cases had discussed for unidirectional converter current (i_{UDC}), bidirectional converter current (i_{BDC}) and DC bus-bar voltage (V_{BUS}) with the constant current and voltage mode stages. The simulation results for four modes of FC hybrid electric vehicle are as follows.

4.1 Case 1: Cold Start Mode

In the cold start mode, FC cannot supply to the UDC converter during starting of vehicle. The FC hybrid electric vehicle under cold start mode is shown in Fig. 10. The unidirectional (i_{UDC}) and bidirectional (i_{BDC}) currents and DC bus-bar voltage (V_{BUS}) of FC vehicle are shown in Fig. 11 under cold start condition.

The UDC current will increase in the later stages of starting process due to cold start mode. During cold start, vehicle will get supply from battery through BDC. Unidirectional converter current (i_{UDC}) is zero at starting due to FC heat up process upto 0.1 s as shown in Fig. 11a. The unidirectional converter current (i_{UDC}) slowly starts increasing and reaches its first peak value and then works in the constant current mode operation. The bidirectional converter current (i_{BDC}) slowly increases and reached its peak value. The i_{BDC} supplies to the FC vehicle until unidirectional converter current reaches its rated value (i_{UDC}) as shown in Fig. 11b.

The DC bus-bar voltage (V_{BUS}) slowly increases and reaches bidirectional converter output voltage (V_{BDC}). DC bus-bar voltage (V_{BUS}) maintains BDC output voltage (V_{BDC}) and it can work under constant voltage mode as shown in Fig. 11c. The i_{BDC} current becomes zero at 0.2 s and then battery is cut down from the vehicle as shown in Fig. 11b.

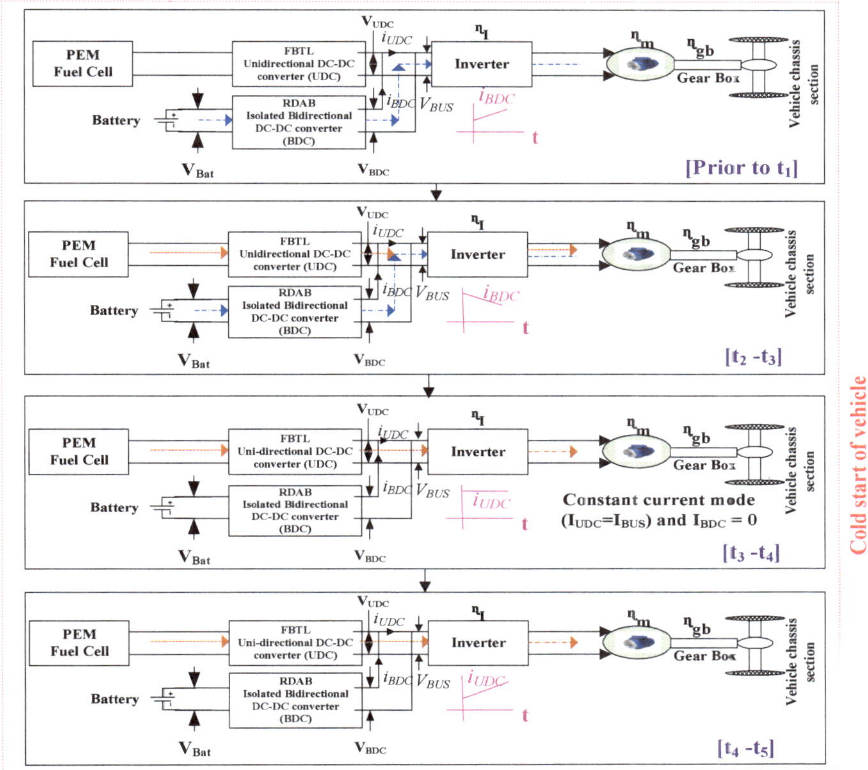

Fig. 10 FC hybrid electric vehicle under cold start mode

4.2 *Case 2: Normal Mode*

In the normal mode, FC alone supplies the vehicle and charges the battery simultaneously. FCHEV under normal mode is shown in Fig. 12. Unidirectional converter (UDC) works in the constant current mode (i_{UDC}) throughout the normal mode as shown in Fig. 13a. The unidirectional and bidirectional currents and DC bus-bar voltage of FC hybrid electric vehicle are shown in Fig. 13 under normal mode.

The current of BDC is negative because battery is being charged. BDC current (3 A) is equal to the value of rating of battery. The i_{BDC} and i_{UDC} are approximately 3.0 and 10.0 A as shown in Fig. 13a, b. The i_{BDC} reaches the rated current of battery 3.0 A and then it works under constant current mode throughout normal mode or until battery gets charged as shown in Fig. 13b. DC bus-bar voltage (V_{BUS}) is equal to the unidirectional converter voltage (V_{UDC}) and works under constant voltage mode as shown in Fig. 13c.

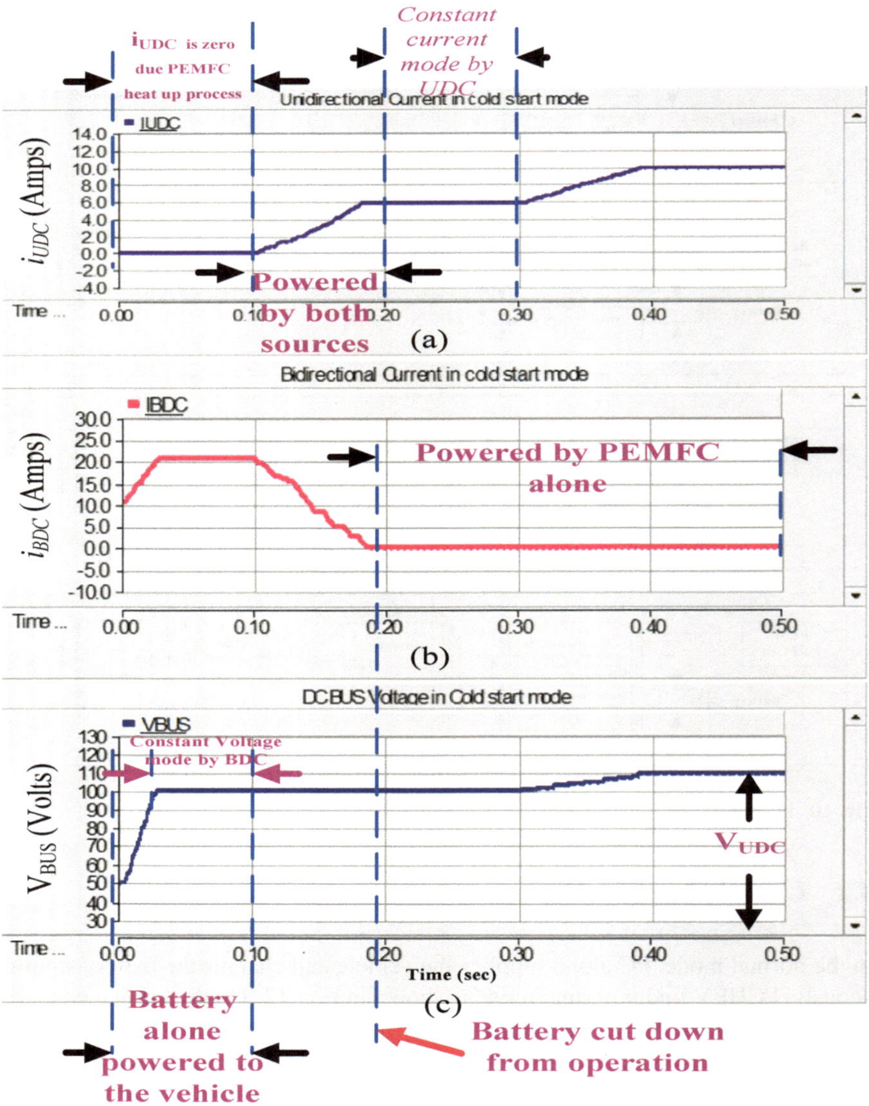

Fig. 11 FC hybrid electric vehicle under cold start mode **a** Unidirectional converter current (i_{UDC}), **b** bidirectional converter current (i_{BDC}), **c** DC bus-bar voltage (V_{BUS})

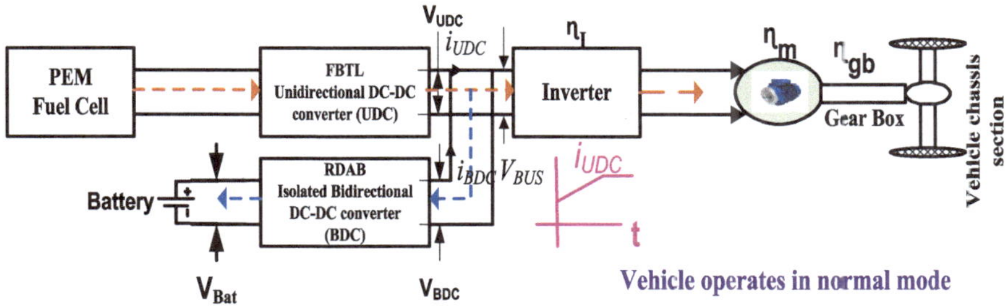

Fig. 12 FC hybrid electric vehicle under normal mode

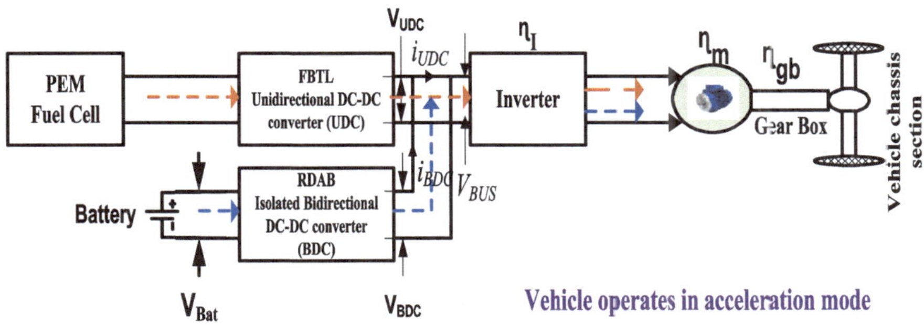

Fig. 13 FC hybrid electric vehicle under acceleration mode

4.3 Case 3: Acceleration Mode

In the acceleration mode, both FC and battery need to assist the vehicle. The FC hybrid electric vehicle under acceleration mode is shown in Fig. 14. The unidirectional and bidirectional converter currents and DC bus-bar voltage of FCHEV are shown in Fig. 15 under acceleration mode.

The unidirectional current (i_{UDC}) is at rated value (10 A) at starting and then slowly increases and reaches the over rated current (i_{Or}) as shown in Fig. 15a. After reaching over rated current (i_{Or}), UDC maintains constant current mode until deceleration mode is completed. The acceleration of FCHEV is accomplished due to support of both FC and battery. The bidirectional current (i_{BDC}) starts increasing linearly and reaches the rated value (26 A) in boost mode. BDC converter works in boost mode in the acceleration mode to assist the vehicle and maintains constant current mode until deceleration mode is completed. The bidirectional converter current (i_{BDC}) becomes zero at 0.35 s as shown in Fig. 15b.

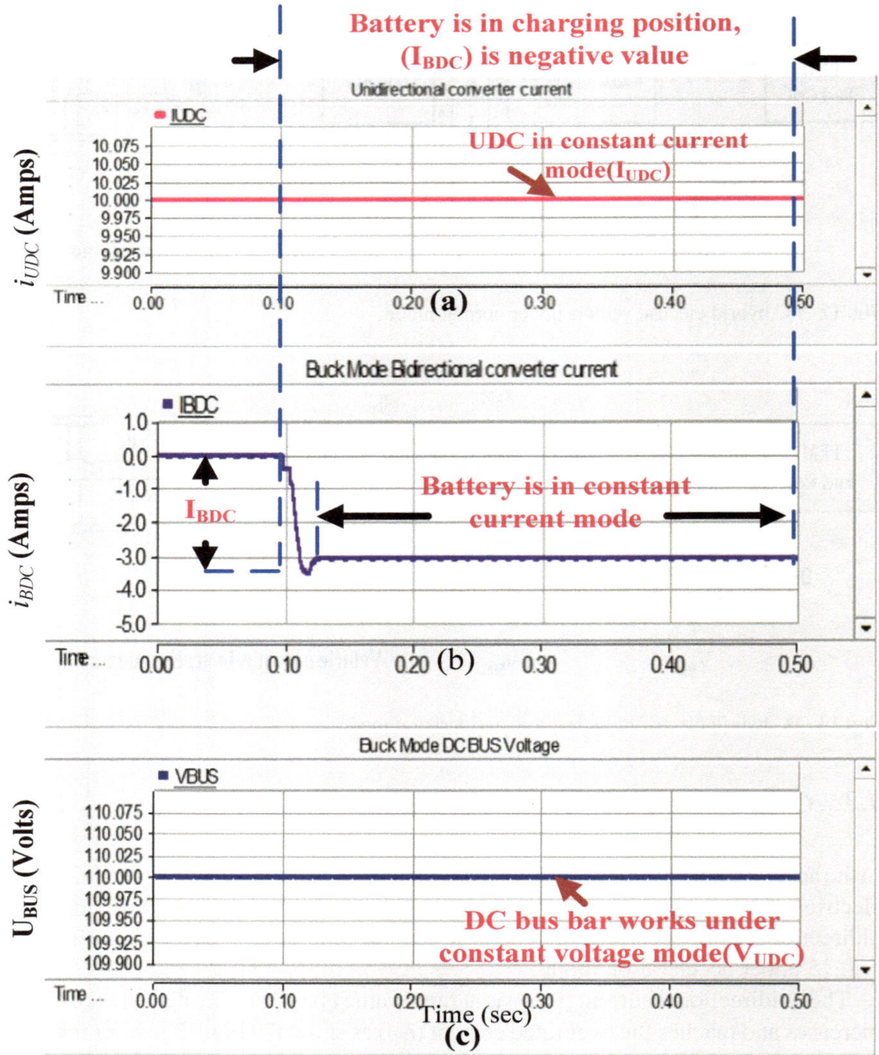

Fig. 14 FC hybrid electric vehicle under normal mode **a** Unidirectional converter current (i_{UDC}), **b** bidirectional converter current (i_{BDC}), **c** the DC bus-bar voltage (V_{BUS})

4.4 Case 4: Deceleration Mode

Battery is in charging in deceleration mode. UDC current (i_{UDC}) of vehicle would come down to zero. FCHEV under deceleration mode is shown in Fig. 16. BDC current (i_{BDC}) can decrease up to rated value of battery under deceleration mode.

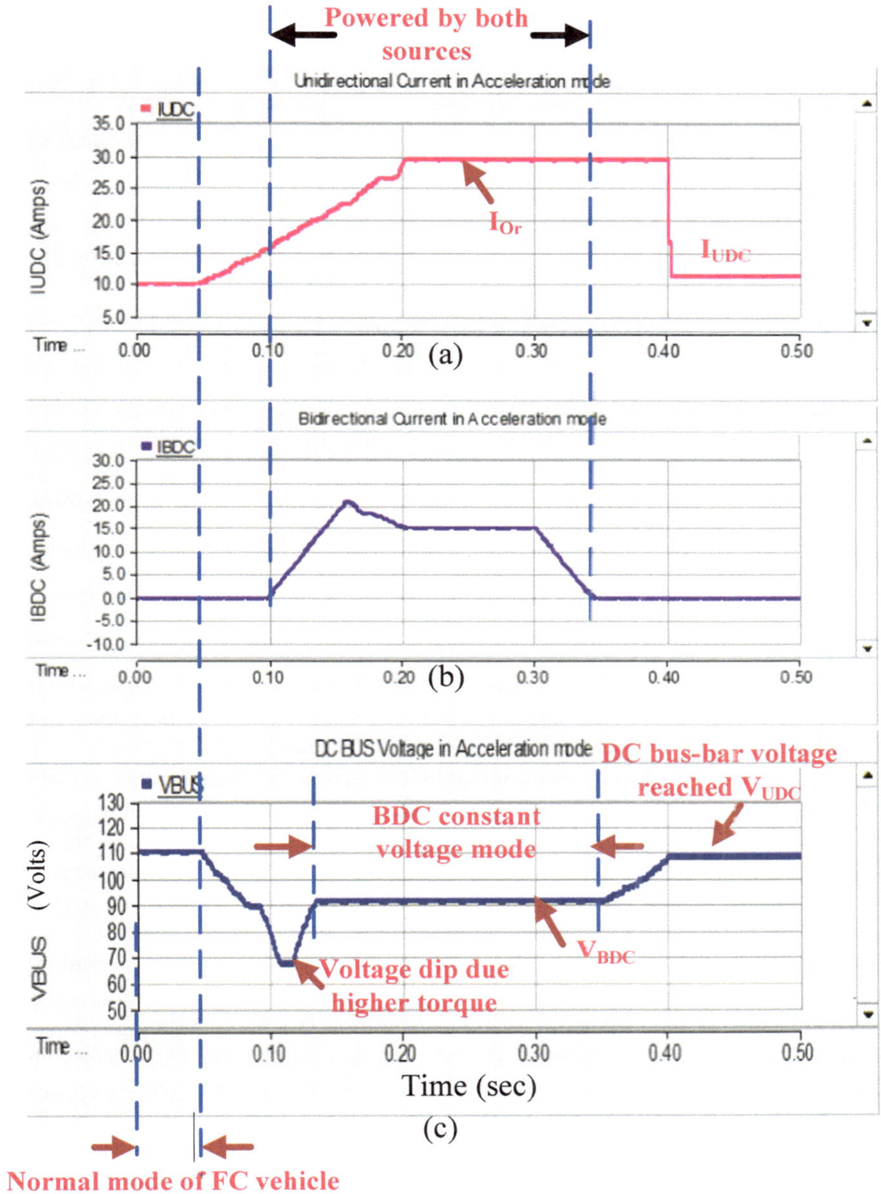

Fig. 15 FC hybrid electric vehicle under acceleration mode **a** Unidirectional converter current (i_{UDC}), **b** bidirectional converter current (i_{BDC}), **c** the DC bus-bar voltage (V_{BUS})

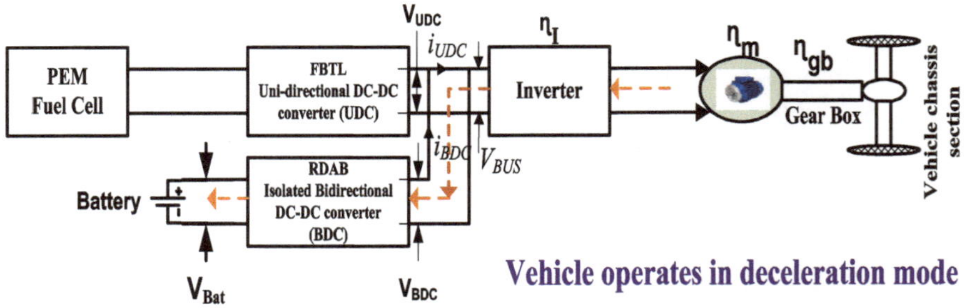

Fig. 16 FC hybrid electric vehicle under deceleration mode

The bidirectional and unidirectional currents and DC bus-bar voltage of vehicle under deceleration mode are shown in Fig. 17. The FCHEV operation divided into four modes due to integration with battery or else FCHEV can operate only in two modes which are cold start and normal modes. Cold start problem has been eliminated due to integration with battery.

5 Conclusion

The architecture and its dynamic modeling of FCHEV has been presented in this article. Accuracy of modeling terminal voltage has been improved by considering fuel and oxidant delay of FC and the design of FCEV have been improved by considering the activation, ohmic and concentration losses while calculating terminal voltage of FC. It has been shown that the FC need not be rated for the acceleration mode in the FC hybrid electric vehicle due to integration with battery as an auxiliary source so that peak power is supplied by the battery in the acceleration mode. When operation shifts from normal mode to acceleration mode as the FC cannot respond quickly, battery supplies power to the vehicle thus improving dynamic performance of FC hybrid electric vehicle. The FCHEV analyzed by using four case studies, which are cold start, normal, acceleration and deceleration modes. Simulation results are validated with theoretical analysis. The advantages of the proposed fuel cell hybrid electric vehicle have been validated with these simulation results. It has been shown that the proposed architecture and modulation methods work satisfactorily in all the modes of operation of the FCHEV.

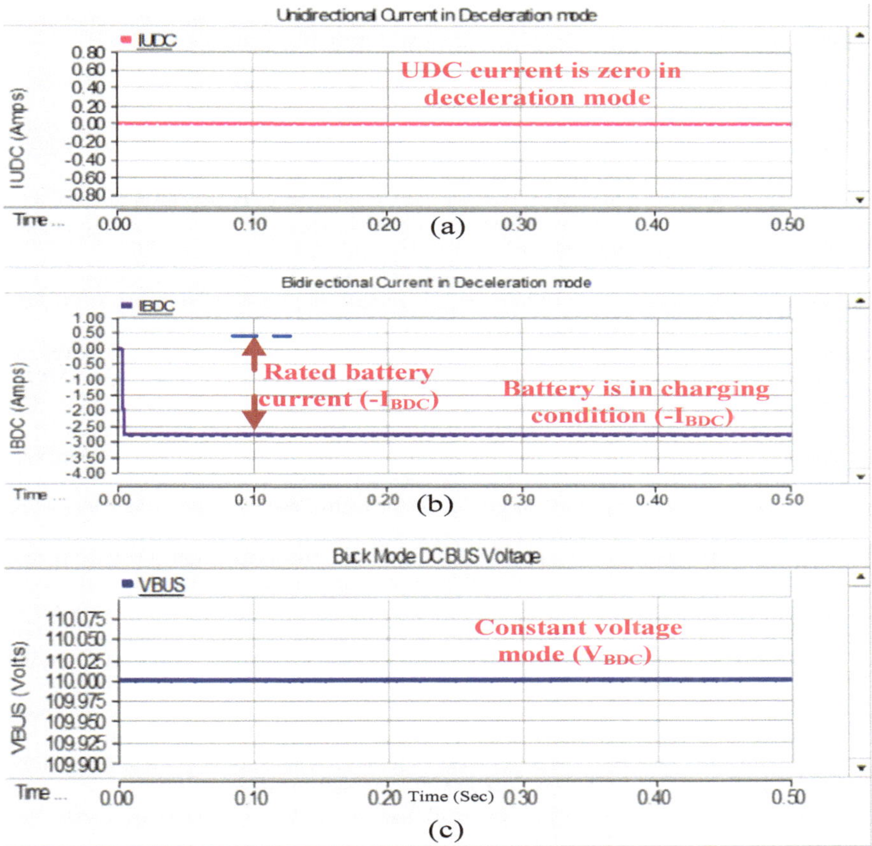

Fig. 17 FC hybrid electric vehicle under deceleration mode **a** Unidirectional converter current (i_{UDC}), **b** bidirectional converter current (i_{BDC}), **c** the DC bus-bar voltage (V_{BUS})

References

1. B. Cook, Introduction to fuel cells and hydrogen technology. Eng. Sci. Educ. J. **11**(6), 205–216 (2002)
2. N.H. Jafri, S. Gupta, An overview of FCs application in transportation, in *IEEE Transportation Electrification Conference and Expo* (Asia-Pacific (ITEC Asia-Pacific), Busan), pp. 129–133 (2016)
3. Fuel Cell Handbook, 7th edn., EG&G Services, Inc., Science Applications International Corporation, DOE, Office of Fossil Energy, National Energy Technology Laboratory (2004)
4. Wikipedia, The Free Encyclopaedia, *List of Fuel Cell Vehicles*. Accessed 24 Mar 2019
5. Wikipedia, The Free Encyclopaedia, *Fuel Cell Vehicle*. Accessed 24 March 2019
6. S. Thomas, M. Zalbowitz, *Fuel Cells-Green Power* (Los Alamos National Laboratory, Los Alamos, New Mexico)
7. F.A. Farret, M.G. Simoes, Power plants with fuel cells, in *Integration of Alternative Sources of Energy* (Wiley-IEEE Press, 2006), pp. 159–197

8. M.H. Nehrir, C. Wang, Principles of operation of fuel cells, in *Modeling and Control of Fuel Cells: Distributed Generation Applications* (Wiley-IEEE Press, 2009), pp. 29–56
9. P.J.H. Wingelaar, J.L. Duarte, M.A.M. Hendrix, Dynamic characteristics of PEM fuel cells, in *IEEE 36th Power Electronics Specialists Conference* (Recife, 2005), pp. 1635–1641
10. J. Snoussi, S.B. Elghali, M. Benbouzid, M.F. Mimouni, Optimal sizing of energy storage systems using frequency-separation-based energy management for fuel cell hybrid electric vehicles. IEEE Trans. Veh. Technol. **67**(10), 9337–9346 (2018)
11. D. Zhou, A. AlDurra, I. Matraji, A. Ravey, F. Gao, Online energy management strategy of fuel cell hybrid electric vehicles: a fractional-order extremum seeking method. IEEE Trans. Industr. Electron. **65**(8), 6787–6799 (2018)
12. Y. Han, Q. Li, T. Wang, W. Chen, L. Ma, Multisource coordination energy management strategy based on SOC consensus for a PEMFC–battery–super capacitor hybrid tramway. IEEE Trans. Veh. Technol. **67**(1), 296–305 (2018)
13. Q. Li, T. Wang, C. Dai, W. Chen, L. Ma, Power management strategy based on adaptive droop control for a fuel cell-battery-super capacitor hybrid tramway. IEEE Trans. Veh. Technol. **67**(7), 5658–5670 (2018)
14. S.S. Yadav, K.S. Sandhu, A grid connected hybrid PV/fuel cell/battery using five level PWM inverter, in *2018 International Conference on Emerging Trends and Innovations In Engineering and Technological Research (ICETIETR)* (Ernakulam, 2018), pp. 1–5
15. A.S. Leger, A multidisciplinary undergraduate alternative energy engineering course. IEEE Trans. Educ. **62**(1), 34–39 (2019)
16. S. Habib, M.M. Khan, F. Abbas, L. Sang, M.U. Shahid, H. Tang, A comprehensive study of implemented international standards, technical challenges, impacts and prospects for electric vehicles. IEEE Access **6**(2), 13866–13890 (2018)
17. D. Zhou, Y. Wu, F. Gao, E. Breaz, A. Ravey, A. Miraoui, Degradation prediction of PEM fuel cell stack based on multi physical aging model with particle filter approach. IEEE Trans. Ind. Appl. **53**(4), 4041–4052 (2017)
18. C.R. Aguiar, G.H.F. Fuzato, R.F.Q. Magossi, R.V.A. Neves, R.Q. Machado, A new method to manage the fuel cell initialization. IEEE Trans. Ind. Appl. **54**(5), 5187–5195 (2018)
19. P.J. Tritschler, E. Rullire, S. Bacha, Emulation of fuel cell systems, in *The XIX International Conference on Electrical Machines—ICEM 2010* (Rome, 2010), pp. 1–5
20. J.C.L. Haj, M.L. Agerskov, M.F. Jensen, P. Lading, Next generation range extension—2 glimpses of the future, in *2013 World Electric Vehicle Symposium and Exhibition (EVS27)* (Barcelona, 2013), pp. 1–8
21. J. Larminie, A. Dicks, *Fuel Cell Systems Explained* (Wiley, New York, 2000)
22. D. Linden, T.B. Reddy, *Battery Handbook* (McGraw Hill Publisher, 2001)
23. L.E. Lesster, Fuel cell power electronics. Fuel Cells Bull. **3**(25), 5–9 (2000)
24. R.W. Erickson, D. Maksimovic, *Fundamentals of Power Electronics*, 2nd edn. (Norwell, MA, USA, Kluwer, 2001)
25. M. Kabalo, B. Blunier, D. Bouquain, A. Miraoui, State-of-the-art of dc/dc converters for fuel cell vehicles, in *IEEE Vehicle Power and Propulsion Conference* (Lille, 2010), pp. 1–6
26. https://www.egr.unlv.edu/~eebag/IEEE_STD_519_1992vs2014.pdf
27. V.F. Pires, R.E. Cadaval, D.I. Roasto, J.F. Martins, Power converter interfaces for electrochemical energy storage systems—a review. Energy Convers. Manage. **86**(2), 453–475 (2014)
28. M.H. Todorovic, L. Palma, P. Enjeti, Design of a wide input range dc/dc converter with a robust power control scheme suitable for fuel cell power conversion, in *Proceedings of Anaheim IEEE Application Power Electronics Conference and Exposition*, pp. 374–379
29. H. Tarzamni, E. Babaei, A.Z. Gharehkoushan, M. Sabahi, Interleaved full ZVZCS dc/dc boost converter: analysis, design, reliability evaluations and experimental results. IET Power Electron. **10**(7), 835–845 (2017)
30. O. Hegazy, J. Van Mierlo, P. Lataire, Analysis, control and implementation of a high-power interleaved boost converter for fuel cell hybrid electric vehicle. Int. Rev. Electr. Eng. **6**(4), 1739–1747 (2011)

31. J. Xie, X. Zhang, C. Zhang, C. Wang, Research on bidirectional dc/dc converter for a standalone photovoltaic hybrid energy storage system, in *Asia-Pacific Power Energy Engineering Conference China* (Chengdu, 2010), pp. 1–4
32. S. Malo, R. Grino, Design, construction, and control of a standalone energy-conditioning system for PEM-type fuel cells. IEEE Trans. Power Electron. **25**(10), 2496–2506 (2010)
33. M.A. Abdullah, A.H.M. Yatim, C.W. Tan, A.S. Samosir, Control of a bidirectional converter to interface ultra-capacitor with renewable energy sources, in *Proceedings of the IEEE International Conference on Industrial Technology* (Cape Town, South Africa, 2013), pp. 673–678
34. V.M. Iyer, S. Gulur, S. Bhattacharya, Small-signal stability assessment and active stabilization of a bidirectional battery charger. IEEE Trans. Ind. Appl. **55**(1), 563–574 (2019)
35. A. Filba, S. Busquets, J. Nicolas, J. Bordonau, Operating principle and performance optimization of a three-level NPC dual-active-bridge dc/dc converter. IEEE Trans. Industr. Electron. **63**(2), 678–690 (2016)
36. B. Zhao, Q. Song, W. Liu, Power characterization of isolated bidirectional dual active bridge dc/dc converter with dual phase shift control. IEEE Trans. Power Electron. **27**(9), 4172–4176 (2012)
37. H. Bai, C. Mi, Eliminate reactive power and increase system efficiency of isolated bidirectional dual active bridge dc/dc converters using novel dual phase shift control. IEEE Trans. Power Electron. **23**(6), 2905–2914 (2008)
38. S. Inoue, H. Akagi, A bidirectional isolated dc/dc converter as a core circuit of the next-generation medium voltage power conversion system. IEEE Trans. Power Electron. **22**(2), 535–542 (2007)
39. H. Bai, C. Wang, S. Gargies, Operation, design and control of dual H-bridge based isolated bidirectional dc/dc converter. IET Power Electron. **1**(4), 507–517 (2008)
40. D. Sha, F. You, X. Wang, A high efficiency current-fed semi dual active bridge dc/dc converter for low input voltage applications. IEEE Trans. Industr. Electron. **63**(4), 2155–2164 (2016)
41. F. Krismer, J. Biela, J.W. Kolar, A comparative evaluation of isolated bidirectional dc/dc converters with wide input and output voltage range, in *Fourtieth IAS Annual Meeting. Conference Record of the 2005 Industry Applications Conference* (Kowloon, 2005), pp. 599–605
42. F. Krismer, J.W. Kolar, Efficiency optimized high current dual active bridge converter for automotive applications. IEEE Trans. Industr. Electron. **59**(7), 2745–2760 (2012)

Light Electric Vehicles and Their Charging Aspects

Chandana Sasidharan, Bhawna Tyagi, and Varun Rajah

1 Introduction

With the advent of electric vehicles (EVs), vehicle market globally are witnessing a massive transition towards clean mobility solutions. EVs are usually considered synonymous with electric cars but electric two and three-wheelers are the largest EV fleet among all modes and expected to dominate till 2030. In 2018, the global stock of electric two-wheeler stands at 260 million followed by electric car at 5.1 million [1]. The penetration of EVs varies across countries depending on supportive policy measures such as fiscal incentives to enable technological advances, economic instruments to increase uptake of EVs, support for deployment of charging infrastructure, among others. The uptake of the electric car remains highest in China followed by Europe and the United States, but the global stock was only 5 million in 2018 [1]. In the case of electric two-wheelers, China had an estimated stock of 250 million units followed by India (0.6 million) and ASEAN countries [1].

Among other segments, light electric vehicles (LEV) is one of the fastest growing segments of the EV industry. There is no single definition for LEV, as this segment includes small vehicles such as electric bikes, electric scooters, electric rickshaws, electric autos, etc. The electric two-wheelers followed by electric three-wheelers constitute the major chunk of LEVs. Resonating the global trend, electric two-wheeler also dominates the EV sales in India with 0.1 million units sold in 2018–19

C. Sasidharan (✉) · B. Tyagi · V. Rajah
Alliance of an Energy Efficient Economy, New Delhi, India
e-mail: chandana.sasidharan@gmail.com

B. Tyagi
e-mail: bhawna.tyagi34@gmail.com

V. Rajah
e-mail: varunrajah.r@gmail.com

N. Patel et al. (eds.), *Electric Vehicles*, Green Energy and Technology,
https://doi.org/10.1007/978-981-15-9251-5_2

and overall fleet size reaching 0.6 million [1, 2]. This is primarily because two-wheelers are a budget-friendly means of personal transport and preferred for last-mile connectivity. This is also evident from the share of conventional two-wheelers (approximately 80%) in the annual automobiles sales for India [3]. Two-wheelers are also preferred for commercial operation due to the ease of driving in congested areas and low maintenance cost [4]. There is also an uptake of electric two-wheelers by fleet operators in various countries such as Germany, Spain and Italy [1] and also for delivery of goods by e-commerce and food industry.

Additionally, in terms of passenger-kilometres travelled, the proportion of public and shared transport vehicles is highest (such as auto-rickshaws, buses, and cabs) [5]. In India, conventional three-wheelers account for about 28% of the total diesel consumed by the transport sector [6]. The similar trend in terms of EV adoption is also observed in the case of electric three-wheelers whose stock stands around 2.38 million, second largest after China (50 million) [1].

As Asian countries, including India and China, are witnessing the continued challenge of air pollution and expected to deteriorate further with rapid urbanization, electric two-wheelers and three-wheelers provide an opportunity to grow the share of cleaner and efficient vehicles. These vehicles will be at the heart of electrification initiative in these countries. As a result, policymakers in India are also pushing the agenda of clean mobility initiative to reduce the vehicular emissions, reduce the carbon footprint and increase energy security by reducing dependence on petroleum import. With the focus on the large-scale adoption of EVs and to maximize the reduction of vehicle emissions, the present chapter focuses attention on charging of electric two-wheelers and three-wheelers as they are used intensively and their electrification will generate a disproportionate reduction in air pollution.

To boost the adoption of the electric vehicles, charging infrastructure is one of the critical components and requires significant consideration. However, most attention has been so far only on standardization of chargers for electric cars and their infrastructure. LEVs are distinct in design from electric cars and is an innovative EV segment which has attracted a lot of entrepreneurs. The resultant of the exercise is the lack of standardization in EV design, which also spills over to the charging related aspects. As the parameters of the batteries are distinct from cars, the suitability of extending the car changing standards for LEVs is rather limited. Even though LEVs are the largest EV segment in numbers, there is hardly any research that is focused on LEV charging. The present chapter is an exclusive study on two of the major LEV components: two-wheelers and three-wheelers and their charging related aspects. To the best of the knowledge of the authors, there are no precedents providing a holistic overview of LEVs in detail. The chapter is designed as a ready reference to LEVs and their charging aspects.

2 Electric Two-Wheelers Market Review

2.1 Regional Overview of the Market

The share of two-wheeler in the automobile segment is dominant primarily in Asian countries. However, electric two-wheelers adoption is led by China and India, among other Asian countries. China dominates the market both in terms of production and stock of vehicles in operation [1]. The penetration is however modest in other Asian countries such as in Indonesia fleet size of electric two-wheelers is around 3000, in Thailand full-size electric two-wheelers stands at 1000 and approximately 6000 in Vietnam [1]. Their presence is also limited in the European countries mainly through the route of rental schemes dominated by two operators Coup and Cityscoot [1].

The uptake of electric two-wheelers in China and India is largely stimulated by fiscal incentives and policy interventions. In the case of China, electric two-wheelers are exempt from vehicle registration which is further extended to 2020 and some cities also introduced a ban on conventional two-wheelers in urban areas [1, 7]. A rebate of USD 23–382 for electric scooters and motorcycles was granted to models with maximum power not exceeding 250 watts under the first phase of Faster Adoption and Manufacture of (Hybrid and) Electric Vehicles (FAME) scheme in India [8]. However, the second phase of FAME limits the demand incentive only to fully electric two-wheelers with advanced battery based on the battery capacity [9].

2.2 Market Segmentation

2.2.1 Types of Electric Two-Wheelers

The electric two-wheeler market is broadly categorized into e-scooters, e-bikes, e-motorcycles, e-foot scooters, mopeds, others. The penetration of e-bikes is primarily concentrated in China and Japan, but remains modest in other Asian countries [1, 10]. EV market is dominated by e-scooters (primarily of two types- e-scooty and e-scooters) in case of India. E-scooty is considered as those electric scooters which are gearless and the majority of variants are available in this category. The price of models varies from USD 265 to 1910 [11]. Presently, more than 25 companies are selling electric two-wheelers in the Indian market [11, 12]. However, there are very few models at present available in the category of e-motorcycles in India.

2.2.2 Battery Type and Capacity

The single biggest component impacting the price of EV is the cost of the battery which primarily depends on the battery type. Based on the battery type, the electric vehicle market can be segmented into sealed lead-acid and lithium-ion battery. Lead-acid batteries are less costly compared to lithium-ion but have a shorter lifespan and require regular maintenance in comparison to lithium-ion. The electric two-wheeler market was initially dominated by models with lead-acid battery due to the low-cost of lead-acid batteries. However, with the restriction of demand incentive to only models with advanced batteries under phase II of FAME scheme, the proportion of models with lithium-ion battery started increasing [9]. Presently, more than 60% of the models offered in the market have lithium-ion batteries [11, 12]. In case of India, the battery capacities of electric two-wheelers, primarily e-scooters and e-motorcycles, varies from 0.31 to 3 kWh with an outlier of 6 kWh as shown in Fig. 1 [11, 12]. There are limited models in the category of e-motorcycles with smaller battery capacity in case of India approximately around 3 kWh. However, globally, e-motorcycles are available with battery size varying from 7.7 to 21.5 kWh [13].

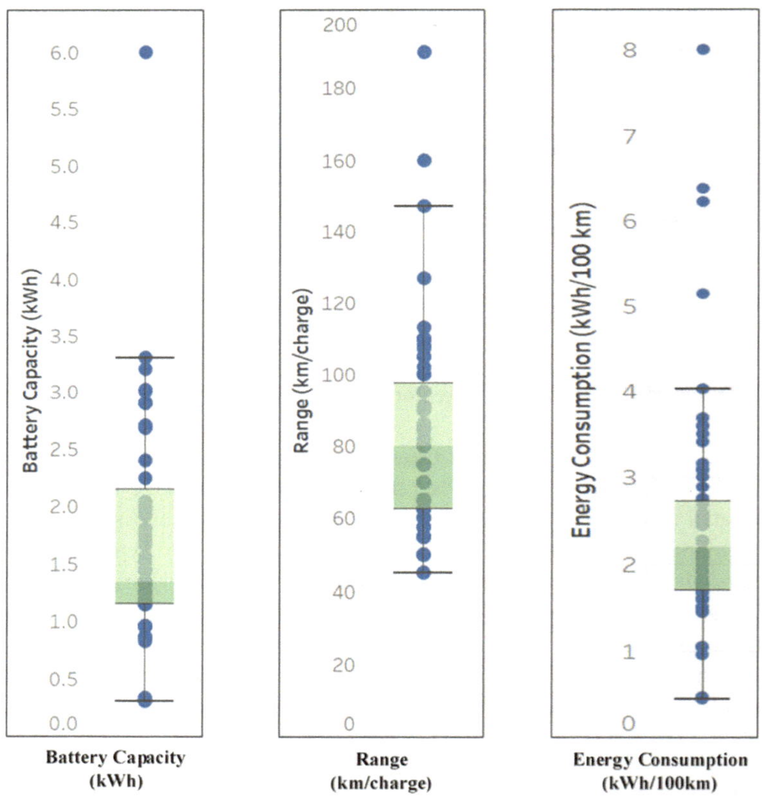

Fig. 1 Battery capacity, range and energy consumption of electric two-wheelers (*Source* Author Analysis)

2.2.3 Speed, Range and Energy Consumption

Range anxiety is one of the deterring factors hampering the purchase of an electric vehicle compared to their conventional counterpart. The range of an electric vehicle largely depends on the battery capacity. Presently, the range of the models offered in the market varies from 45 to 190 km as shown in Fig. 1 [11, 12]. However, the actual range varies depending on factors such as ambient temperature, traffic conditions, driving style, etc. Initially, the majority of the models offered were low speed and short-range vehicles. The market has started shifting towards vehicles that are comparable with the conventional vehicles particularly in the motorcycle segment with the range more than 100 km [1, 11, 12]. On an average 85%, two-wheeler owners in rural areas and 75% two-wheeler owners in urban drive their vehicles within 20 km distance [14]. Thus, the range offered by models available in the market is sufficient to meet the requirements of the customer. The speed of the models varies from 25 to 80 kmph [11, 12]. Unlike the fuel efficiency in conventional vehicles, energy efficiency is estimated for electric vehicles. The ratio between battery capacity and range is considered as a proxy indicator for energy efficiency. Based on our estimates, energy efficiency varies from 0.45 kWh/100 km to 8 kWh/100 km as shown in Fig. 1 [11, 12].

3 Electric Three-wheelers

3.1 Regional Overview of the Market

Three-wheelers are used as one of the major modes of public transport primarily in Asian countries [15–17]. They referred by different names in different countries such as auto-rickshaws, tuk-tuks, trishaws, autos, rickshaws, ricks, tricycles, among others [16]. Following the increasing use of three-wheelers for public transport and their contribution to pollution (as have two or four-stroke gasoline engines), they have been converted into LPG, and CNG such as autos in Delhi run on CNG, tuk-tuks in Bangkok run on LPG [15, 16]. Due to low speed and less distance covered in a day, three-wheelers became the ideal candidate for electrification. Resonating the trend of two-wheelers, the penetration of electric three-wheelers is highest in China and India, among other Asia-Pacific countries. In 2018, stock of electric three-wheelers exceeded 50 million in China and about 2.38 million in India [1]. A study by CRISIL forecasted that by 2024 approximately 50% of the three-wheelers sold in India will be electric [18]. The Asia-Pacific electric three-wheeler market is projected to reach USD 11,935.1 million by 2023 owing to low ownership cost, the decline in battery prices, and favourable government policies [19].

In the case of India, the adoption of electric three-wheelers was also stimulated by the demand incentive provided by the government under FAME scheme. The scheme granted a rebate of USD 45–800 for mild hybrids, plug-in and battery electric three-wheelers [8]. However, the second phase of FAME limits the demand incentive only to fully electric three-wheelers with advanced battery [9].

3.2 *Market Segmentation*

3.2.1 Types of Electric Three-Wheelers

The electric three-wheelers are broadly of two types- e-rickshaw and e-auto. The basic difference between the two is that e-autos have regenerative braking system compared to e-rickshaw. Based on their usage, they can be further classified into-passenger carrier and a load/goods carrier. There is another category of the electric vehicle, called quadricycle, which is seen as an alternative to three-wheelers primarily owing to less stringent requirements. However, its usage is limited in India. In the electric three-wheeler segment, the penetration of e-rickshaw is highest as compared to e-auto in India. They are mainly preferred for intra-city public transport for shorter distances and resolved the issue of drudgery for rickshaw pullers. At present, there are more than 80 variants available in the category of e-rickshaw both for passenger and goods carrier category [1]. However, the number of models for e-autos are limited in number and costly compared to their conventional counterparts. The price of e-rickshaw varies from USD 650 to 2000, while in the case of e-autos it varies from USD 2500 to 3500 [11].

3.2.2 Battery Type and Capacity

Based on battery type, electric three-wheelers are mainly available into two variants a sealed lead-acid and lithium-ion battery. More than 65% of the models in the market are presently available with lead-acid batteries as they are less costly compared to lithium-ion [11, 12]. However, variants of e-autos are primarily available with lithium-ion batteries, thus, making them costly to their conventional counterparts. Restriction on demand incentive only for models with advanced batteries under phase II of FAME scheme stimulated the market towards lithium-ion batteries [9]. Presently, more than 60% of the models offered in the market have lithium-ion batteries [11, 12]. The battery capacities of e-rickshaws vary from 1.2 to 8 kWh for passenger and the good carrier category as shown in Fig. 2 [11, 12]. In the case of e-auto, the battery capacities vary from 3.84 to 7.37 kWh for passenger and good carrier category [11, 12].

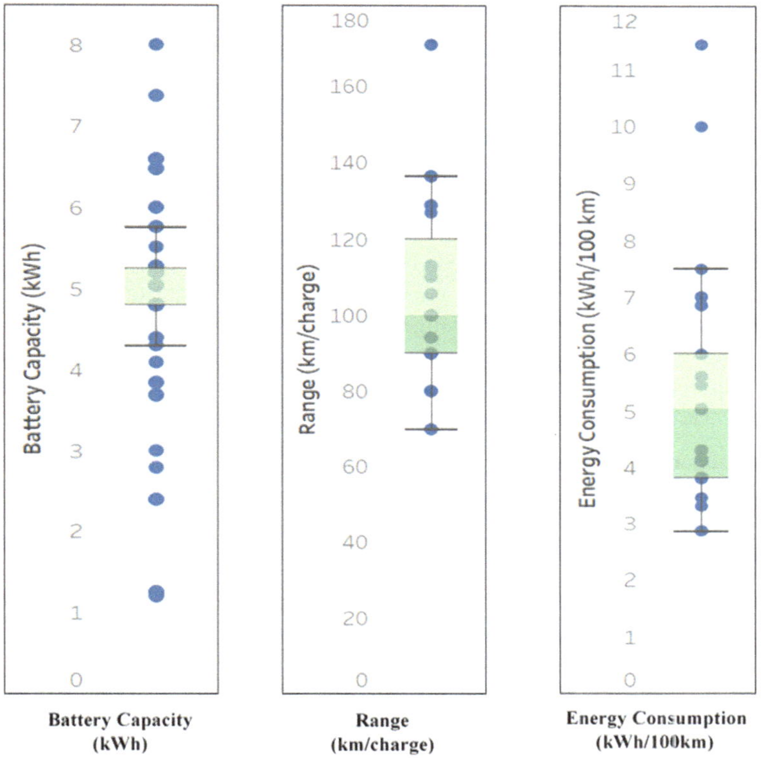

Fig. 2 Battery capacity, range and energy consumption of electric three-wheelers (*Source* Author Analysis)

3.2.3 Speed, Range and Energy Consumption

As electric three-wheelers are used for shared public transport and to carry goods, the range per charge becomes a critical factor. Time spent in charging during the day results in revenue loss for the driver. The range of an electric vehicle largely depends on the battery capacity, but could be impacted by factors such as ambient temperature, traffic conditions, driving style, etc. Presently, the range of the e-rickshaw varies from 70 to 140 km and for e-autos from 90 to 170 km as shown in Fig. 2 [11, 12]. In the case of India, e-rickshaw has a maximum speed of 25 kmph and e-autos with speed around 40–55 kmph. In terms of energy efficiency, calculated as the ratio between battery capacity and the range varies from 2.86 kWh/100 km to 11.43 kWh/100 km as shown in Fig. 2 [11, 12].

4 LEV Charging

4.1 Distinct Features of LEV Charging

EV charging is replenishment of energy inside the battery. All EVs have re-chargeable batteries or ultra-capacitors for energy storage. More often LEVs use lithium-ion or lead-acid batteries. LEVs are diverse in their design from electric cars, especially with respect to smaller battery sizes. Though LEV batteries are smaller in capacity than electric car, they are still the most expensive component on the vehicle. The distinct design features cause challenges and provides opportunities for charging of these vehicles. The first advantage with LEV batteries is that the individual charging power requirement is small. The following characteristics make LEVs unique in their charging needs:

- Battery packs are small with voltages under 120 V
- No active cooling mechanism for batteries.

The challenges for LEV batteries are that the chargers need to be designed to suit the battery voltage. There is no standardization of the battery pack voltage, and it could be anywhere between 48 and 72 V. The established charging standards for EVs are designed for battery packs with high voltages. Hence those standards cannot be applied for LEV batteries. As LEV batteries do not have active cooling mechanisms, charging has both shock and fire hazard. In China, there were cases of fire due to the charging of LEV batteries inside apartments. This led to a policy initiative to discourage home charging and promote public charging with safety regulations. Efforts have been made to develop safety standards EN 50604 and ISO 18243. The next sub-section examines battery parameters in detail [20–22].

4.2 Battery Parameters

A battery is an electrochemical storage and the set of parameters that has an impact on its selection, charging, and thermal management are explained in detail in Table 1.

Table 1 Summary of major battery parameters (*Source* Author Analysis)

Battery chemistry	Nominal voltage (V)	Gravimetric density (Wh/kg)	Volumetric density (Wh/1)	Efficiency (%)	Charging temperature °C
Lead acid	2	30–40	80–100	50–85	−20 to +50
Nickel metal hydride	1.2	70–95	180–220	65–95	0 to +45
Lithium Ion	3–3.7	120–250	200–400	95–99	0 to +45

- **Battery chemistry**: The common chemistries observed in EV batteries are Lead-Acid, Nickel Metal Hydride and Lithium-ion chemistries. Globally, electric cars use advanced lithium batteries, but within the LEV segment, Lead-Acid battery chemistries are also common [20].
- **Battery configuration**: Three design variants of lithium-ion batteries are—cylindrical, pouch and prismatic configurations design, which has an implication on the thermal management of the battery. Battery pack configuration, i.e. the number of cells in series and parallel, has an implication on the voltage and current capacity of the battery pack [20].
- **C-rate**: This is the rate at which a battery is discharged or charged with respect to its maximum capacity. For every chemistry and design type, battery packs have maximum c-rate. High c-rate implies that the batteries can be charged faster. It is calculated as the ratio of battery capacity (in Ah) to charging current. 1C, 0.5C and 2C rate means to discharge or charge the battery in 1 h, two hours and half an hour respectively [23].
- **Battery Capacity**: Battery capacity is generally specified as the total ampere-hours (Ah) available at a specific c-rate. Capacity can also be specified in Watt-hour (Wh) by multiplying Ah capacity with the nominal voltage. It should be noted that battery capacity decreases as c-rate increases [23].
- **Energy Efficiency**: The energy efficiency of a battery is calculated as the ratio of output energy discharged from the battery and to the input energy used for charging. It is also possible to determine the coulombic efficiency of a battery as the ratio between the discharge and charge capacity [24].
- **Battery Voltage**: Battery voltage is dependent on the electrochemistry of the battery pack, and the battery configuration. When a battery is getting discharged its terminal voltage decreases, and the decrease depends on the chemistry of the battery. Similarly, the terminal voltage of cells increases when it gets charged [23–25].
- **State of Charge**: SOC indicates the remaining capacity of a battery, and is expressed as a percentage of the maximum capacity. SOC determines the amount of charging energy needed and is impacted by operating conditions, especially temperature. For example, typically the charging of batteries is undertaken from 20% SOC to 90% SOC, which implies that the energy needed is 70% of the battery capacity [23, 24].
- **Depth of Discharge (DOD)**: DOD is the inverse of SOC, and is used to indicate the percentage of the total battery capacity that has been discharged. For deep-cycle batteries which are commonly used in EV, the DOD is 80% or more [23, 24].
- **Specific Energy**: The specific energy of a battery can be expressed as gravimetric density: per weight, or volumetric density: per volume. Both are important for LEVs, as space and weight are dependent on vehicle design. The specific energy of a battery determines the size of a battery need for the propulsion of the vehicle. In LEVs, the battery size and weight also have an implication on removing batteries for charging. Lithium chemistries help in the reduction of weight and space for batteries [25].

- **Charging temperature**: The temperature at which the batteries can be charged depends on the battery chemistry. This is an essential consideration for LEVs as there are no active cooling mechanisms in these batteries [25–30].

4.3 Battery Charging Basics

The basic charging mechanism is the same for LEVs similar to other EVs. Charging happens in two steps, first the rectification of grid AC power, and then battery current regulation. So, every charger has an AC/DC rectifier with power factor correction (PFC), and a variable DC/DC converter as shown in Fig. 3 [25, 26].

Based on the size and corresponding weight, the charger units could be either fixed or portable. In the case of LEVs, the charger is quite small that it is often part of a portable cable connector like a laptop charger. While most low power chargers are installed on the EV itself, the high-power chargers are part of the charging power outlet. Charging power outlets, called Electric Vehicle Service Equipment (EVSE), also houses the communication and protection circuitry for charging process apart from the power connection. Based on the charger configuration and type of output, the EVSE could be an AC EVSE or a DC EVSE. Power electronic components are a part of both the charging circuitry and the power train of a LEV. The drive train typically consists of a bi-directional DC/DC converter (shown in Fig. 4) for the battery which can be used to charge batteries. Using the bi-directional converter as the charger helps to reduce the weight and space required for the charger in LEVs. Many researchers have also proposed solutions using the motor converter during the regenerative breaking for charging batteries, but it is not a common solution for LEVs [25–28].

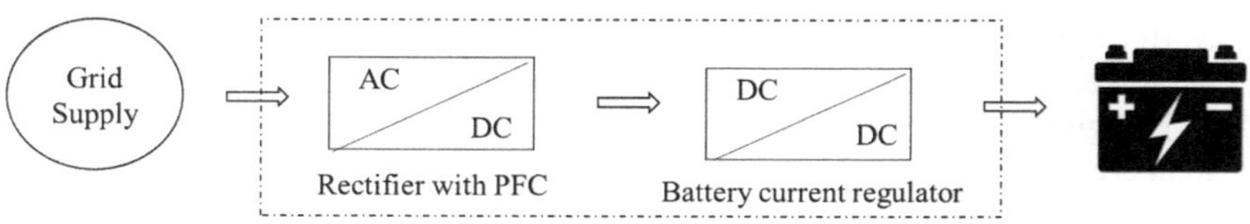

Fig. 3 Block diagram of a basic EV charger (*Source* Author Analysis)

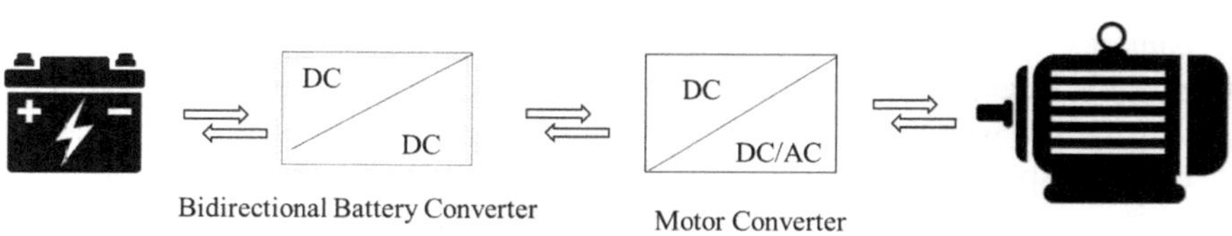

Fig. 4 Block diagram of an EV drive train (*Source* Author Analysis)

4.4 Charging Control Methods

The two most common charge control methods used in EV charging are constant current (CC), constant voltage (CV) and constant power (CP) as shown in Fig. 5. The most common charging for LEVs is slow charging, performed in the CC mode, where a low current level is set such that the battery can be charged fully in an extended period of time. These current levels could be set as low as 10–30% of the rated current of the battery. It is essential to ensure that charging gets cut-off timely, as overcharging does not lead to overheating of batteries [25–28].

In the case of fast charging, either a combination of CC and CV modes or pulse charging is used. In the combined CC/CV mode, first CC charging is performed and after the voltage of the battery pack reaches a certain level the mode is switched to CV. In the pulse charging mode, the battery is charged in a series of current pulses with rest periods as shown in Fig. 6. The speed of charging can be controlled by the width of the pulse. The rest period allows time for stabilization of batteries and at times, negative pulses are applied during the rest period to aid in the depolarization. Other methods of charging are taper charging and trickle charging. Taper charging is a variant of CV charging where charging current reduces as cell voltage increases. Trickle charging is when a very small current is done to compensate self-discharge and is applicable to lead-acid batteries [25–28].

Researchers are also focusing on the advanced closed loop charging technique called the constant temperature (CT) charging. This charging technique offers a lot of advantages in comparison with the CC charging, especially because it uses

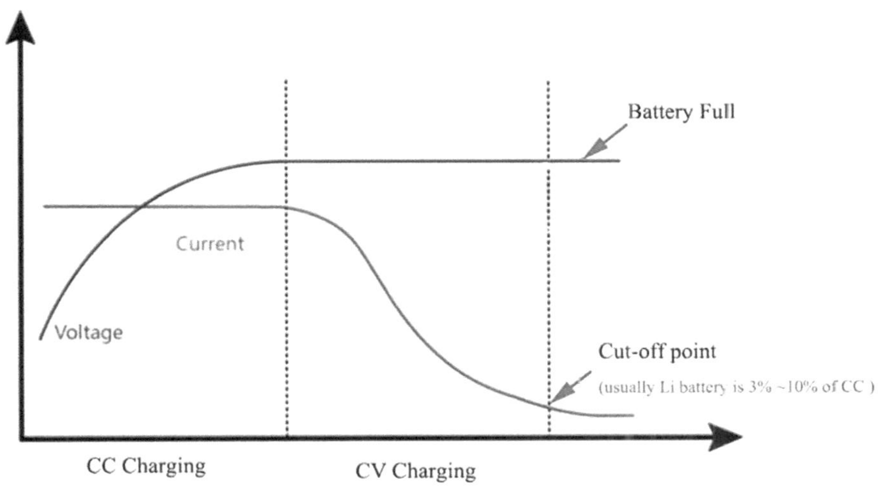

Fig. 5 CC and CV charging methods for EV batteries (*Source* Author Analysis)

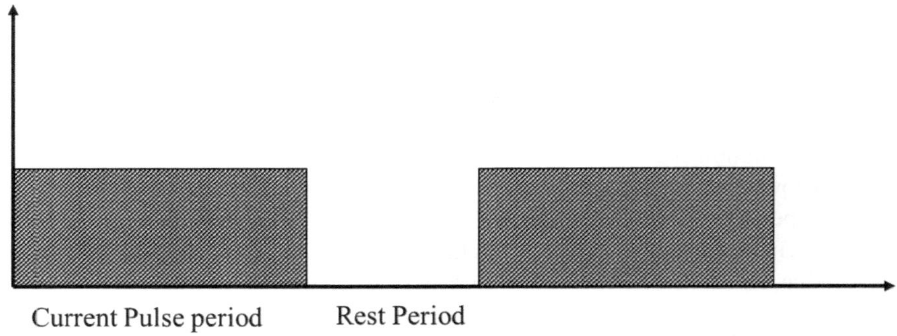

Fig. 6 Pulse charging for EV batteries (*Source* Author Analysis)

temperature to modulate charging. In this charging method, the magnitude of the charging current can be adjusted with response to the environment the charging is happening. This makes it highly relevant for LEVs, as it can provide faster charging within temperature constraints [25].

4.5 Charging Technologies for LEVs

Typically, all batteries are charged via a hardwired connection between the battery, charger and the EVSE. This type of charging can be performed for LEVs while the batteries are inside EV or by removing batteries from the EV as shown in Fig. 7. The

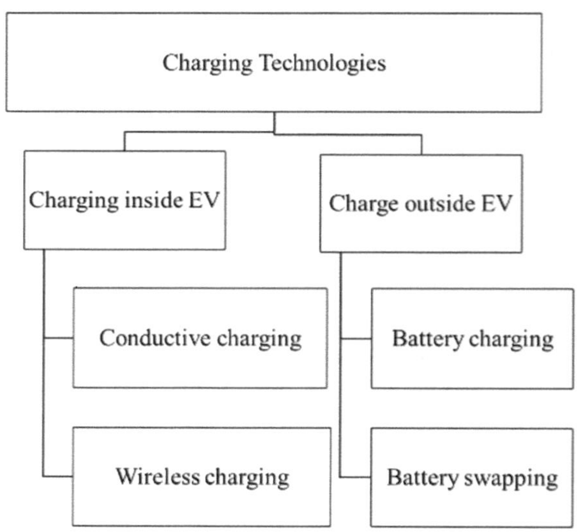

Fig. 7 Charging technologies for LEV batteries (*Source* Author Analysis)

first type of charging, with batteries inside the EVs with a hardwired connection is called as a plug-in or conductive charging. Conductive charging is the most common and popular method for charging LEVs [26–31].

The second option is critical when it comes to LEVs because thermal management of batteries is an important consideration during charging. The charging process is a chemical reaction that generates a lot of heat as by-product, and the heat generated is proportional to the rate of charging. Faster the charging process, more the heat is generated. In the case of an electric car getting charged, its HVAC system comes on and assists in the thermal management of the battery. The ability to charge faster is limited by the thermal management capability for the LEVs without on-board cooling mechanisms [27–31].

But as their battery packs are smaller and lighter, it is easier to remove them from the vehicle for charging. This is particularly important for EV deployment in sub-tropical and tropical regions, with relatively high ambient temperature and humidity. Once the batteries are removed from the vehicles, it is possible to charge them faster in an air-conditioned environment, provided the battery chemistry allows charging at a faster rate. The batteries can be charged individually or can be replaced with a charged battery. The second type of operation where a discharged battery is replaced with a charged spare battery is called battery swapping. The EV batteries can also be charged inside the vehicle wirelessly via induction. However, these systems require EVs to have receiver coils for receiving the energy transferred wirelessly. There are no reported commercial applications for wireless charging for LEVs, researchers have considered this as an option [26–31].

4.5.1 Plug-in Charging

The easiest and most popular way of charging a LEV is through a wired connection to an EVSE. This type of charging is commonly AC charging performed with the help of portable or on-board chargers. The AC EVSEs needed for this type of charging are relatively simpler and could be done with normal domestic or industrial plug points of the appropriate capacity. The charging could be slow or fast charging based on the c-rate. The charging time for fast charging is typically not less than two hours, and is performed with the help of advanced chargers with CC/CV functionality. The slow charging time is anywhere between 4 and 10 h, and is commonly CC or taper CV charging. It should be noted that the charging time for lead-acid batteries are at least twice of the charging time for lithium-ion batteries, as the charging rates are typically slow [26–28].

It is also possible to do DC charging of LEVs, in case charging is done through off-board chargers. This is not a common technology, however, more and more LEV manufacturers are trying to perform DC charging for two-wheelers. However, the standard for LEV and conductive charging of LEV—IEC 61851-3 are under development. Only a few manufacturers, including Harley Davidson, use the combined charging system protocol for DC charging [32, 33].

Table 2 Sample calculation of charging power and energy required at different c-rates (*Source* Author Analysis)

Battery capacity (Ah)	Battery voltage (V)	C-rate	DC current required (A)	Power required (kW)	Grid AC voltage (V)	Grid AC current (A)
50	48	0.25	12.5	0.7	230	4
50	48	0.5	25	1.5	230	7
50	48	1	50	3	230	14
50	72	0.25	12.5	1.1	230	5
50	72	0.5	25	2	230	11
50	72	1	50	4	230	21
100	48	0.25	25	1.5	230	7
100	48	0.5	50	5	230	22
100	48	1	100	6	230	28
100	72	0.25	25	4.5	230	22
100	72	0.5	50	4	230	21
100	72	1	100	17.5	415	26

For plug-in charging the power of the charger is dependent on the Ah capacity of the battery, the rate of charging, and the voltage of the battery pack. A sample calculation is illustrated and results for different battery Ah and voltages are presented in Table 2. The steps for a 48 V battery bank of 50 Ah capacity are presented below.

- The DC voltage rating for the charger will be closer to the battery voltage. It can be estimated using a $\pm$ 10% window around the battery pack voltage and in this case, it would be around 43–53 V.
- The next step is to calculate the charging current based on the c-rate. The charging current can be estimated by multiplying the capacity with the charging current. For 0.25 C rate, DC charging current will be 12.5 A.
- The charging power can be estimated using the upper limit of charging voltage and the charging current. Using a power factor of 0.9, the charging estimated for this case is 0.7 kW.
- The grid supply needed for this charger from a 230 V AC supply is 4 A.
- If the battery bank needed to be charged at 1C rate, the charging voltage remains the same, but the DC charging current and power needed will be 50 A and 3 kW respectively. The grid connection needed for this is 230 V, 16 A.
- If the battery bank was of 72 V, 50 Ah, then the charging current needed would remain the same at the corresponding charging c-rate. But the charger power will increase to 1.1 kW.

4.5.2 Battery Swapping

In case, the LEVs batteries are removable, their batteries can be swapped with other batteries. The batteries removed can be charged individually or in a centralized operation in a swapping facility. The same charging methods described in sections above are applicable to the charging aspect of battery swapping. However, charging can take place in a controlled environment, and hence faster charging can be achieved in swapping. Sun mobility has established battery swapping facilities in India where batteries can be charged within an hour in an air-conditioned environment, which is twice as fast as other swapping service providers. This is the best option to fast charge any battery while ensuring monitoring of the health of the battery [34].

In terms of economics, the cost of chargers of swapping and plug-in charging are comparable. Battery swapping has two other important aspects, which has a significant implication on the cost of service, the cost for the battery swapping mechanism and the cost of the spare batteries. In the first aspect of the swapping, there is an advantage for LEVs. Manual swapping is possible for LEVs if the battery packs are light enough to be lifted by a person. Gogoro operates swapping facilities for electric two-wheelers with specifically designed batteries of the right weight [35].

Hence for LEVs, the choice of battery chemistry with the right specific energy becomes important. Lithium-ion batteries are available with specific energies between 120 and 250 Wh/kg. A 2 kWh lithium-ion battery could weigh anywhere between 8 and 16 kg, i.e. one could be twice the weight of another. However, the cost of batteries increases with specific density, and the choice will depend on the optimization of both cost and weight. Three-wheeler batteries are of higher capacity and are heavier, and might require the assistance of hydraulic tools to help in the lifting process. One way to tackle the weight problem is to split the battery bank into multiple battery packs of optimal weight [34].

The next most important cost for the swapping aspect is the number of spare batteries that are needed. Batteries are the most expensive component of the electric vehicle and the cost of the spare batteries is substantial. Optimal case of swapping station operation is with the least amount of spare batteries, and for that, the charging time of the batteries need to be as low as possible. There is a lot of research ongoing on battery swapping operation with the focus on optimization of costs. Battery swapping stations are potential grid storage facilities, and many researchers have focused on the grid services battery swapping stations can offer [36–39].

5 Challenges, Best Practices and Recommendations

In India, China & other South-East Asian countries, many challenges are prevailing with the existing LEV models such us poor pick up, low top speed, old battery technology and recharging continue to be top barriers in the uptake of LEVs. Globally, governments have adopted a variety of policy instruments for the promotion of EVs. However, LEV is a segment that has often received less attention in comparison with electric cars and buses. The key focus of this section is to draw recommendations and highlight the best practices from the analysis done in the earlier sections.

5.1 Safety and Thermal Management of Batteries

Enabling fast charging and reducing recharging time is what manufactures and consumers aspire, however, for LEVs battery thermal management is a limitation. At present, there are no active thermal management systems in the majority of LEVs. If the thermal management system fails, the temperature of the cells could rise and potentially send the cells into thermal runaway. So far, there is no formal standard that has been developed for safety aspects of LEV operation and charging in extreme temperature. These battery thermal considerations need to be addressed before the wider adoption, especially in tropical climate conditions where temperature and humidity play a crucial role in the performance of the batteries. Phase changing materials has a huge role to play in the thermal management of LEVs [39, 40].

5.2 Battery Swapping

Battery swapping technology offers the best alternative for fast charging especially for the LEV segment. For instance, Sun mobility offers quick interchange station for LEVs wherein batteries are charged at 1C rate in an air-conditioned environment. Taiwanese company Gogoro also offers fast charging for two-wheeler batteries. With faster charging the number of spare batteries needed, and the cost of the swapping operation can be brought down. The swapping stations can also offer grid support services.

5.3 Charging Techniques

Currently, most of the chargers for LEVs uses charging techniques which are not considerate of the temperature. As thermal management is the most important consideration for LEVs, advanced charging techniques with closed feedback for temperature monitoring becomes important. Alternatively, charging modes which enable over-voltage protection also becomes important.

5.4 Solar Charging

The charging requirements of LEVs are such that it can be effectively integrated into Solar Photovoltaic power. In the workplace and other public charging, chargers for LEVs can be integrated with the rooftop generation. Locally generated DC power from solar PV systems can be effectively used to charge LEVs. This also helps to reduce the losses in the rectification of AC power that is incurred in other chargers [40].

5.5 Vehicle to Grid Integration

LEVs are not considered as the ideal candidates for smart charging or feeding back to the grid. Firstly, the charging of LEVs was not controllable, and the chargers were not bi-directional. However, using smarter chargers, and aggregation LEVs also deserve consideration as grid support reserves. On the other hand, battery swapping facilities can be integrated easily as distributed grid storage systems.

References

1. IEA, *Global EV Outlook 2019: Scaling-up the Transition to Electric Mobility* (International Energy Agency (IEA), France, 2019), p. 2019
2. JMK Research & Analytics: Monthly EV update-April 2020. Haryana (2020)
3. SIAM, *Automobile Domestic Sales Trend in India* (Society of Indian Automobile Manufacturers (SIAM), New Delhi, 2020)
4. MECA, *Emission Control of Two- and Three-Wheel Vehicles* (Manufacturers of Emission Controls Association (MECA), 2014)
5. Delhi EV policy
6. PPAC and Nielsen, *All India Study on Sectoral Demand of Diesel & Petrol* (Petroleum Planning & Analysis Cell (PPAC), Ministry of Petroleum & Natural Gas, Government of India, New Delhi, 2013)
7. GlobeNewswire, Asia Light Electric Vehicle Market Set to Reach USD 52.8 billion by 2024, Observing a CAGR of 7.0%: VynZ Research. (2019, April 22). Retrieved from https://www.globenewswire.com/news-release/2019/04/22/1807410/0/en/Asia-Light-Electric-Vehicle-Market-Set-to-Reach-USD-52-8-billion-by-2024-Observing-a-CAGR-of-7-0-VynZ-Research.html
8. Ministry of Heavy Industry and Public Enterprises, *Notification—Scheme for Faster Adoption and Manufacturing of (Hybrid &) Electric Vehicles in India* (FAME India. New Delhi, 2015)
9. Ministry of Heavy Industries and Public Enterprises, *Publication of notification in Gazette of India (Extraordinary) regarding Phase-II of FAME India Scheme* (The Gazette of India. New Delhi, 8 Mar 2019)
10. ADB and SIDA, *Electric Two-Wheelers in India and Viet Nam—Market Analysis and Environmental Impacts* (Asian Development Bank, Philippines, 2009)
11. Primary data collection based on the automobile manufacturers websites selling in India
12. FAME Dashboard, FAME India Scheme Phase II (National Mission on Electric Mobility). Website https://fame2.heavyindustry.gov.in/

13. Primary data collection based on the automobile manufacturers websites selling globally
14. TERI, *Faster Adoption of Electric Vehicles in India: Perspective of Consumers and Industry* (The Energy and Resources Institute (TERI), New Delhi, 2019)
15. S.M. Lukic, P. Mulhall, G. Choi, M. Naviwala, S. Nimmagadda, A. Emadi, Usage Pattern Development for Three-Wheel Auto Rickshaw Taxis in India (IEEE Vehicle Power and Propulsion Conference, Arlington, 2007), pp. 610–616
16. ITPL, *Two and Three Wheelers in India* (Innovative Transport Solutions (iTrans) Pvt. Ltd., TBIU, IIT Delhi, New Delhi, 2009)
17. Climate Technology Centre & Network, Motorised three-wheeler taxis. Retrieved from https://www.ctc-n.org/technologies/motorised-three-wheeler-taxis
18. CRISIL, By 2024, nearly half of new 3-wheelers sold will be e-autos (11 Feb 2020). Retrieved from https://www.crisil.com/en/home/newsroom/press-releases/2020/02/by-2024-nearly-half-of-new-3-wheelers-sold-will-be-e-autos.html
19. Businesswire, Asia-Pacific Electric Three-Wheeler Market Outlook to 2023 (27 Feb 2019). Retrieved from https://www.businesswire.com/news/home/20190227005487/en/Asia-Pacific-Electric-Three-Wheeler-Market-Outlook-2023
20. LEVA, Light Electric Vehicle Technician Training (2019). Retrieved from www.LEVAssociation.com
21. IEA, *Hybrid and Electric Vehicles: The Electric Drive Hauls* (Technology Collaboration Programme on Hybrid and Electric Vehicles, 2019)
22. IEA, *Hybrid and Electric Vehicles: The Electric Drive Automates* (Technology Collaboration Programme on Hybrid and Electric Vehicles, 2018)
23. MIT, A Guide to Understanding Battery Specifications (2008)
24. M. Wagemaker, EV battery parameters, TU Delft lecture series on electric cars (2018)
25. L. Patnaik, A.V.J.S. Praneeth, S.S. Williamson, A closed-loop constant-temperature constant-voltage charging technique to reduce charge time of lithium-ion batteries. IEEE Trans. Industr. Electron. **66**(2), 1059–1067 (2018)
26. M. Yilmaz, P.T. Krein, Review of battery charger topologies, charging power levels, and infrastructure for plug-in electric and hybrid vehicles. IEEE Trans. Power Electr. **28**(5), 2151–2169 (2012)
27. H. Shareef, M.M. Islam, A. Mohamed, A review of the stage-of-the-art charging technologies, placement methodologies, and impacts of electric vehicles. Renew. Sustain. Energy Rev. **64**, 403–420 (2016)
28. K. Morrow, D. Karner, J. Francfort, *Plug-in Hybrid Electric Vehicle Charging Infrastructure Review* (US Department of Energy-Vehicle Technologies Program, vol. 34, 2008)
29. W. Wu, S. Wang, W. Wu, K. Chen, S. Hong, Y. Lai, A critical review of battery thermal performance and liquid based battery thermal management. Energy Convers. Manag. **182**, 262–281 (2019)
30. A. Khaligh, S. Dusmez, Comprehensive topological analysis of conductive and inductive charging solutions for plug-in electric vehicles. IEEE Trans. Veh. Technol. **61**(8), 3475–3489 (2012)
31. A.A. Pesaran, Battery thermal management in EV and HEVs: issues and solutions. Battery Man **43**(5), 34–49 (2001)
32. Ather, Ather Grid Fast DC Charging (2019). Retrieved from https://www.atherenergy.com/grid
33. Harley Davidson, *Specs and Pricing 2020 Livewire* (2020). Retrieved from https://www.harley-davidson.com/gb/en/motorcycles/livewire/detailed-specs-and-pricing.html
34. C. Sasidharan, S. Das, *Assessment of Charging Technologies Currently Used to Charge Electric Two and Three Wheelers in India* (India Smart Utility Week, 2020)
35. Gogoro, *Gogoro Network Battery Swapping Platform*. Retrieved from https://www.gogoro.com/gogoro-network/
36. H.Y. Mak, Y. Rong, Z.J.M. Shen, Infrastructure planning for electric vehicles with battery swapping. Manag. Sci. **59**(7), 1557–1575 (2013)
37. M.R. Sarker, H. Pandžić, M.A. Ortega-Vazquez, Optimal operation and services scheduling for an electric vehicle battery swapping station. IEEE Trans. Power Syst. **30**(2), 901–910 (2014)

38. P. You, S.H. Low, L. Zhang, R. Deng, G.B. Giannakis, Y. Sun, Z. Yang, Scheduling of EV battery swapping-Part II: distributed solutions. IEEE Trans. Control Netw. Syst. **5**(4), 1920–1930 (2017)
39. X. Tan, G. Qu, B. Sun, N. Li, D.H. Tsang, Optimal scheduling of battery charging station serving electric vehicles based on battery swapping. IEEE Trans. Smart Grid **10**(2), 1372–1384 (2017)
40. M. Keyser, A. Pesaran, Q. Li, S. Santhanagopalan, K. Smith, E. Wood, S. Ahmed, I. Bloom, E. Dufek, M. Shirk, A. Meintz, A. Markel, Enabling fast charging—battery thermal considerations. J. Power Sources **367**, 228–236 (2017). https://doi.org/10.1016/j.jpowsour.2017.07.009

Plug-In Hybrid Electric Vehicles (PHEVs)

Himanshu Singh, A. Ambikapathy, K. Logavani, G. Arun Prasad, and Saravanan Thangavel

1 Introduction

1.1 About Plug-In Hybrid Vehicles (PHEVs)

Plug-in hybrid electric vehicles are simply a fusion of an electric vehicle and a conventional vehicle (fossil fuel) vehicle whose battery can be charged externally. It has the benefits of both worlds. It is not only energy conserving but also efficient. A plug-in hybrid electric vehicle (PHEV) is a hybrid electric vehicle whose battery can be recharged via external power source as well as by the onboard electric generator. Generally, all PHEVs are made for passenger use, but nowadays, PHEVs are also being implemented in other fields like commercial vehicles, military vehicles, scooter, medical vehicles, etc. Some of the PHEVs range are shown in Fig. 1.

Apart from the features, it incorporates from a conventional car, it also takes up what is best about electric vehicles. A PHEV cuts down the operating cost of the vehicle. It also brings down the emission of harmful pollutants; therefore, making it

H. Singh · A. Ambikapathy (✉) · K. Logavani · G. Arun Prasad · S. Thangavel
Galgotias College of Engineering and Technology, Greater Noida, UP, India
e-mail: ami_ty21@yahoo.com

H. Singh
e-mail: theahmadfaraz@gmail.com

K. Logavani
e-mail: Vani.tulips@gmail.com

G. Arun Prasad
e-mail: arun.prasad@galgotiacollege.edu

S. Thangavel
e-mail: tsaravcse@gmail.com

K. Logavani
Government College of Engineering—Salem, Salem, Tamil Nadu, India

N. Patel et al. (eds.), *Electric Vehicles*, Green Energy and Technology,
https://doi.org/10.1007/978-981-15-9251-5_3

Fig. 1 Compact BMW X models in PHEV range [2]. *Source* electrive.com

environment friendly. In a study, it was found that a normal driver drives for about 50 miles a day on an average. Keeping this study in mind, PHEVs are the best alternative to the conventional vehicles which emit a lot of harmful pollutants in the environment. A PHEV has an engine and also a battery. These two components work in perfect sync to provide the best output [1].

1.2 Terminology

A plug-in hybrid vehicle is made with specific terminology which have a special meaning defining the ability of the vehicle. It is represented with PHEV (km s), and the number in front of PHEV represents the distance vehicle can travel on battery power alone.

Plug-in Hybrid Vehicles are also known as:

I. GO-HEV (Gas Optional Hybrid Electric Vehicle).

II. Extended Range Electric vehicle [1].

2 The History of PHEVs

To know about plug-in hybrid vehicles, we must have an idea behind it and how it was coined; the history of electric vehicles goes back to the late eighteenth and the early nineteenth century. The first model electric vehicle was attributed to the success of various people from all around the world; in the year 1828, Anyos Jedlik invented an electric motor and created a model car. This was an important and crucial breakthrough as it served as the base of the idea that electricity can also be used as fuel to operate vehicles. Thomas Parker built the first practical electric car in the year 1895. Although the history of PHEVs is span over hundred years; however, the commercial developments took place after the year 2002. All this happened in reference to electric vehicles.

Specifically, talking about PHEVs, Lohner-Porsche Mixte Hybrid as shown in Fig. 2, produced in the year 1899, the first hybrid electric car. Early hybrids could

Fig. 2 Lohner-PorsheMixteHybrid [1]. *Source* Wikipedia

be charged only before the use and they worked until the battery was drained out. Hybrid word was used in reference to the fact that these vehicles could be charged from the wall socket as well. Andrew Frank, a professor UC Davis is called as a "Father of Modern Plug-in Hybrid" [1].

2.1 The Demise of PHEVs

By the invention of electric starters by Charles Kettering, the PHEVs saw their rapid demise. The price of PHEV was very high, their top speed was very low, and they had very limited range. Henry ford was a big name in the industry around that time and he initiated mass production of gas powered vehicle. He brought the prices of the gas powered vehicles down which ensured the demise of all-electric vehicles. At this time, the major concern was not the environment so the common people gave priority to the low initial cost, above normal top speed, low maintenance, above normal powered gas/fuel vehicles [1].

2.2 Revival of Interest

The PHEVs started reviving when Renault started selling Elect'road, which was the plug-in variant of the popular Kangoo. It not only had an unmatched engine, it could also be plugged into electric outlet for recharging purpose. It could be charged up to 95% in 4 h. The car was initially priced at €25,000 and sold in France. It was further upgraded and redesigned in the year 2007 [1].

The second reason for revival of interest in the plug-in hybrid is the increase in gas prices in the USA in the year 2004. For this very reason, some of the plug-in hybrids were converted from the existing hybrids. By the use of Pb-acid battery, the car could go 15 km s in one charge only on electric battery.

Fig. 3 Basic electric vehicle

After this, every company wanted to work on plug-in hybrid projects. Several big players like General Motors, Toyota, etc., invested in this which revived the interest of people back in the PHEVs [1].

3 Comparison

3.1 Basic Electrical Vehicles (BEVs)

Basic electrical vehicles are those which solely run on electricity. They use only electricity as fuel. The emission from tailpipe is zero. They use a special process to recover battery by regenerative braking technique. It can be charged via household socket.

BEVs have high initial cost which is their major drawback. But the positive point is this that the maintenance cost is low. Yet it is still a trouble for common people. Another drawback is the distance covered. BEVs can only travel up to 20–25 km s, once full charge. That too when operated at constant low speed. It has low acceleration; therefore, no above normal initial torque. But they are most environment friendly, as they do not produce any noise pollution or air pollution.

For example: Tesla Model-3, Nissan Leaf, Kia e-niroetc [3] (Fig. 3).

3.2 Hybrid Electrical Vehicles (HEVs)

HEVs are hybrid vehicles that have both an internal combustion engine as well as a battery. It primarily runs on the internal combustion engine when the power requirement is more and in the economy mode, it swiftly shifts to the electric battery. Hybrid vehicles have a very small battery that serves as an add-on to the internal combustion engine and adds some all-electric range to the vehicle. Due to this, the tailpipe emission of the car is brought down and finally aids in reducing air pollution. The battery that is incorporated in the hybrid vehicle can not be charged externally. It can only be charged through a special technique of regenerative braking. Since the battery can not be charged externally by the user; therefore, we can not use the

Fig. 4 Hybrid electric vehicle

battery at its full potential. HEVs majorly depend on the gas/fuel to run the car's engine. Although the emission from the vehicle is reduced by the use of battery, it is significantly high. Hybrid vehicles are also known as self-charging vehicles.

Some HEVs are named as follows: Toyota Corolla Hybrid, Lexus 450 h, Toyota Yaris Hybrid [3] (Fig. 4).

3.3 Plug-In Hybrid Electrical Vehicles (PHEVs)

The "plug-in" word is self explanatory, meaning you just need to plug the car into a charging point so that the battery could be charged. It is a combination of Hybrid Electric Vehicles and Basic Electric Vehicles. It has benefits of both the worlds. Just like BEVs, it can be plugged into an external power source to charge the battery and like HEVs, it has an internal combustion engine with the powerful set of batteries. The PHEVs are designed for several modes of operations keeping in mind the requirement of the customer. In PHEVs, the internal combustion engine and motor/generator works in perfect harmony.

A Plug-in hybrid generally goes about 40–45 km s solely on the battery, and then the internal combustion engine will take over. PHEVs are designed for the people who want to move away from the use of petrol/diesel and also want to stay away from constrains of BEVs. For short trips, you can use only electric power in a PHEV. For bigger trips, you are using electric energy in the start in a PHEV, but then have the luxury of knowing that you can refill in no time up at any gas/petrol station and move ahead with your journey. However, this means that if you drive more than about 45 km s a day on a regular basis, your PHEV will be acting like a normal fuel operated vehicle for all of the distance after 45 km s, and will be polluting the environment.

Among all the PHEVs present today, some are listed as: Mitsubishi Outlander, Volvo XC60 Twin Engine, Toyota Prius plug-in, BMW 225xe, Mercedes-Benz E350 e SE [3] (Fig. 5).

Fig. 5 Plug-in hybrid electric vehicle

4 Construction of PHEVs

A typical plug-in hybrid electric vehicle consists of many small electrical and mechanical components as shown in Fig. 6. Some are discussed in this segment. First and the major component that acts as the heart of PHEVs is the battery. In a plug-in vehicle, the auxiliary battery gives power to drive the vehicle, then the main battery is locked in which controls the vehicle. If there are batteries, then they must be charged. In addition to batteries, there is a charging port. The port permits the car to associate with an external power supply so as to charge the main battery. Thirdly, there is a DC/DC converter. This gadget changes high voltage power to a low voltage DC power from the main pack of batteries to the other components and the auxiliary battery which work in perfect synchronism to run the vehicle smoothly.

Then is the electric generator which generates power from the turning wheel of the vehicle while slowing down, moving the kinetic energy back to the main pack of batteries. This technique is also known as regenerative braking. Few car utilizes the engine that enacts both the functions, i.e., drive and recovery capacities. Then is the electric main motor. This motor uses DC power from the main pack of batteries and

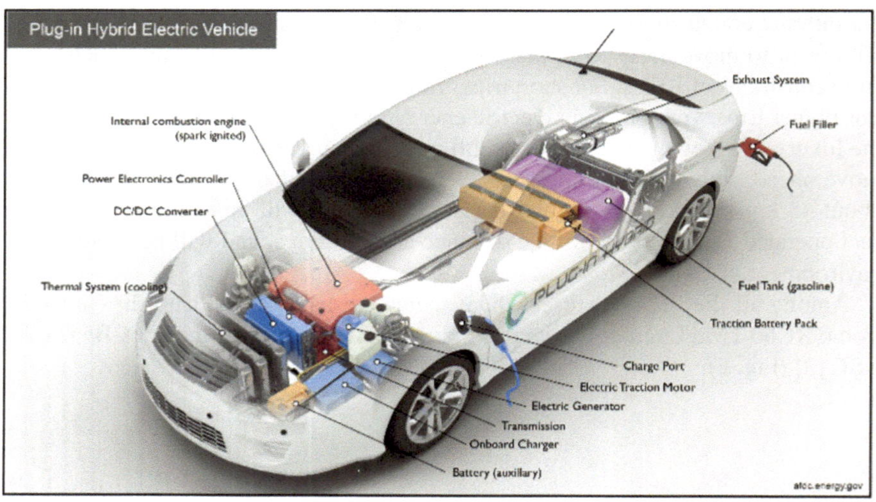

Fig. 6 Basic components of a PHEV [4]. *Source* afdc.energy.gov

transfers it to the transmission control which in turn moves the wheels. This motor is designed in such a way that it works in perfect synchronism with the conventional ICE to provide the desired torque. Then comes the exhaust system. This system improves the pollution reduction mechanism and also improves the efficiency of the car as it emits all the gases coming from the engine through the attached tailpipe. Then is the fuel filler. This is a "spout" used to refill fuel for the next journey of the PHEV. Fuel tank acts as a reserve for fuel that is to be used on demand of the PHEV. The fuel is used only when the main battery level is below a certain point and is not able to keep up with the vehicle's requirement. Then is the internal combustion engine. This design has gases infused in either the admission complex also called as the burning region, here it combines with air, and the air/fuel blend which is ignited from the flash of a spark plug. Following this is the onboard charger which processes the approaching alternating current provided by means of the charging point and changes to direct current power for charging the main battery. It screens battery attributes, for example, V, I, temp, and the remaining percentage of charging. Then is the power electronics controller. These units deal with the progression of energy conveyed by the main battery, commanding the rpm of the main engine, and the torque produced by it. Then is the thermal framework (cooling). This framework keeps up an appropriate working temperature scope of the motor, electric engine, power gadgets, and different segments. Traction battery pack reserves power that is to be used by the electric main engine. Lastly is the transmission. It transfers the power generated in the ICE or electric motor to the wheels of the vehicle [4, 11].

5 Working of PHEVs

Figure 7 shows a basic model of a plug-in hybrid electric vehicle. A PHEV combines an internal combustion engine with an electric motor and an enormous powered battery. In contrast to customary cross breeds, PHEVs can be connected and energized externally, permitting them to work purely on the electric motor. When the battery is finished, the traditional motor comes into play and car works as a regular breed. Since they can run on power from the grid—and on the grounds that power is regular energy source than fuel or diesel. PHEVs do not pollute the environment when running on all-electric mode. Since they have an electric motor and a battery, they add to its eco-friendliness. They utilize less fuel in comparison to the conventional vehicles, which reduces the running cost of vehicles. Although, this cost is not significant when calculated for a day but in the long run, this reduces the overall expenditure on the vehicle. A plug-in hybrid electric vehicle requires a 120 V outlet which is normally found in residential houses which adds to the luxury of the owner.

The cost advantage on the basis of fuel consumption of PHEVs over conventional vehicle attracts the consumers for the adoption of PHEVs. These models add to the luxury of the owner by giving option between all-electric range and hybrid mode. In the hybrid mode, both the electricity and the fuel are used in perfect harmony depending upon the power requirement.

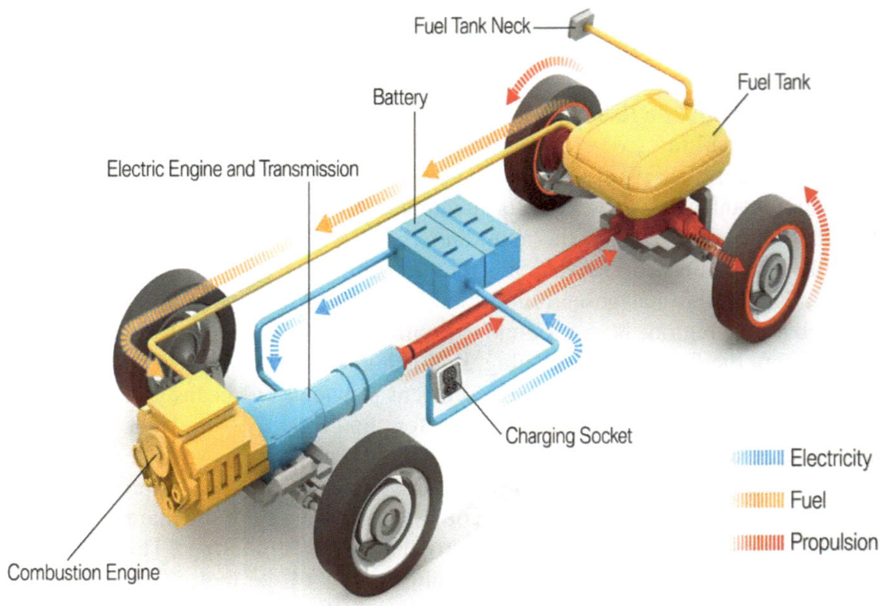

Fig. 7 Model of a PHEV [5]. *Source* bmw.com/en/innovation

Apart from all the above-mentioned advantages of a PHEV over conventional vehicles, there is one more technology that adds to the feature of a PHEVs. This technology is called regenerative braking. The regenerative braking is the terminology used when the battery is recharged while the brakes are being applied.

Various models of module cross-breed electric vehicles may likewise have diverse drivetrains, the mechanical parts that convey capacity to the driving wheels [6, 12].

5.1 Block Diagram of Plug-In Hybrid

See Fig. 8.

5.2 What is a Drive Train?

A drivetrain is the assortment of segments that convey power from a vehicle's motor or engine to the vehicle's wheels. In plug-in hybrid electric vehicles, the drivetrain's plan decides how the electric engine functions related to the regular motor. The drivetrain influences the vehicle's mechanical proficiency, fuel utilization, and running cost. In cross breeds with equal drivetrains, the electric engine and the ignition motor can

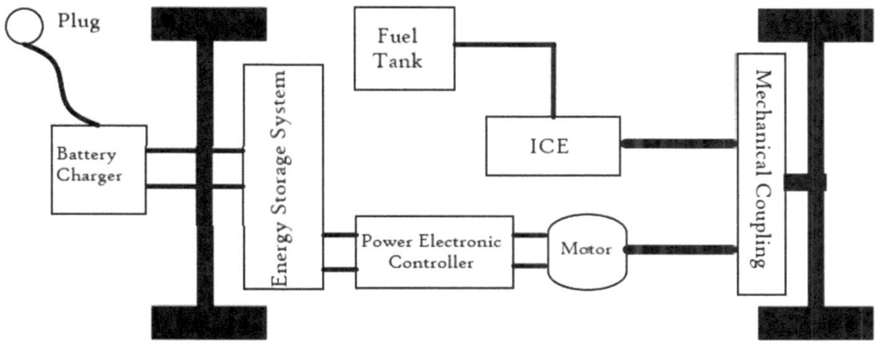

Fig. 8 Block diagram of PHEV

give mechanical force at the same time. Arrangement/equal drivetrains empower the motor and electric engine to furnish power freely or related to each other [6].

5.2.1 Series Drive Train

Arrangement drive trains are the most straightforward half and half setup. In an arrangement crossover, the electric engine is the main source of giving power to the wheels. The electric engine gets this power from two sources. From the main pack of batteries and the generator present in the PHEV. This energy consumption by the electric motor depends directly on the torque requirement. If the torque requirement is more, then ICE comes into play. A PC decides the amount of force that originates from the battery or the motor/generator. Both the motor/generator and the utilization of regenerative slowing down to revive the battery pack.

Arrangement half and halves perform at their best during unpredictable traffic, where fuel and diesel motors are wasteful. The vehicle's PC can select to control the engine with the battery pack just, sparing the motor for circumstances where it is progressively effective [6].

5.2.2 Parallel Drive Train

In vehicles with equal half and half drive trains, the motor and electric engine work couple to create the force that drives the wheels. Equal mixtures will in general utilize a littler battery pack than arrangement drive trains, depending on regenerative slowing down to keep it revived. At the point when force requests are low, equal half breeds additionally use the engine as a generator for supplemental reviving, much like an alternator in ordinary vehicles.

Since the motor is associated straightforwardly to the wheels in equal drive trains, the energy lost due to the mechanical coupling of the wheels with the transmission system increases. This increase in the lost energy, however, does not take out the productivity of the electric engine. An electric engine is highly efficient in go back and forth kind of traffic [6].

5.2.3 Series/Parallel Drive Train

Series/parallel drive trains have both the points of interest and confusion of a series and a parallel drive trains. By consolidating the two plans, the motor can both drive the wheels straightforwardly (as in the equal drive train), and be adequately detached, with just the electric engine giving force (as in the arrangement drive train). The Toyota Prius helped make arrangement/equal drive trains a mainstream plan.

With gas-just and electric-just alternatives, the motor works at close to ideal effectiveness all the more regularly. At lower speeds, it works more as an arrangement vehicle, while at above normal speeds, where the arrangement drive train is less productive, the motor dominates and energy misfortune is limited.

This framework acquires greater expenses than an unadulterated equal half and half since it requires a generator, a bigger battery pack, and additionally processing capacity to control the double framework. However, its efficiency imply that the arrangement/equal drive train can perform better—and utilize less fuel—than either the arrangement or equal frameworks alone [6].

6 Heart of PHEVs-Battery

Battery science has made some amazing progress since in 1800 Alessandro Volta invalidated the normal hypothesis that power must be made by living creatures. The batteries in the past had so small capacity that running an electric vehicle was just a dream. Due to continuous technological advancements, the capacity of the batteries are improved to such a level that electric vehicles are no longer a dream. The batteries are continuously being reduced in size, weight, and discharging rate. Now depending upon the requirement, we have batteries that can suffice a high load for short time or medium load for a longer period of time. To understand this concept of batteries following are some type.

6.1 Sorts of Battery

6.1.1 Li-Ion Battery

Li-particle battery is presently utilized in most versatile customer gadgets, for example, PDAs and workstations in view of their above normal energy per unit mass comparative with other electrical energy stockpiling frameworks. They additionally

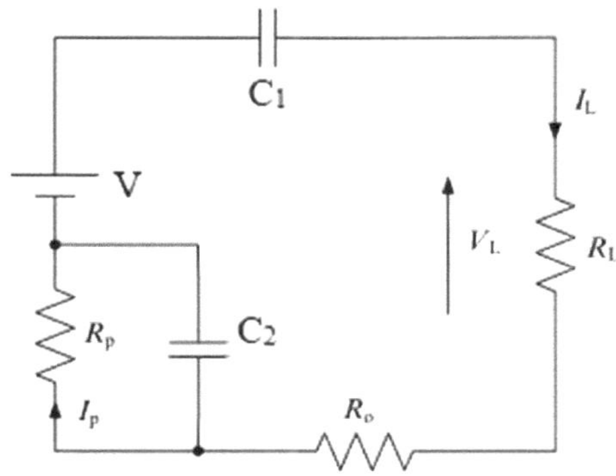

Fig. 9 Li-ion lumped parameter model (LPM)

have a high capacity to-weight proportion, above normal energy proficiency, great above normal-temperature execution, and low self-release. Most parts of Li-particle battery can be reused, yet the expense of material recuperation stays a test for the business. The vast majority of the present module half breed electric vehicles and every single electric vehicle use Li-particle battery. Innovative work is continuous to diminish cost and expand their helpful life cycle [4].

Reactions Involved in a Li-Ion Battery:

- $CoO_2 + Li^+ + \text{electron} \leftrightarrow LiCoO_2$
- $LiC_6 \leftrightarrow C_6 + Li^+ + \text{electron}$
- $LiC_6 + CoO_2 \leftrightarrow C_6 + LiCoO_2$
- $Li^+ + e^- + LiCoO_2 \rightarrow LiO_2 + CoO$
- $LiCoO_2 \rightarrow Li^+ + CoO_2 + \text{electron}$

Lumped Model Of**Li-ion Battery**: Lumped model of a Li-ion battery is shown in Fig. 9.

6.1.2 Ni-Metal Hydride Battery

Ni-metal hydride battery is utilized routinely in clinical hardware and digital cameras. It offers sufficient explicit energy and explicit power capacities. Ni-metal hydride battery is more efficient than Pb-acid battery. These batteries are generally utilized in cross-breed electric vehicles. The primary difficulties with Ni-metal hydride battery are their significant expense, above normal decay and heating age at above normal temperature, and the need to control the lost hydrogen [4].

Reactions involved in a Ni-metal hydride battery:

$H_2O + \text{Metal} + \text{electron} \leftrightarrow OH^- + MH$
$Ni(OH)_2 + OH^- \leftrightarrow NiO(OH) + H_2O + \text{electron}$

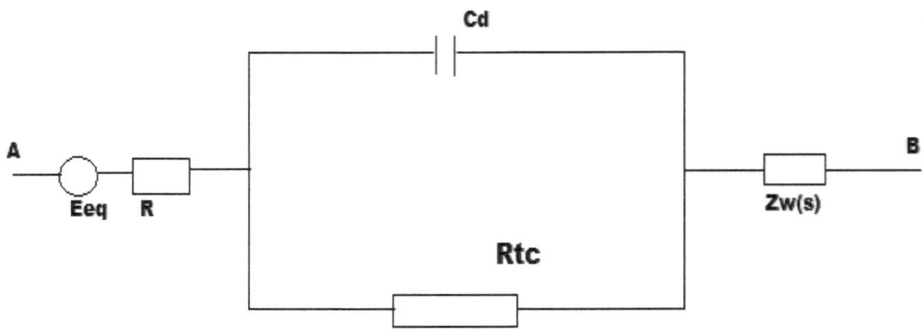

Fig. 10 Electric circuit of Ni-MH cell (equivalent)

Equivalent Circuit Model Ni-Metal Hydride Battery: The equivalent circuit is shown in Fig. 10.

6.1.3 Pb-Acid Battery

Pb-corrosive battery can be intended to be above normal force and are modest, safe, and solid. Be that as it may, low explicit energy, poor cold-temperature execution, and short schedule and cycle life hinder their utilization. Propelled above normal-power Pb-corrosive batteries are being grown, yet these batteries are just utilized in monetarily accessible electrically driven vehicles for auxiliary burdens [4].

Reactions involved in a Ni-metal hydride battery:

- $Pb^{2+} + SO_4^{2-} \rightarrow PbSO_4$
- $PbO_2 + H_2SO_4 \rightarrow PbSO_4$

Equivalent Circuit of Pb-Acid Battery: The equivalent circuit of Pb-acid battery is shown in Fig. 11 (Fig. 12).

7 Charging

7.1 How Would I Charge My Plug-In Hybrid?

The least and the basic requirement for the charging of plug-in hybrid electric vehicle is the household 120 V supply. On the off chance, when the vehicle is parked, it can be parked near an electric outlet or a small electric outlet can be setup near the parking region.

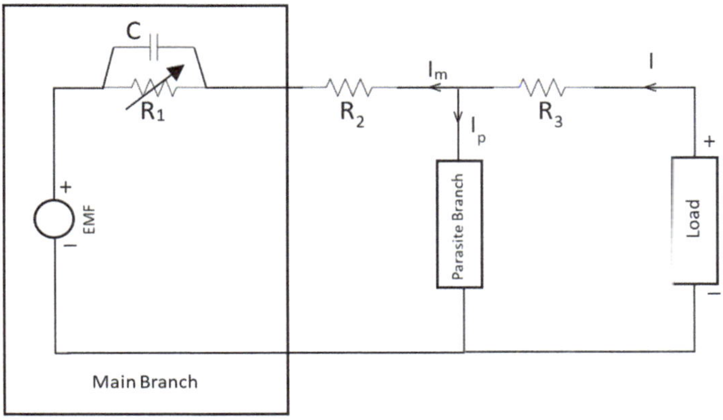

Fig. 11 Equivalent circuit of Pb-acid battery

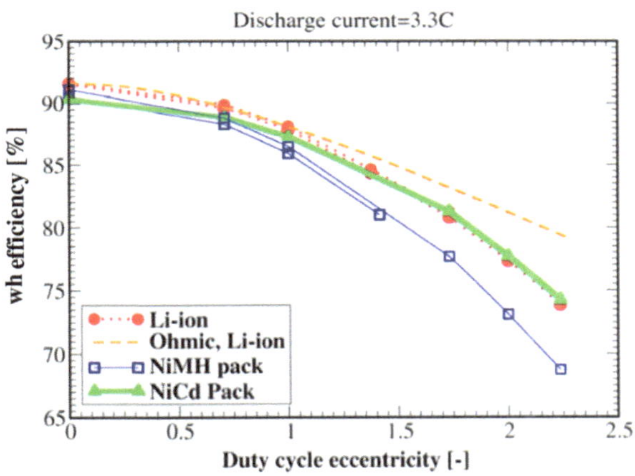

Fig. 12 Discharge cycle [7]. *Source* umanitoba.ca/outreach

In present time, several universities and MNCs have setup electric outlets for the employees in the parking area in order to promote the use of electric/plug-in hybrid electric vehicles.

Due to technological advancements in the charging field, there are several options for the users to charge an electric vehicle depending upon their requirements. For example, all home charging choices utilize a standard fitting for the vehicle. Depending upon the time required to charge an electric vehicle, it can utilize either a 120 V or 240 V power supply. It can also be subdivided into several types, depending upon the type of electric power supply and fitting of the vehicle.

Various types of charging port are shown in Figs. 13, 14, 15 and 16 [8].

Fig. 13 CHAdeMO [4]. *Source* afdc.energy.gov

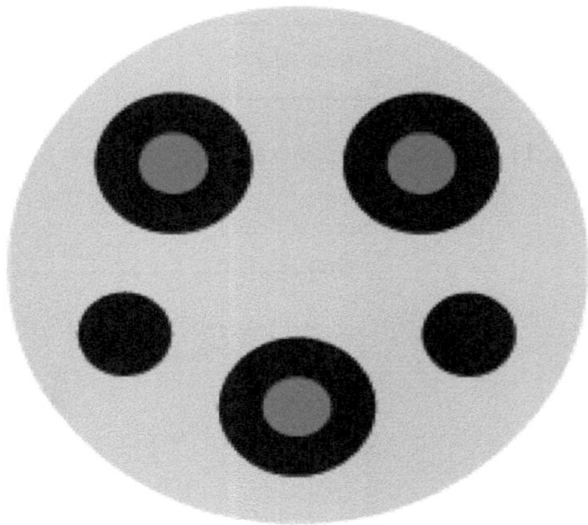

Fig. 14 J1772 charge port [4]. *Source* afdc.energy.gov

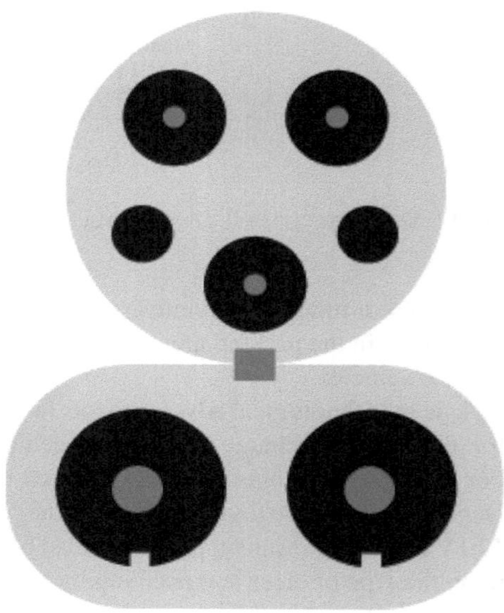

Fig. 15 J1772 Combo [4]. *Source* afdc.energy.gov

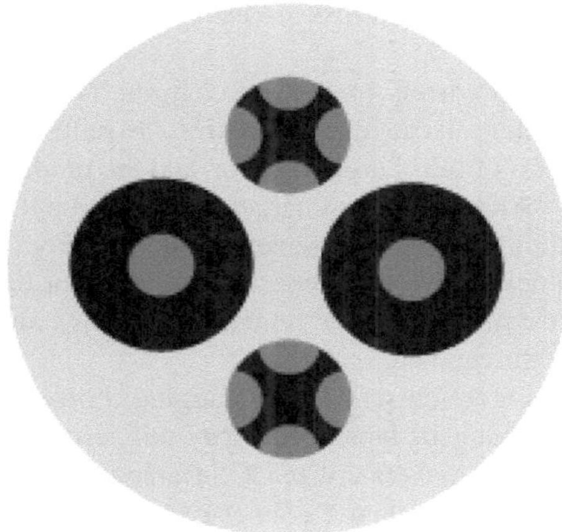

Fig. 16 Tesla supercharger [4]. *Source* afdc.energy.gov

7.2 Types

7.2.1 Level One

These are just like the normal hardware on another electric vehicle. In this type of charging, we just need a 120 V outlet which is grounded. In a period of 8 h (Short Duration Charge), it can charge the PHEV to go around 40 miles of range. It becomes reasonable for hybrid vehicles that are used for either a short or medium duration of time to charge them overnight because there is a hefty amount of battery already remaining from the last charge [8].

7.2.2 Level Two

In this type of charging, it generally requires a 240 V charging system, one that is incorporated in electrical dryers. The rate of charging relies upon the PHEVs acknowledgment and the greatest accessibility of current. It requires a run of the mill thirty ampere circuit. Once it is charged for eight hours, it can travel up to a distance of 180 miles. Level two chargers are widely recognized as open chargers, and they can be found at workplaces, supermarkets, and similar structures. With the exception of Tesla that requires a connector, open level two chargers having standard Electric Vehicles association plug for every single current vehicle [8].

7.2.3 Level Three or Fast Direct Current Charging

Direct current method of charging is the quickest right now accessible energizing strategy. It can regularly add 50–90 miles in a short time, contingent upon the capacity limit of the charging station and the model ofelectric vehicles.

Although superchargers developed by Tesla are considerably quicker, providing about one hundred seventy miles in only charging of thirty minutes. Direct current quick charging is generally valuable for bigger excursions, vehicles being used the vast majority of the time (ex: taxicabs), and the one's who have restricted access to home reviving.

Direct current quick charging utilizes three diverse fitting sorts which can not be changed. Japan ordinarily utilize the CHAdeMO model; generally, Europe and US automakers utilize CCS framework. Supercharging stations established by Tesla utilize an exclusive connector that is specifically for the Tesla built vehicles [8].

7.3 Charge Time

The period of time required to revive a discharged PHEV relies upon two variables: firstly, it depends on the amount of energy or battery remaining in the PHEV andsecondly, it depends on the type of charger and charging techniques, i.e., level 1, level 2 or fastcharging. If you choose to charge with a level 1 type charger, then it adds approximately four to five miles of range to PHEV per hour until fully charged. Module half and half vehicles frequently have scope of twenty-fifty miles; therefore, it should not take more than eight hours to completely charge a battery. While if we prefer to use a level two charger, they are based on 240 V charging technique and are quicker, conveying around fifteen-twenty five miles of scope per hour. If we are to charge a long-extend battery with the help of level two charger, then it will take not more than 8 h to charge it completely. And if we use a direct current quick charging option (which by far is the best), it will add fifty to seventy miles of range to the PHEV in a very short duration of time.

It is essential to understand there are only a few drivers that will go over fifty miles every day, so even a moderately moderate type one charger will be sufficient for the day-to-day energizing needs of a great many people needing to change from gas to an electric vehicle [8].

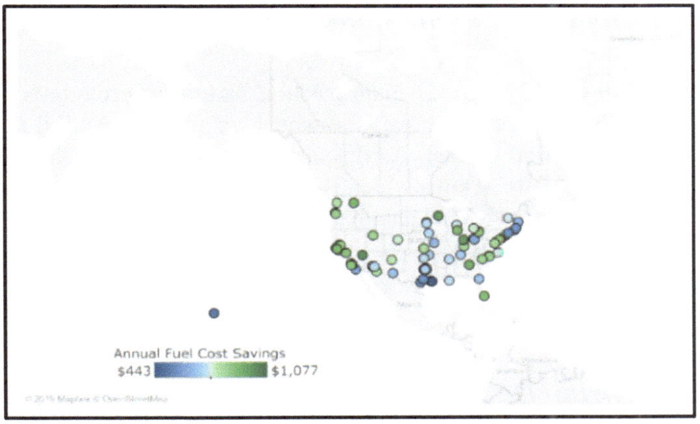

Fig. 17 Figure with average fuel cost savings [9]. *Source* Infoplease.com/atlas

8 Why Choose Plug-In Hybrid Electric Vehicles?

8.1 Plug-In Hybrids Are Cost Saving

According to the data collected by an USA-based agency, an individual worker on a working day travels for about 60–80 km s each day. Depending upon the data collected, an average individual spends for 60–80 km s of fuel. It costs near about 300 rupees for a daily commute of 60–80 km s for an ICE vehicle, which might not be significant on daily basis but if we look at the end of the year, this cost sums up to 80,000 rupees, which now is a significant amount. On the other hand, a PHEV comes with an all-electric range of 60–80 km s, which means no money is spent for the first 80 km s which suffice the commute of a working individual. Therefore, at the end of the day no money is spent on the fuel and thus a plug-in hybrid electric vehicle is cost saving.

The average fuel saving is shown in Fig. 17 [8].

8.2 PHEVs Enhances the Driving Experience

Torque is produced by the incorporated electric motor in the PHEV, which implies that PHEVs give smooth, responsive increasing speed and deceleration. They additionally have a low focus of gravity, which improves taking care of, responsiveness, and ride comfort. The driving execution of electric vehicles has been perceived in race arrangement like Formula One neighborhood drag strips and bore witness to by a huge number of PHEV drivers the nation over [8].

8.3 PHEVs Reduce the Fuel Consumption by a Lot

It is as clear as day, a PHEV uses main traction battery to provide power to the electrical motor which in return moves the wheels of the vehicle. All this is done without even using a drop of fuel. As discussed above, everyday commute of an individual is not more than 80 km s which is easily sufficed by the all-electric mode of a PHEV. This in other sense means that the car is running for 80 km s without even using a single drop of fuel. Due to this very reason, the cost as well as the fuel consumption is reduced significantly, which serves as a better approach [8, 14, 15].

8.4 PHEVs Bring Down the Rate of Emission (Environment Friendly)

A plug-in hybrid electric vehicle is known for its two operating modes—all-electric mode and hybrid mode. As far as the all-electric mode of PHEV is concerned, there is no CO_2 emission. When the PHEV is working in the hybrid mode, then only CO_2 emission takes place which is much less than that of ICE vehicles. The discharges execution of PHEVs is bound to just improve as wind and sun-oriented force uproot coal-terminated power age. Numerous EV proprietors are likewise deciding to combine their EV with housetop sunlight-based boards and home energy stockpiling units. At the point, when controlled solely by sustainable power source, an EV can work about emanations free. The carbon emission of various EVs is shown in Fig. 18 [8, 13, 16].

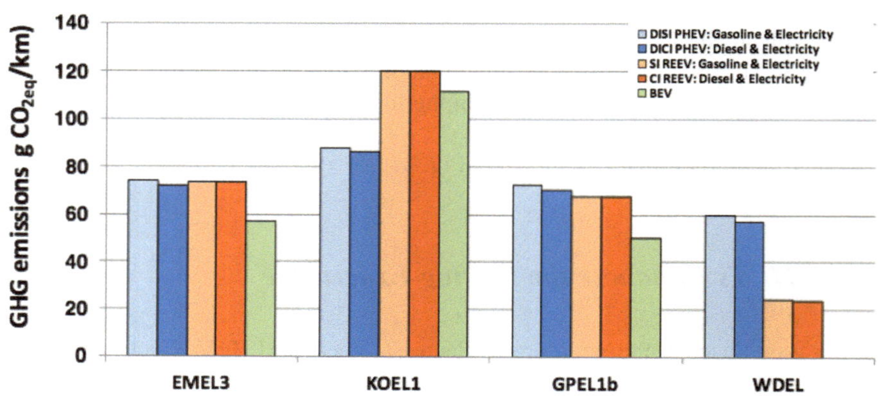

Fig. 18 Graph with CO_2 emission data [10]. *Source* Green Car Congress

9 Conclusion

The idea of a plug-in hybrid electric vehicle boomed with the growth and advancement of electric vehicles. Its idea way back to late eighteenth and early nineteenth century. A plug-in hybrid electric vehicle in-cooperates the use of internal combustion engine and electric motor in perfect harmony. A plug-in hybrid electric vehicle is every bit of an electric vehicle and more. It provides several modes of operation depending upon the power requirement of the vehicle. In short, a plug-in hybrid electric vehicle if we answer to the challenges faced by a BEV.

PHEVs come with a lot of features that makes it exclusive. Features like, different modes of operation, different modes of charging, independent choice of fuel (electricity or gasoline). PHEVs are not only environment friendly due to less carbon dioxide emission but also cost efficient. An electric vehicle limits the user to a certain range of distance in a single charge while a PHEV is independent of the choice of distance. At first, it is designed to operate under all-electric mode and then to the hybrid mode.

For the growth of plug-in hybrid electric vehicles, charging substations play a crucial role. They are directly related as it adds to the luxury of the owner. With the further advancement in the battery technology, the all-electric mode of the PHEV can be enhanced. The initial cost of a PHEV is somewhat high, which can be a disadvantage but as the market grows, the cost will eventually come down. For now, consider this as a day-to-day LED; expenses in the beginning but helpful, efficient, and cost friendly in the long run.

References

1. J.J. Fialka, Coalition turns on to 'plug-in hybrids': utilities, localities, DaimlerChrysler give traction to professor's drive for high mileage. Wall Street J. (2015). Retrieved on 21 June 2011
2. Carrie, Compact BMW X models to come in PHEV versions, *Electrive Industry Service for Electric Mobility* (2020)
3. H. Wang, w. Yang, *Overview of Hybrid Electric Vehicle Trend* (World Wide Science—The Global Science Gateway, 2018)
4. Plug-ins progress, in *Green Car Congress*, 29 Sept 2006. Retrieved 30 Dec 2017
5. J. Gonder, T. Markel, Energy management strategies for plug-in hybrid electric vehicles, in *SAE World Congress*, 25 Sept 2007
6. Working of plug-in hybrid electric vehicles. Int. J. Eng. Technol. (2018)
7. L.O. Valoen, M.I. Shoesmith, *The Effect of PHEV and HEV Duty Cycles on Battery and Battery Pack Performance* (Norway, Canada)
8. *Electric Vehicle Charging Types, Time, Cost and Savings* (Union of Concerned Scientists, 2018)
9. *State of Charge Electric Vehicles' and Fuel-Cost Savings Across the United States* (Union of Concerned Scientists, 2012)
10. EEA report: EVs are better for climate and air quality, in *Green Car Congress Energy, Technologies, Issues and Policies for Sustainable Mobility* (2018)
11. A. Anand et al., Optimal selection of electric motor for E-rickshaw application using MCDM tools, *Cognitive Informatics and Soft Computing* (Springer, Singapore, 2020), pp. 501–509

12. K.C. Manjunatha, A.K. Bhoi, K.S. Sherpa, Design and development of buck-boost regulator for DC motor used in electric vehicle for the application of renewable energy, *Advances in Smart Grid and Renewable Energy* (Springer, Singapore, 2018), pp. 33–37
13. N. Priyadarshi, F. Azam, A.K. Bhoi, A.K. Sharma, A multilevel inverter-controlled photovoltaic generation, in *Advances in Greener Energy Technologies.* (Springer, Singapore, 2020), pp. 149–155
14. N. Priyadarshi et al., A closed-loop control of fixed pattern rectifier for renewable energy applications, *Advances in Greener Energy Technologies* (Springer, Singapore, 2020), pp. 451–461
15. A. Sahu et al., Design of permanent magnet synchronous generator for wind energy conversion system, *Advances in Smart Grid and Renewable Energy* (Springer, Singapore, 2018), pp. 23–32
16. S. SenGupta, A.F. Zobaa, K.S. Sherpa, A.K. Bhoi, *Advances in Smart Grid and Renewable Energy* (2018)

Power Electronics—EV Battery Charging

Biswamoy Pal, Shib Sankar Saha, and Papun Biswas

1 Introduction

Transportation sector demands for major share of worldwide production of fossil fuels, like petrol, diesel, natural gas, etc. The internal combustion (IC) engine-driven vehicles contribute more than two-third of global carbon monoxide (CO) production due to inefficient and incomplete combustion of fossil fuels and about one-third of the total volatile organic compound emissions [1–3]. Electric vehicles (EV) driven by electric motors have therefore emerged as a viable solution towards mitigation of environment issues [4]. Launching of EVs by General Motors in 1996 and subsequently by Ford, Toyota and Honda quickly became very popular. In all versions of EVs, batteries constitute an integral storage device of electrical energy [5–9]. Batteries used in EVs are intended to have high energy density to provide long driving distance and high power density to get proper acceleration with less space requirement and loading to EV [10]. The lithium-ion (Li-ion) batteries with high energy density (140 Wh/kg) available in the market, low self-discharge rate of nickel–metal hydride (Ni-MH) and lead acid batteries gradually being replaced for use in EVs. However, the Li-ion batteries are strictly required to be operated within two safe limits of the state of charge (SOC) to prevent premature end of life [11] and explosion due to thermal runaway. Power electronics-based battery management systems (BMS) are therefore used to estimate and equalize the SOCs of battery cells [12–17]. Charging of EV batteries is done from AC or DC sources. AC charging system uses

B. Pal (✉) · P. Biswas
JIS College of Engineering, Kalyani, West Bengal, India
e-mail: biswamoy.paul@jiscollege.ac.in

P. Biswas
e-mail: papun.biswas@jiscollege.ac.in

S. S. Saha
Kalyani Government Engineering College, Kalyani, West Bengal, India
e-mail: sahashib@hotmail.com

N. Patel et al. (eds.), *Electric Vehicles*, Green Energy and Technology,
https://doi.org/10.1007/978-981-15-9251-5_4

AC–DC converter to charge EV batteries at different power levels. Level 1 (110 V, 12–16 A, 1.5 kW) chargers are generally used to charge the batteries of small EVs within a time range between 0.5 to 12.5 h. These are mostly preferred for overnight charging [11, 18]. Level 2 (240 V, 16–80A, 14.4 kW) charging systems are used for EVs with on-board charger and the batteries get charged through direct connection to the utility grid [11, 19]. Level 3 (240/480/600 V, 150–400A, > 14.4 kW) charging systems use permanently wired chargers, dedicated only for EV battery charging. Level 3 chargers are usually fast chargers and do not take more than 30 min to charge the EV batteries. DC chargers require dedicated wiring and installation, normally mounted in charging stations and usually provide faster charging than AC chargers. Only DC supply voltage needs to be regulated for such systems to match the battery requirement. DC chargers can also be classified into the categories of Level 1 (200–450 V, 80A, 36 kW), Level 2 (200–450 V, 200A, 90 kW) and Level 3 (200–600 V, 400A, 240 kW).

Efficiency of EV battery charging primarily depends on the power electronic converter topologies, used in the chargers. Converter topologies presented in [20–22] use single-stage AC–DC power conversion for EV battery charging. Two-stage conversion systems use an AC–DC converter followed by an active power factor correction (PFC) and a DC–DC converter for tight regulation of output voltage against line and load fluctuations. Unidirectional power converters are mostly used for charging the batteries of small EVs and interfacing renewable energy sources to the charging systems [23–26]. Bidirectional power converters can provide both the features of EV charging from the grid, i.e. grid-to-vehicle (G2V) mode operation and feeding the battery stored electrical energy back to the grid, i.e. vehicle-to-grid (V2G) mode operation [27–35]. However, EV battery charging systems, being high power load, may sometimes have the adverse effects of protection relay tripping and overloading of distribution transformers, fuses, cables, etc. [36–38], if the charging is done at peak hours. Charging at off-peak hours and proper coordination among different charging stations are therefore recommended to be adopted to minimize the adverse effects of load imbalance, energy shortage, voltage instability, power quality degradation, etc. [39, 40].

2 Typical Power System Layouts of EV Battery Charging Systems

Grid tied EV battery charging systems have two stages for power conversions; a front-end AC–DC converter with PFC followed by a regulated DC–DC converter. These two stages are connected through a bulky electrolytic DC link capacitor. The two-stage conversion system may have unidirectional/bidirectional power flow capability. Figure 1a shows the block diagram of an EV charger suitable for G2V operation. It consists of a conventional diode bridge rectifier (DBR) followed by a PFC converter. The DBR and PFC converters together form the AC–DC conversion stage. The

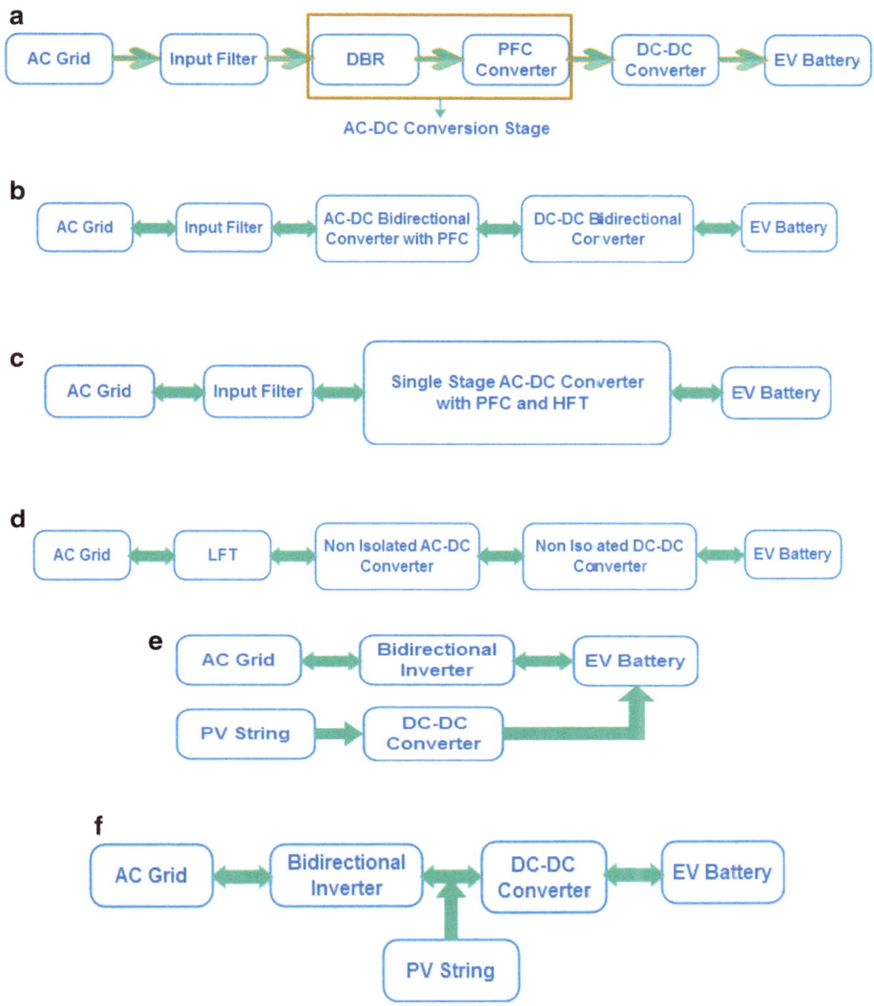

Fig. 1 **a** Power system layout of two-stage unidirectional EV battery charger. **b** Power system layout of two-stage bidirectional EV battery charger. **c** Power system layout of single-stage bidirectional EV battery charger. **d** Power system layout of bidirectional EV battery charging system for public charger. **e** Power system layout of grid tied PV powered EV battery charger. **f** Power system layout of grid tied PV powered EV battery charger

PFC stage is generally a boost converter, output of which is fed to another DC–DC converter for tight regulation of output voltage. The DC–DC stage may also provide galvanic isolation through a high-frequency transformer (HFT). This stage involves combination of controlled switches and diodes at both sides of HFT. Figure 1b shows the block diagram of a EV battery charger performing two-stage conversion of power with bidirectional power flow control capability, which employs a front-end bidirectional AC–DC converter, capable of operating in both G2V and V2G mode with active power factor correction. This converter is cascaded with a bidirectional DC–DC converter through a DC link capacitor. The two-stage converter structure has the drawbacks of reduced efficiency, more component count and presence of bulky electrolytic DC link capacitor. Figure 1c shows a single-stage EV battery charging scheme, in which the grid frequency AC voltage is first converted to a high-frequency AC voltage and then converted to a regulated DC voltage for charging the EV battery [21, 25, 31]. All the converter topologies shown in Fig. 1a–c are most suitable for residential charging. The public charging stations, shown in Fig. 1d, use a large step-down transformer and high-frequency non-isolated converters with increased power density and high efficiency. Figure 1e–f show grid tied PV (photovoltaic) powered charging schemes. The scheme shown in Fig. 1e produces high ripple charging current for the battery. Figure 1f shows an arrangement using a common DC–DC converter for both the inputs from AC grid and PV array.

2.1 AC Charging System

EV batteries can be charged from AC or DC power supply at different voltage or current levels depending on the charging speed requirement. AC charging systems need AC–DC converters to charge the batteries. According to Society of Automotive Engineers (SAE) standards, AC charging systems can be categorized into three different power levels [11].

Level 1: The voltage level is 120/240 V and the current can be in between 12 A to 16 A depending on the charger rating. This system does not require any special arrangement and hence can be used at household outlets. Charging the batteries of small EVs with this arrangement can take from 0.5 to 12.5 h. This feature makes the system suitable for overnight charging [11, 18].

Advantages

Low cost of installation.
Less peak hour impact on electric utility.

Disadvantages

Charging is slow. EVs can run maximum 3–5 miles against each hour of charging.

Level 2: In this system of AC charging, the EV battery is directly connected to the utility grid through on-board charger. Maximum rating of Level 2 systems are 240 V, 60 A and 14.4 kW [11, 18].

Advantages

- Charge speed is significantly higher than Level 1. EVs can run 10 to 20 miles per hour of charge.

Table 1 AC charging system parameters

Operating level	Input voltage (V)	Maximum current (A)	Output power (kW)	Miles of range per hour of charging	Typical duration of charge event (h)
Level 1	120	12–16	1.08–1.44	5	6–10
Level 2	208–240	16–80	3.3–14.4	10–20	1–3
Level 3	208/480/600	150–400	>14.4	50–60	0.5–1

- More energy efficient than Level 1.

Disadvantages

- Higher cost of installation than Level 1.
- Potentially higher peak hour impact on electric utility.

Level 3: In this system, a dedicated EV battery charger is permanently wired to the utility grid. Such charging systems are used for power rating above 14.4 kW. These are fast chargers and can charge a typical battery pack used in EVs in less than half an hour. Table 1 summarizes the parameters of SAE defined AC charging systems of different categories [11].

2.2 DC Charging System

DC charging systems use dedicated wiring and installations with a permanently wired charger, usually mounted at parking area or public charging stations. DC chargers are faster than AC chargers. The output voltage level in modern DC charging stations is automatically changed to meet the demand of the battery packs. DC charging systems are also categorized into three power levels.

Level 1: The rated voltage, current and power ratings of such systems are 450 V, 80 A and 36 kW, respectively.
Level 2: Its voltage rating is same to that of Level 1, but the current and power ratings are 200 A and 90 kW, respectively.
Level 3: The voltage, current and power ratings of this category are 600 V, 400 A and 240 kW, respectively.

Table 2 summarizes the parameters of SAE defined DC charging systems of different categories [11].

Table 2 DC charging system parameters

DC charging system	Supply voltage (V)	Maximum current (A)	Output power level (kW)
Level 1	200450	80	1.08–1.44
Level 2	200–450	200	3.3–14.4
Level 3	200–600	400	>14.4

3 Power Electronic Converters for EV Battery Charging

Power electronic converter is the power processing unit of an EV battery charger. Performance and rating of a charger are primarily decided by the converter topology and its rating. Different converter configurations used in battery charging systems for electric vehicles are categorized in Fig. 2. The simplest form of power electronic converter used in EV battery charger is the AC–DC converter, which connects the EV battery to the utility grid or microgrid. The AC–DC converters can have the features of unidirectional or bidirectional power flow capability. Two-stage power conversions (AC–DC followed by DC–DC conversion) are the most common power converter topology, used for EV battery charging. DC–DC converter-based EV chargers are employed to extract the energy obtained from solar photovoltaic sources, fuel cells, etc. Single-stage AC–DC power conversion is also not very uncommon nowadays [41]. Both AC–DC and DC–DC power converters might have isolated and non-isolated features with the capability of unidirectional/bidirectional power flow. AC–DC stage mainly employs a non-isolated converter to produce DC output voltage from input AC grid voltage with PFC operation. The output of AC–DC converter is fed to a DC–DC converter through DC link capacitor. The DC–DC stage uses an isolated/non-isolated DC–DC converter for controlling the rate of EV battery charging. For safety reasons, isolated DC–DC converters are used for EV battery

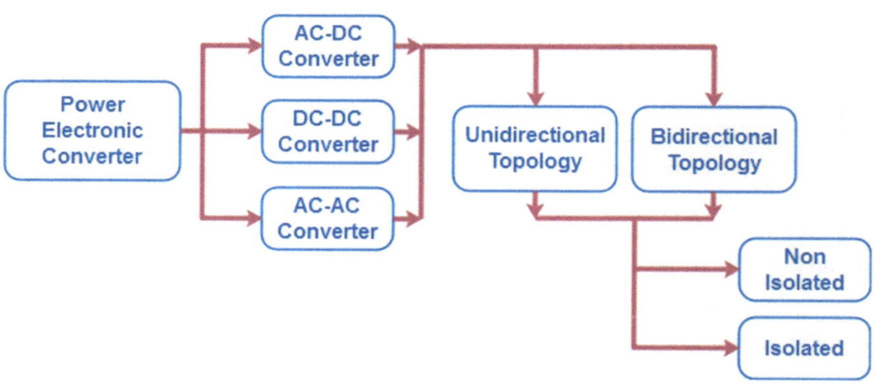

Fig. 2 Power converter topologies used in EVs batteries charging systems

charging applications. In the following section, operating principles of different basic AC–DC and DC–DC converter topologies used for EV charging are discussed, in brief.

3.1 AC–DC Converter Topologies

AC–DC converters mainly convert grid AC voltage to DC voltage. Some AC–DC converters also perform power factor correction. The current drawn by conventional AC–DC converter is highly distorted resulting in high total harmonic distortion (THD) of line current and poor power factor operation. Thus, to improve input power quality, all AC–DC converters are intended to include power factor correction (PFC) circuits, which force the input line currents to follow the line voltage waveforms. Different PFC converter topologies have been developed in recent years [42–49]. Figure 3 shows a diode bridge rectifier (DBR) and a boost PFC circuit to minimize line current THD and improve the input PF.

When the boost switch (S_b) is closed, the boost inductor L_b stores the energy from the supply through DBR and the output capacitor (C_o) drives the load. The inductor current increases during this time. When S_b is opened, the inductor (L_b) acts as a source of energy and its stored energy is transferred to the output. In this duration, the inductor current decreases. The switching operation of S_b is done at high frequency, in the order of tens of kHz. The duty cycle of the switch (S_b) is regulated in such a way that, the average inductor current tracks the waveform of input line voltage to get improved power factor operation and simultaneously, the output voltage is maintained to the desired level. Both voltage mode control and current mode control techniques are employed to vary the duty ratio of the boost switch (S_b). In voltage mode control, the output voltage is fed to controller as feedback signal to control the duty ratio. In current mode control, an inner current loop is added to get inductor current as feedback signal, in addition to the output voltage signal. However, boost converter topologies suffer from the drawback of high output voltage ripple. Size of boost inductor also becomes very large in high-power applications.

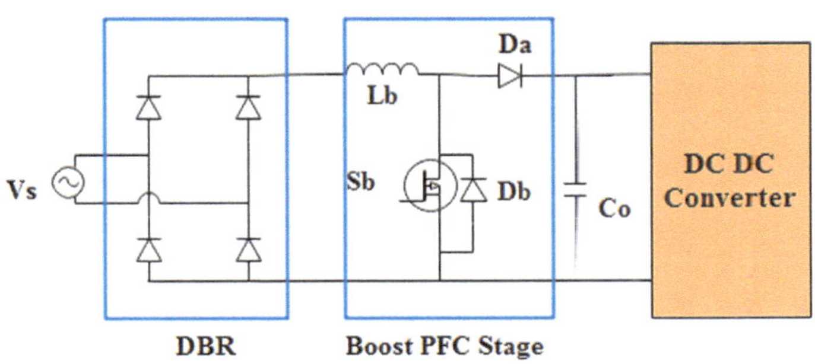

Fig. 3 Conventional front-end AC–DC converter with boost PFC circuit

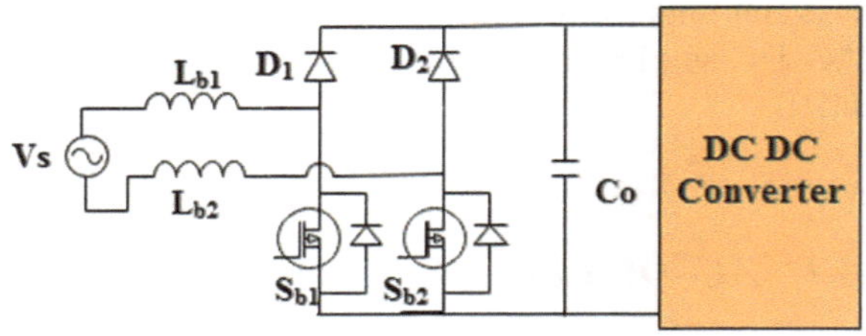

Fig. 4 Front-end bridgeless AC–DC converter with boost PFC circuit

Bridgeless boost converter topologies are an attractive solution for AC–DC power factor conversion in high-power applications. However, these converters suffer from the drawbacks that, complex sensing circuitry is required for closed-loop operation and large electromagnetic interference (EMI) disturbances are introduced into the system. Other boost converter-based PFC topologies include interleaved boost converter, bridgeless interleaved boost converter, etc.

The bridgeless topology as shown in Fig. 4 comprises of two controlled switches (S_{b1} and S_{b2}), two diodes (D_1 and D_2) and two boost inductors (L_{b1} and L_{b2}). During positive half-cycle, the boost inductor (L_{b1}) stores energy through L_{b1}, S_{b1}, and the body diode of S_{b2}, when $S_{b1}$is gated. As S_{b1} is turned off, stored energy of L_{b1} is released through D_1 and body diode of S_{b2} to the load. Similarly, during negative half-cycle of supply voltage, S_{b2} is turned on and the boost inductor L_{b2} stores energy. Later during the off-state of S_{b2}, L_{b2} releases its energy to the load. The above AC–DC converter topologies are capable of unidirectional power flow and their application is restricted to only grid-to-vehicle (G2V) mode operation.

3.2 DC–DC Converter Topologies

The basic function of DC–DC stage is to regulate voltage and current as per battery requirement. Both constant voltage (CV) and constant current (CC) methods are applied for charge control of EV batteries. Different parameters, such as voltage, current and state of charge (SOC) of the batteries are sensed and fed to the charge controller to vary the duty ratio of power switches of the DC–DC converter. The DC–DC converters used in battery charging systems can be isolated or non-isolated type. Based on the direction of power flow, these may be classified as unidirectional and bidirectional. In this section, different non-isolated/isolated topologies are discussed with the capability of unidirectional/bidirectional power flow.

3.2.1 Buck Converter

Figure 5a shows non-isolated DC–DC buck converter used in EV battery charger. It consists of a controlled switch (S_1), a diode (D_1) and an LC filter at the output. The switch (S_1) is usually operated at a high switching frequency. When, S_1 is turned on, current through the battery increases gradually due to the presence of inductance in the current flow path and consequently, L stores energy. During this time, the diode D_1 remains reverse-biased. When S_1 is turned off, the battery is disconnected from the source and the inductor L supplies its stored energy to the load (battery) through D_1. Now, the current through the battery gradually decreases. The value of the inductor decides whether operation of the converter will be under continuous conduction mode (CCM) or under discontinuous conduction mode (DCM). By proper duty ratio control, power flow to the battery is regulated. For such converters, output voltage is always less than the input voltage. This low-cost converter is simple in structure and operation. However, large input current ripple and sluggish transient response are the main drawbacks of this converter.

3.2.2 Boost Converter

Boost converters are mostly used in AC–DC PFC stage along with DBR. Figure 5b shows circuit diagram of a conventional boost converter, consisting of a switch (S_1), a diode (D_1), an inductor (L) and filter capacitor (C_o). By turning on the controlled switch, the boost inductor L starts storing energy from input supply (AC–DC converter). During this time, D_1 does not conduct and the load (battery) is supplied by C_o. During turn off mode of the converter, D_1 gets forward-biased, the input supply in series with the boost inductor supplies the load. Hence, the load voltage gets boosted up above the input voltage level.

3.2.3 Buck–Boost Converter

A buck–boost configuration is shown in Fig. 5c. It uses the same components, as used by a buck or a boost converter. When S_1 is turned on, inductor L stores energy. During this time, the diode (D_1) does not conduct and the load (battery) is supplied by C_o. As S_1 is turned off, L reverses its polarity and supplies the battery through D_1. Depending on duty ratio of controlled signal, the converter can act as buck or boost converter. Large output current ripple and reverse polarity output voltage are the main drawbacks of this converter.

3.2.4 Cuk Converter

Cuk converter is derived from the basic buck–boost converter. It has combined all the advantages of basic DC–DC converter topologies, like continuous input and output current and capability of working both in buck and boost mode. A cuk converter, as shown in Fig. 5d, consists of a switch (S_1), two inductors ($L_1$and L_2), two capacitors (C_1 and C_o) and a diode (D_1). The capacitor (C_1) is the main energy transfer element, which gets connected to input and output alternatively through D_1 and S_1,

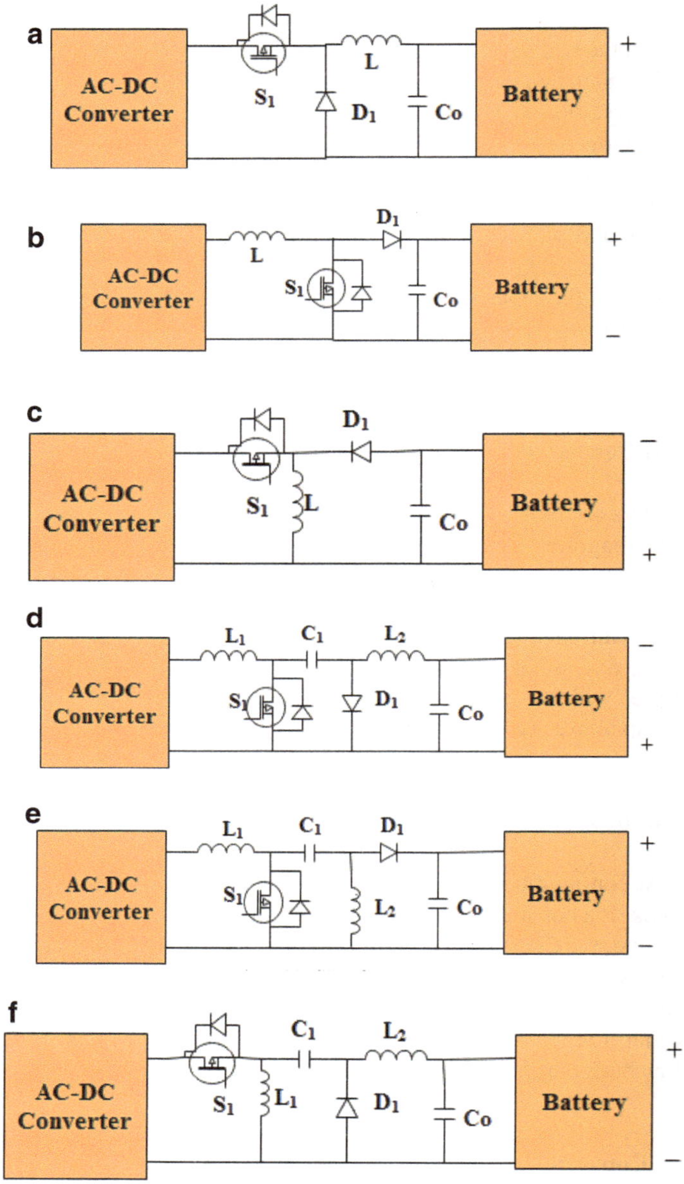

Fig. 5 **a** Buck converter. **b** Boost converter. **c** Buck–boost converter. **d** Cuk converter. **e** SEPIC converter. **f** Zeta converter

respectively. During off-state of S_1, the capacitor C_1 is charged from the input supply through L_1 and D_1. During the on-state of S_1, the capacitor (C_1) discharges its stored energy to the load (battery) through L_2. The inductors (L_1 and L_2) convert the voltage sources to equivalent current sources. This conversion is necessary to restrict excessive current flow through capacitor C_1. The main disadvantage of Cuk converter is reversal of output voltage polarity with respect to the input voltage polarity.

3.2.5 SEPIC Converter

A SEPIC converter, shown in Fig. 5e, is derived from the basic buck–boost topology. This converter is essentially a combination of a boost converter and a buck–boost converter. This converter can operate in buck as well as boost mode without any reversal of output voltage polarity. By controlling the duty ratio of S_1, the load (battery) terminal voltage can be regulated. When S_1 is turned on, L_1 stores energy from input, the capacitor C_1 sends energy to L_2 and C_o supplies the load. Soon as S_1 is opened, L_1 delivers its stored energy to C_1 and the load through D_1. In addition, L_2 also sends its stored energy to the load through D_2. The drawbacks of SEPIC converter are large output current ripple and the requirement of large capacitor (C_1) to transfer energy between the source and the load.

3.2.6 Zeta Converter

It is another popular DC–DC converter topology, obtained by rearranging the circuit elements of SEPIC converter. This converter has power transfer capability both in buck or boost mode and produces continuous output current. This converter also produces output voltage of same polarity to that of input voltage. Compared to SEPIC converter, zeta converter produces low output voltage ripple. But its input current ripple is high requiring large input capacitor. When the switch S_1 is turned on, both L_1 and L_2 are energized by the input supply. The capacitor (C_1) now transfers its energy to L_2 and the load (battery). As S_1 is turned off, two current loops are formed. In one loop, the capacitor (C_1) gets charged by the stored energy of L_1 and in another loop, L_2 discharges to the battery through D_1.

3.2.7 Bidirectional DC–DC Converters

All the topologies of basic DC–DC converter discussed so far are capable of only unidirectional power flow, i.e. from input supply (AC–DC converter) to the load (battery). These converters are thus suitable for grid-to-vehicle (G2V) mode operation. In recent yeras, demand of vehicle-to-grid (V2G) mode operation to feed back the stored battery energy to the grid to meet the peak energy demand are getting prime attention. In Fig. 6(a–d), different non-isolated bidirectional DC–DC converter topologies are shown. In Fig. 6a, the topology works in buck mode during

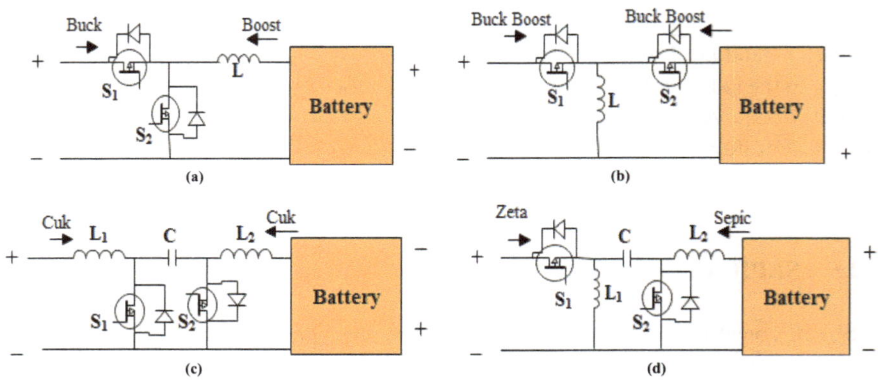

Fig. 6 a–d Non-isolated DC–DC bidirectional converter topologies

G2V operation, whereas it operates in boost mode during V2G operation. The buck operation is controlled by switch S_1 and the boost operation is controlled by the switch S_2. Figure 6b shows a bidirectional buck–boost converter, capable of working in both G2V and V2G mode. In G2V and V2G mode operations, the output voltage is controlled by the switches S_1 and S_2, respectively. Similarly, the bidirectional Cuk–Cuk and Zeta–SEPIC converters are shown in Fig. 6c–d, respectively.

The non-isolated converters discussed above enjoy the advantages of less component count, high power density and high efficiency, but suffer from the drawback that the load (EV battery) is not electrically isolated from the grid. In order to ensure safety to the user, it is mandatory to electrically isolate the load (battery) from the utility grid. Several isolated DC–DC converters like flyback converter, forward converter, half-bridge converter and full-bridge converter are finding wide applications in battery charging systems.

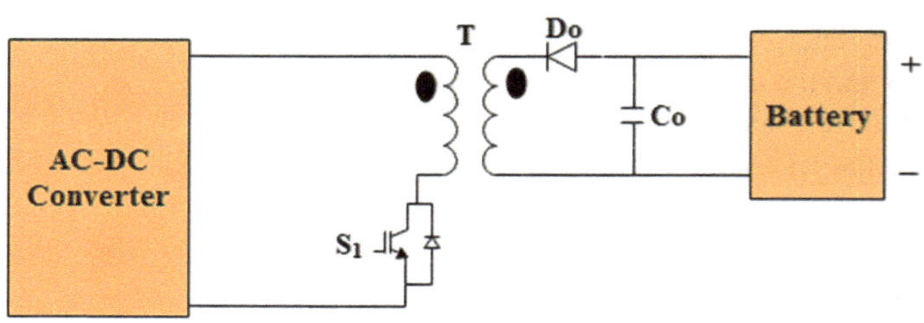

Fig. 7 Flyback converter

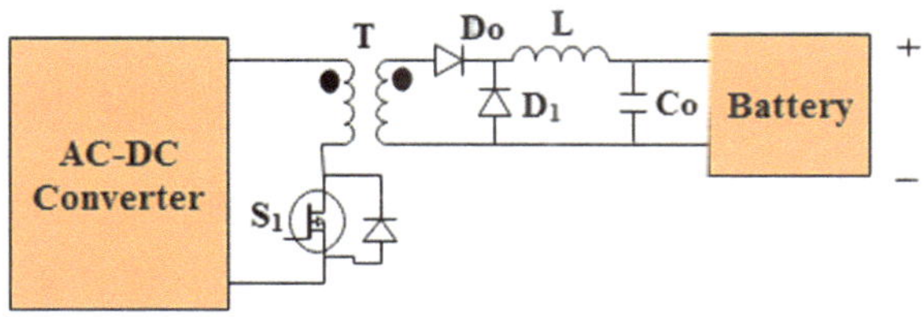

Fig. 8 Forward converter

3.2.8 Flyback Converter

The flyback topology, as shown in Fig. 7, is a simple and low-cost isolated DC–DC converter with minimum component count. This converter has been derived from the conventional buck–boost converter with the inductor replaced by a coupled inductor (transformer). It uses one controlled switch S_1, a high-frequency transformer (HFT) and a diode (D_o). The switch is operated at high switching frequency to minimize the size and weight of the flyback transformer. The transformer provides electrical isolation between the grid and the load (battery). When S_1 is turned on, the magnetizing inductor of the transformer stores energy, as the output diode (D_o) is reverse-biased. As S_1 is turned off, the energy stored in the magnetizing inductance is delivered to the load through D_o. By controlling the duty ratio of S_1, a regulated output voltage is achieved. Voltage gain of this converter depends on the duty ratio and the turns ratio of the HFT. However, flyback converters produce huge voltage spikes across the switch (S_1) due to the leakage inductance of the HFT. Snubber circuits are therefore required to reduce this voltage spike. Flyback converters are popularly used in low-power applications.

3.2.9 Forward Converter

The forward converter topology, as shown in Fig. 8, is almost similar to that of a flyback converter with one extra diode connected in antiparallel with the LC filter. Another difference in circuit configuration is the output diode (D_o) of forward converter, connected in reverse polarity to that of flyback converter. When the switch S_1 is turned on, current flows through the primary winding of the HFT and power is transferred to the load (battery) through D_o. When the switch is turned off, the freewheeling diode D_1 is forward-biased and starts conduction.

3.2.10 Half-Bridge Converter (HBC)

The half-bridge and full-bridge converters have large power handling capability, but require more number of active and passive components. Both half-bridge and full-bridge converters are capable of transferring power in either direction. A half-bridge bidirectional converter as shown in Fig. 9 uses two semiconductor switches (S_1, S_2)

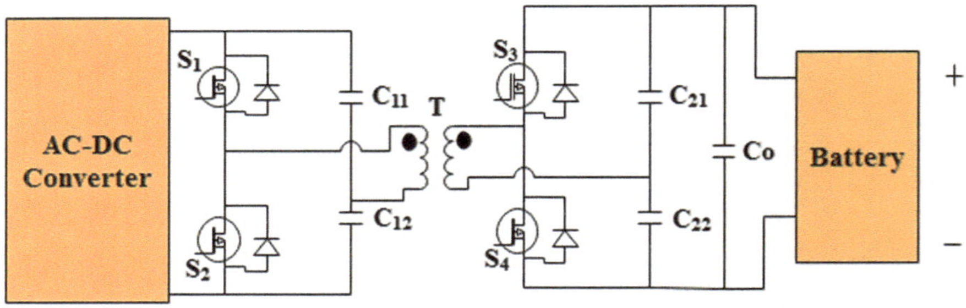

Fig. 9 Half-bridge bidirectional converter

with body diodes and two capacitors (C_{11} and C_{12}) at the primary side of the HFT. The secondary side is also comprised of similar components. When S_1 is in on-state, half of input voltage is impressed across the primary of the HFT. The capacitor (C_{21}) now gets charged through body diode of S_3. When S_2 is in on-state, reverse polarity voltage is impressed across the primary of the HFT and the capacitor (C_{22}) now gets charged through the body diode of S_4. Thus, the battery connected across the series combination of C_{21} and C_{22} gets charged. In this mode, the switches at the primary side act in inverter mode and the diodes of the secondary side act as rectifier. However, if the battery energy is to be fed back to the input side, then the role of primary and secondary circuits are reversed. Now, the switches (S_3 and S_4) of the secondary side act in inverter mode and the body diodes of S_1 and S_2 act as rectifier.

3.2.11 Full-Bridge Converter (FBC)

Operation of full-bridge converter as shown in Fig. 10 is similar to that of half-bridge converter, only with the exception that, two diagonal switches (S_1, S_2 or S_3, S_4) are kept in on-state to impress supply voltage across the primary of the HFT. The secondary induced voltage charges the output capacitor (C_o) through two diodes (D_1, D_2 or D_3, D_4). Here, the primary switches operate in inverter mode and the secondary

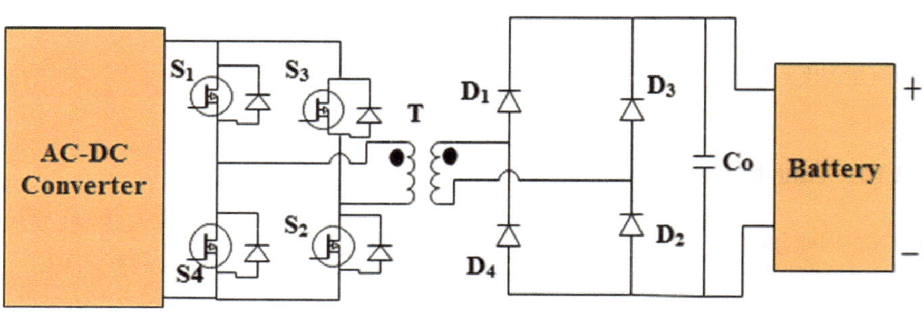

Fig. 10 Full-bridge unidirectional converter

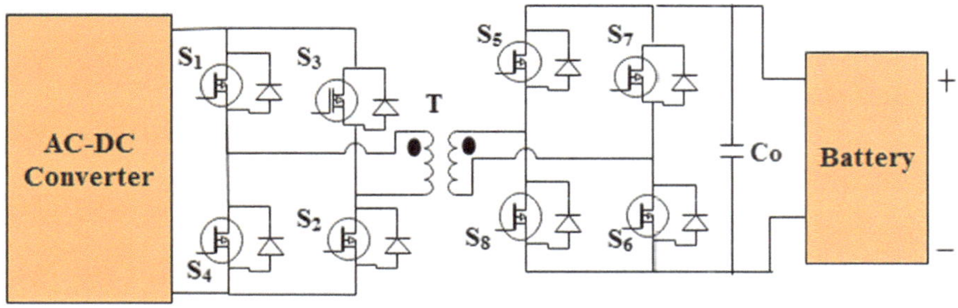

Fig. 11 Full-bridge bidirectional converter

diodes act as rectifier. The bidirectional full-bridge converter shown in Fig. 11 allows power flow from the load (battery) to the input by operating the secondary switches in inverter mode and the primary diodes as rectifier.

4 Recent Development in Power Electronic Converter Topologies for EV Battery Charging

Apart from the converters discussed above, several new topologies of AC–DC and DC–DC converters have been suggested by power electronic researchers for performance improvement of EV battery chargers [41, 50–53]. Recent trend in power converter design is the reduction of size and weight of the converter and simultaneously increasing the efficiency. Size and weight on power converters can be reduced in great extend with high switching frequency operation. However, high switching frequency operation may also lead to increased switching loss, high voltage and current stress to the switches and large EMI, if the switches are operated in hard switching mode. Therefore, to reduce the switching loss and voltage and current stress of the switches even under high-frequency operation, semiconductor switches of modern power electronic converters are operated either with zero voltage switching (ZVS) and/or with zero current switching (ZCS).

4.1 Single-Phase Single-Stage ZCS Boost PFC Converter

The single-phase single-stage ZCS boost PFC converter [53] shown in Fig. 12 uses two semiconductor switches in totem–pole configuration. The resonating circuit, comprising of L_r and C_r, facilitates ZCS operation of the switches and minimized reverse recovery losses of the diodes. The semiconductor switches are controlled in such a way that, the input power factor gets improved.

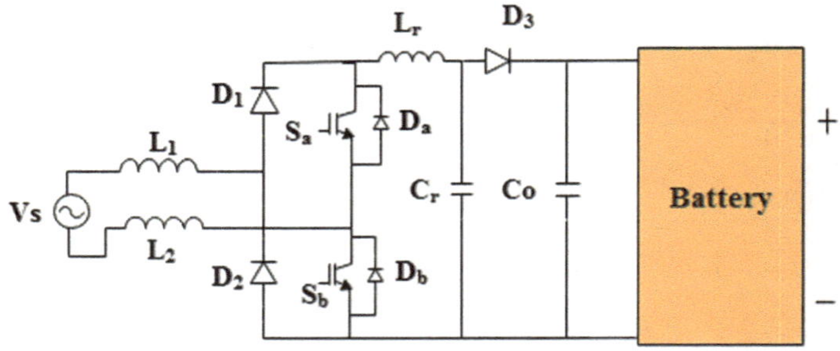

Fig. 12 Single-phase single-stage ZCS boost PFC rectifier

Assuming the supply voltage (V_s) to be positive and both the switches (S_a and S_b) are in off condition, the input current now flows through D_1, L_r, D_3, the load and D_b. When S_a is turned on, due to the presence of resonant inductance (L_r) the current through S_a starts rising from zero and thus the switch (S_a) turns on under ZCS condition. When the current through S_a is increased to the input current, the resonant inductor current is reduced to zero. Thus, the diodes (D_3 and D_b) turn off softly. Again, as the switch S_b is turned on, C_r starts discharging in resonance with L_r through the switches, S_a and S_b. After half resonant cycle, the resonant capacitor (C_r) is charged to opposite polarity and then starts discharging in opposite direction through D_b and S_a. As, the resonant inductor current increases to the input current, the current through the switch (S_a) is reduced to zero. After that, the excess resonant inductor current passes through the antiparallel body diode (D_a), impressing zero voltage across S_a. The switch (S_a) can now be turned off under ZV-ZCS condition. As the supply voltage (V_s) goes negative, the converter operation is similar, only with the exception that the roles of the switches S_a and S_b get interchanged.

4.2 *ZCS-PWM Flyback Converter*

The ZCS-PWM flyback converter [52] as shown in Fig. 13 operates in CCM. Initially, with both the switches in off-state, the energy stored in the magnetizing inductance (L_m) is transferred through the diode (D_o) to the output. As the switch (S_m) is turned on, its current starts rising resonantly from zero, due to the presence of leakage inductance. Thus, S_m turns on under ZCS condition. As the current of S_m is reached to the magnetizing current, the current through D_o is reduced to zero in the same fashion. Thus, D_o commutates softly. Now, the magnetizing inductance gets energized from the input supply (V_s), through the switch (S_m). The auxiliary switch (S_a) can now is turned on under ZCS condition, as its current starts increasing from zero due to the resonance between L_r and C_r. After half resonant period, as the capacitor gets charged to twice the supply voltage, the switch S_a is turned off under ZCS condition and the diode (D_a) is commutated softly. The resonant capacitor (C_r) now starts discharging

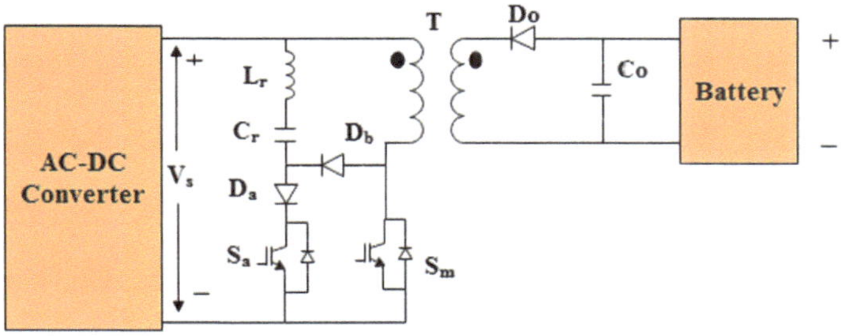

Fig. 13 ZCS-PWM flyback converter

resonantly through L_r, V_s, S_m, D_b and the switch (S_a). Thus, the current through S_a also reduces in the same fashion. As the resonant inductor current increases to the magnetizing current, the current through the switch (S_a) is reduced to zero. The excess resonant inductor current now passes through the antiparallel body diode of S_m, impressing zero voltage across S_a. The switch (S_a) can now be turned off under ZV-ZCS condition. The resonant capacitor continues to discharge through L_r, V_s, body diode of S_m and D_b until the resonant inductor current is reduced to zero and the diode (D_b and body diode of S_m) commutate softly. Thus, all switching devices of this converter operate under soft-switched condition and the diodes commutate softly.

4.3 Bidirectional Isolated AC–DC Converter with PFC

A bidirectional isolated AC–DC converter [41], as shown in Fig. 14, is suitable for both G2V and V2G mode operation of EV battery charging. The AC side of the converter operates as a current fed half-bridge converter (HBC) to feed power to

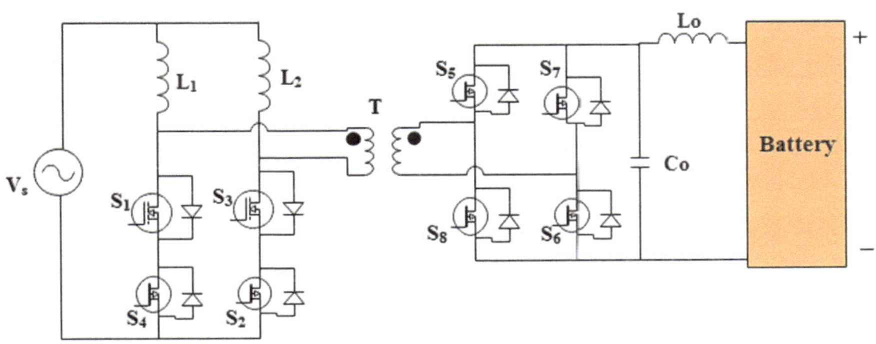

Fig. 14 Bidirectional isolated AC-DC converter with PFC

the load (battery) through a HF transformer (T), followed by a full-bridge converter (FBC). The HBC, at the AC side, regulates the input current to follow the line voltage, whereas the FBC at the output side of the HFT regulates the load (battery) voltage and current. All switches of the converter operate under soft-switched condition. The converter can operate in four V-I quadrants to facilitate G2V and V2G mode operation.

In first quadrant G2V mode operation, during positive half-cycle of supply voltage (V_s), considering that the switch (S_1) is in on-state and the switch (S_3) is in off-state. Now, the inductor (L_1) gets energized from the input supply and the stored energy of L_2 is transferred to the load (battery) through the HFT and antiparallel body diodes of S_6 and S_7. When S_3 is turned on, its current starts increasing gradually and current through S_1 decreases. When the current of S_3 becomes equal to that of S_1, the load current through the antiparallel body diodes of S_6 and S_7 becomes zero and the diodes commutate softly. In order to turn off the primary switch (S_1), the secondary side switches (S_5 and S_8) are turned on simultaneously. This causes the negative output voltage to appear across the secondary of HFT and the reflected voltage at the primary reduces the current through the switch (S_1) and ultimately the switch (S_1) turns off under ZCS condition. The second quadrant operation of the converter or the operation in negative half-cycle of the supply voltage is similar to that of first quadrant operation. In third quadrant operation, the switches S_1 and S_3 are kept in off-state, whereas the switches S_2 and S_4 are kept in on-state. In this mode, the converter acts as full-bridge voltage-fed converter with current doubler at the load (battery) side. Similarly, in fourth quadrant operation, power flows from battery to grid side.

5 Conclusion

This chapter briefly presented the power system layouts of EV battery chargers and discussed the working principles of different power electronic converter topologies suitable for EV battery charging applications. The basic isolated and non-isolated converter topologies have been discussed with unidirectional/bidirectional power transfer capabilities. A thorough discussion has also been made highlighting the modifications needed in existing converter topologies to make them fit for modern EV battery charging systems. A number of recently developed soft-switched power converter topologies have also been explained in brief.

References

1. G.J. Henry, G.W. Heinke, Environmental Science and Engineering London (Prentice-Hall International 1996)

2. P.D. Bailey, *modelling future vehicle exhaust emissions in Europe, Water, Air and Soil Pollution*, vol. 85 (Kluwer Academic Publishers, printed in the Netherlands, 1995), pp. 1879–1884
3. Lilley, *A New Approach to Emissions Inventory Modelling; Assessing Fuel And Vehicle Impacts on Air Quality, Air Pollution*, vol. 3, pp. 389–398 (2000)
4. U.M. Netravati, V.C. Patil, T.M. Veeragangadhar Swamy, S. Goud, vehicular pollution in india and solution. Int. J. Disaster Recovery Bus. Continuity **11**(1), 1010–1014 (2020)
5. J.Y. Yong, V.K. Ramachandaramurthy, K.M. Tan, N. Mithulananthan, A review on the state-of-the-art Technologies of electric vehicle, its impacts and prospects. Renew. Sustain. Energy Rev. **49**, 365–385 (2015)
6. C.C. Chan, The state of the art of electric and hybrid vehicles. Proc. IEEE **90**, 247–275 (2002)
7. O.M.F. Camacho, P.B. Nørgård, N. Rao, L. Mihet-Popa, Electrical vehicle batteries testing in a distribution network using sustainable energy. IEEE Trans. Smart Grid. **5**, 1033–1042 (2014)
8. A. Khaligh, Z. Li, Battery, ultracapacitor, fuel cell, and hybrid energy storage systems for electric, hybrid electric, fuel cell, and plug-in hybrid electric vehicles: State of the art. IEEE Trans. Veh. Technol. **59**, 2806–2814 (2010)
9. H.J. Chiu, L.W. Lin, A bidirectional DC–DC converter for fuel cell electric vehicle driving system. IEEE Trans. Power Electron. **21**, 950–958 (2006)
10. Y. Gao, M. Ehsani, J.M. Mille, Hybrid electric vehicle: overview and state of the art, in *Proceedings of the IEEE International Symposium on Industrial Electronics, Dubrovnik, Croatia*, 20–23 June 2005, pp. 307–316
11. S.S. Williamson, A.K. Rathore, F. Musavi, Industrial electronics for electric transportation: Current state of- The-art and future challenges. IEEE Trans. Ind. Electron. **62**, 3021–3032 (2015)
12. P.A. Cassani, S.S. Williamson, Feasibility analysis of a novel cell equalizer topology for plug-in hybrid electric vehicle energy-storage systems. IEEE Trans. Veh. Technol **58**, 3938–3946 (2009)
13. A.C. Baughman, M. Ferdowsi, Double-tiered switched-capacitor battery charge equalization technique. IEEE Trans. Ind. Electron. **55**, 2277–2285 (2008)
14. K. Nishijima, H. Sakamoto, K. Harada, A PWM controlled simple and high performance battery balancing System, in *Proceedings of the IEEE Power Electronics Specialists Conference, Galway, Ireland*, vol. 1 (2000), pp. 517–520
15. P.A. Cassani, S.S. Williamson, Design, testing, and validation of a simplified control scheme for a novel plug-in hybrid electric vehicle battery cell equalizer. IEEE Trans. Ind. Electron. **57**, 3956–3962 (2010)
16. Y.S. Lee, M.W. Cheng, Intelligent control battery equalization for series connected lithium-ion battery Strings. IEEE Trans. Ind. Electron. **52**, 1297–1307 (2005)
17. Y.S. Lee, M.W. Cheng, S.C. Yang, C.L. Hsu, Individual cell equalization for series connected lithium-ion batteries. IEICE Trans. Commun. **E89-B**, 2596–2607 (2006)
18. E.A. Grunditz, T. Thiringer, Performance analysis of current BEVs based on a comprehensive review of specifications. IEEE Trans. Transp. Electr. **2**, 270–289 (2016)
19. H. Shareef, M.M. Islam, A. Mohamed, A review of the stage-of-the-art charging technologies, placement methodologies, and impacts of electric vehicles. Renew. Sustain. Energy Rev. **64**, 403–420 (2016)
20. K. Itani, A.D. Bernardinis, Z. Khatir, A. Jammal, M. Oueidat, Regenerative braking modeling, control, and simulation of a hybrid energy storage system for an electric vehicle in extreme conditions. IEEE Trans. Trans. Electrification. **2** (2016)
21. C. Li, Y. Zhang, Z. Cao, D. XU, Single-phase single-stage isolated ZCS current-fed full-bridge converter for high power AC/DC applications. IEEE Trans. Power Electron. (2016)
22. C.H. Chang, C.A. Cheng, E.C. Chang, H.L. Cheng, B.E Yang, An integrated high-power-factor converter with ZVS transition. IEEE Trans. Power Electron. (2015)
23. H.N. Kutkut, D.M. Divan, D.W. Novotny, R.H. Marion, Design Considerations and Topology Selection for a 120-kW IGBT Converter for EV Fast Charging. IEEE Trans. on power Electron. **13** (1998)

24. S. Ratanapanachote, H.J. Cha, P.N. Enjeti, A Digitally Controlled Switch Mode Power Supply Based on Matrix Converter. IEEE Trans. on power Electron. **21** (2006)
25. T. Mishima, K. Akamatsu, M. Nakaoka, A high frequency-link secondary-side phase-shifted full-range soft-switching PWM DC–DC converter with ZCS active rectifier for EV battery chargers. IEEE Trans. on power Electron. **28** (2013)
26. J, Deng, S. Li, S. Hu, C.C. Mi, R.Ma, Design methodology of LLC resonant converters for electric vehicle battery chargers. IEEE Trans. vehicular Technol. **63** (2014)
27. C. Wang, M. Xu, F.C. Lee, B. Lu, EMI Study for the Interleaved Multi-Channel PFC, in *Proceedings of the IEEE Power Electronics Specialists Conference (PESC)*, (Orlando, FL, USA, 2007), pp. 1336–1342
28. D.S. Gautam, F. Musavi, W. Eberle, W.G. Dunford, A zero voltage switching full-bridge dc-dc converter with capacitive output filter for a plug-in-hybrid electric vehicle battery charger, in *Proceedings of the IEEE Applied Power Electronics Conference and Exposition* (Orlando, FL, USA, 2012), pp. 1381–1386
29. F. Musavi, M. Craciun, D.S. Gautam, W. Eberle, W.G. Dunford, An LLC resonant DC-DC Converter for wide output voltage range battery charging applications. IEEE Trans. Power Electron. **28**, 5437–5445 (2013)
30. Y. Gurkaynak, Z.M. Li, A. Khaligh, A novel grid-tied, solar powered residential home with plug-in hybrid electric vehicle (PHEV) loads, in *Proceedings of the 5th Annual IEEE Vehicle Power and Propulsion Conference*, (Dearborn, MI, USA, 2009), pp. 813–816
31. O. Onar, *Bi-Directional Rectifier/Inverter and Bi-Directional DC/DC Converters Integration for Plug-in Hybrid Electric Vehicles with Hybrid Battery/Ultra-capacitors Energy Storage Systems* (Illinois Institute of Technology, Chicago, IL, USA, 2009)
32. M.H. Rashid, *Power Electronics Handbook: Devices, Circuits and Applications* (Elsevier, Amsterdam, The Netherlands, 2010)
33. B. Koushki, A. Safaee, P. Jain, A. Bakhshai, Review and comparison of bi-directional AC-DC converters with V2G capability for on-board EV and HEV, in *Proceedings of the 2014 IEEE Transportation Electrification Conference and Expo (ITEC)* (Dearborn, MI, USA, 2014)
34. J. G. Pinto, Vítor Monteiro, H. Gonçalves, J.L. Afonso, Onboard Reconfigurable Battery Charger for Electric Vehicles With Traction-to-Auxiliary Mode, IEEE trans. Vehicular Technol. **63** (2014)
35. H.S. Krishnamoorthy, P. Garg, P.N. Enjeti, A matrix converter-based topology for high power electric vehicle battery charging and V2G application
36. L. Yao, W.H. Lim, T.S. Tsai, A real-time charging scheme for demand response in electric vehicle parking station. IEEE Trans. Smart Grid **8**, 52–62 (2017)
37. L. Kütt, E. Saarijärvi, M. Lehtonen, H. Mõlder, J. Niitsoo, A review of the harmonic and unbalance effects in electrical distribution networks due to EV charging, in *Proceedings of the 2013 12th International Conference on Environment and Electrical Engineering (EEEIC)* (Wroclaw, Poland, 2013)
38. P. Richardson, D. Flynn, A. Keane, Optimal charging of electric vehicles in low-voltage distribution systems. IEEE Trans. Power Syst. **27**, 268–279 (2012)
39. F. Mwasilu, J.J. Justo, E.K. Kim, T.D. Do, J.W. Jung, Electric vehicles and smart grid interaction: a review on vehicle to grid and renewable energy sources integration. Renew. Sustain. Energy Rev. **34**, 501–516 (2014)
40. B. Geng, J.K. Mills, D. Sun, Two-stage charging strategy for plug-in electric vehicles at the residential transformer level. IEEE Trans. Smart Grid **4**, 1442–1452 (2013)
41. U.R. Prasanna, A.K. Singh, K. Rajashekara, Novel bidirectional single-phase single-stageisolated AC-DC converter with PFC forcharging of electric vehicles. IEEE Trans. Transp. Electrification (2017)
42. L. Yang, B. Lu, W. Dong, Z. Lu, M. Xu, F.C. Lee, W.G. Odendaal, Modeling and characterization of a 1KW CCM PFC converter for conducted EMI Prediction, in *IEEE Applied Power Electronics Conference and Exposition, APEC,* vol. 2, (2004), pp. 763–69
43. C. L. Sirio, Analysis of line current harmonics for single-phase PFC converters, in *IEEE Power Electronics Specialists Conference*, (2007), pp. 1291–1297

44. D. Xu, J. Zhang, W. Chen, J. Lin, F.C. Lee, Evaluation of output filter capacitor current ripples in single phase PFC converters, in *Proceedings of the Power Conversion Conference, PCC*. vol. 3 (Osaka, Japan, 2002), pp. 1226–1231
45. D. Mitchell, AC–DC converter having an improved power factor, United States Patent # 4,412,277 (1983)
46. U. Moriconi, A bridgeless PFC configuration based on L4981 PFC Controller. STMicroelectronics Application Note AN1606, (2002)
47. W.Y. Choi, J.M. Kwon, E.H. Kim, J.J. Lee, B.H. Kwon, Bridgeless boost rectifier with low conduction losses and reduced diode reverse-recovery problems. IEEE Trans. Ind. Electron. **54**, 769–780 (2007)
48. J. M. Hancock, Bridgeless PFC boosts low-line efficiency, infineon technologies (2008)
49. L. Huber, Y. Jang, M.M. Jovanovic, Performance evaluation of bridgeless PFC boost rectifiers. IEEE Trans. Power Electron. **23**, 1381–1390 (2008)
50. F. Musavi, W. Eberle, W.G. Dunford, A high-performance singlephase ac-dc power factor corrected boost converter for plug in hybrid electric vehicle battery chargers, in *IEEE Energy Conversion Congress and Exposition*, (ECCE, Atlanta, Georgia, 2010)
51. P. Nayak, S.K. Pramanick, K. Rajashekara, Isolated single stage AC-DC converter topologies with regenerative snubber circuit for EV application, in *IECON 2018 - 44th Annual Conference of the IEEE Industrial Electronics Society* (2018)
52. C.M. Wang, A Novel ZCS-PWM flyback converter with a simple ZCS-PWM commutation cell. IEEE Trans. Ind. Electron. **55** (2008)
53. K.S. Muhammad, D.D.C Lu, Single-phase single-stage ZCS boost PFC rectifier with reduced switch count, in *Australasian Universities Power Engineering Conference, AUPEC 2014*, (Curtin University, Perth, Australia, 2014)

An Overview of Solar-Powered Electric Vehicle Charging in Vehicular Adhoc Network

Farooque Azam, Neeraj Priyadarshi, Harish Nagar, Sunil Kumar, and Akash Kumar Bhoi

1 Introduction

Due to the rise in pollution and increase in the CO_2 level, different measures are being considered. To use electrical vehicle instead of petrol/diesel vehicle is the new promotions these days. Electrical vehicle is eco-friendly transportation system [1]. According to [2], EVs depend on rechargeable battery, produce zero emissions, and may bring down the emission from the transport sector too. Because of the mentioned benefits, electrical vehicle is going to have huge market in the transport. Depending on the assessment from Electric Power Research Institute (EPRI) [3], the electric vehicle demand level can reach 35%, 51%, and 62% by 2020, 2030, and 2050, respectively. Renewable energy sources are clean, eco-friendly, and abundant in nature [4–13]. As the penetration of EVs in market grows, load on the grid during peak hour must be supplemented by solar energy to provide uninterrupted services.

F. Azam
Sangam University, Bhilwara, Rajasthan 311001, India
e-mail: farooque53786@gmail.com

N. Priyadarshi (✉)
Department of Energy Technology, Aalborg University, 6700 Esbjerg, Denmark
e-mail: neerajrjd@gmail.com

H. Nagar · S. Kumar
Department of Computer Science, Sangam University, Bhilwara, Rajasthan 311001, India
e-mail: harish.nagar@sangamuniversity.ac.in

S. Kumar
e-mail: sunil.kumar@sangamuniversity.ac.in

A. K. Bhoi
Department of Electrical and Electronics Engineering, Sikkim Manipal Institute of Technology, Sikkim Manipal University, Sikkim 737136, India
e-mail: akashkrbhoi@gmail.com

N. Patel et al. (eds.), *Electric Vehicles*, Green Energy and Technology,
https://doi.org/10.1007/978-981-15-9251-5_5

The increase in the adoption of EVs highly depends on recharging battery that poses the requirement of charging station at different location and the real-time traffic information. Apart from this, information of electric load, EV location is also required [1]. Vehicular ad hoc network provides promising solution to the problem mentioned above. VANET is a self-organizing network. VANET provides two types of communication, viz. v2v and v2I for communication and exchanging critical information [3, 14]. Lu et al. [15] elaborate a scheme for car parking where it guides vehicles about the selected space from entry to the designated space. Addition to this, integrating the information of various electric vehicle supply equipment (EVSE) and electric vehicles in an area for charging enhances the user satisfaction. Dedicated short-range communication (DRSC) [16] and wireless access in vehicular environment (WAVE) are standards for V2V and V2I communications in VANETs.

Because of the high mobility and volatility, VANET suffers from both insider and outsider attack [17]. These attacks pose three challenges among researchers which are security, privacy, and trust. In this research work, an overview of security issues in VANET is being discussed.

The paper is organized as: Sect. 2 discusses the VANET system models and communication standards. Section 3 elaborates an interaction models for EV charging/discharging. Section 4 outlines the overview of solar-powered EV charging in VANET. Section 5 presents the conclusions and future work.

2 Overview of VANET

2.1 VANET Architecture

Figure 1 shows the VANET architecture which consists of electric vehicle equipped with on-board unit (OBU), electric vehicle supply equipment (EVSE) point, road-side unit (RSU), and their communication technology, viz. DSRC and 4G/5G. The on-board unit embedded in the electrical vehicle comprises of sensors to share vehicle information with other vehicles' on-board unit and the road-side units [18]. The road-side units are mounted alongside the road. In Fig. 1, for EV charging–discharging message, RSU exhibits as a service provider. RSUs facilitate the charging/discharging services for the electric vehicles. Road-side units are connected to the EV as well as other road-side units and infrastructure. EVSE is deployed within the communication range of dedicate RSU so that the EV can avail the services as and when required.

2.2 Standards

The main standards dealing with VANETs include (a) dedicated short DSRC and (b) WAVE.

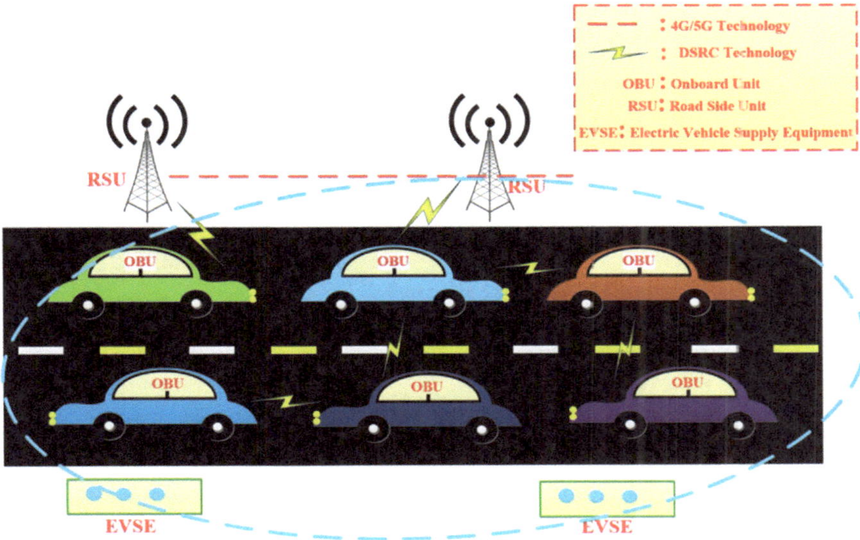

Fig. 1 VANET architecture

(a) DSRC
The United States' Federal Communications Commission (FCC) has allotted the spectrum of 75 MHz band wide from 5.850-5.952 GHz for dedicated short-range communication DSRC [19]. Figure 2 shows the seven channels of DSRC [20].

(b) WAVE
The wireless access in vehicular environments (WAVE) standard provides the communication in VANET. The smart grid communication standard is IEC 61850 [21]. The communication among EVs takes place through WAVE. Figure 3 shows the WAVE architecture. The WAVE model is a layered architecture comprises of IEEE 802.11p, 1609.4, 1609.3, 1609.2, 1609.1 standards. Table 1 shows the purpose of each protocol.

WAVE comprises of OBU which is equipped on the EV and RSU which are normally kept at the road or intersection. RSU acts as a service provider for relaying and disseminating the information.

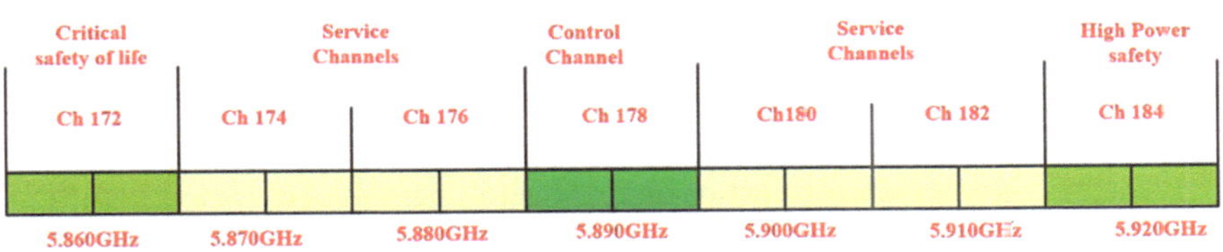

Fig. 2 DSRC

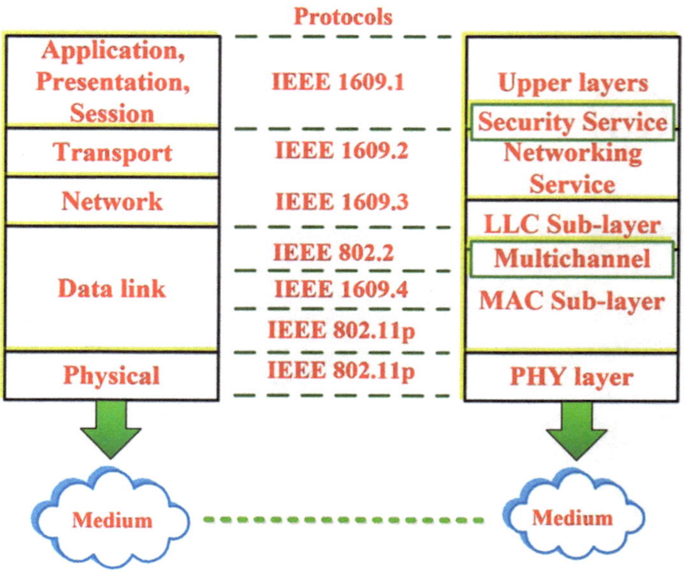

Fig. 3 WAVE architecture

Table 1 IEEE standards in WAVE

Standards	Purpose
IEEE 802.11p	Single-channel operations
IEEE 1609.4	Multi-channel operations
IEEE1609.3	Networking services
IEEE 1609.2	Security services
IEEE1609.1	Resource management

2.3 *Communication Pattern in VANET*

There are four types of communication pattern, viz. (a) warning message propagation among vehicles, (b) group communication among vehicles, (c) beaconing among vehicles, and (d) warnings between infrastructure and vehicles [22].

(a) Warning Message Propagation among Vehicles
Warning messages are sent to the designated EV in case of catastrophic situation. The algorithm must adhere to the strict time constraints in VANET.
(b) Group Communication among Vehicles
VANET is characterized by highly volatile decentralized self-organizing infrastructure. Group communication must adhere to its nature while establishing communication a particular set of EVs.
(c) Beaconing among Vehicles

RSU regularly send messages to all nearby EVs regarding traffic information, charging information, etc. This is sometimes called as beacon message.

(d) Warnings between Infrastructure and Vehicles
Alert messages must be sent by RSUs to all EVs in its range in case of danger expected or detected especially for complex road intersection.

3 EV Charging/Discharging Interaction Model

In this section, we introduce an overview of system model for charging and discharging.

In Fig. 4, RSU and EVSE are deployed in the different places. All electrical vehicles equipped with OBU are linked to nearby RSUs. We suppose that the EVs are able to communicate through VANET infrastructure their position, EV battery percentage status, etc., to the nearby road-side units. Here, RSU acts as service provider. The EV can avail the nob-safety services from the RSUs while passing by. Here, RSUs can send beacon messages about the charging/discharging services to nearby EV within its range. The interaction between the EV and RSU can be understood by Fig. 5.

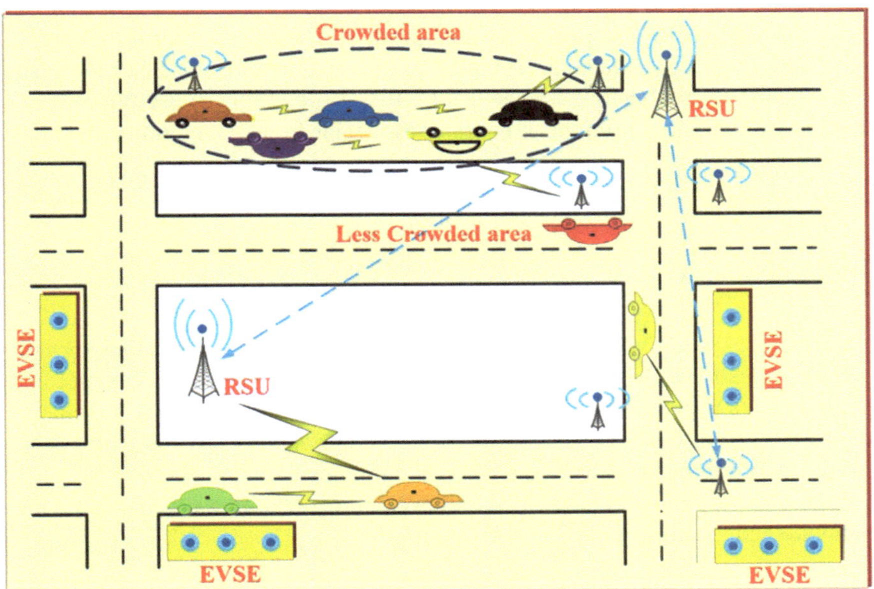

Fig. 4 Interaction model for EVs and RSU

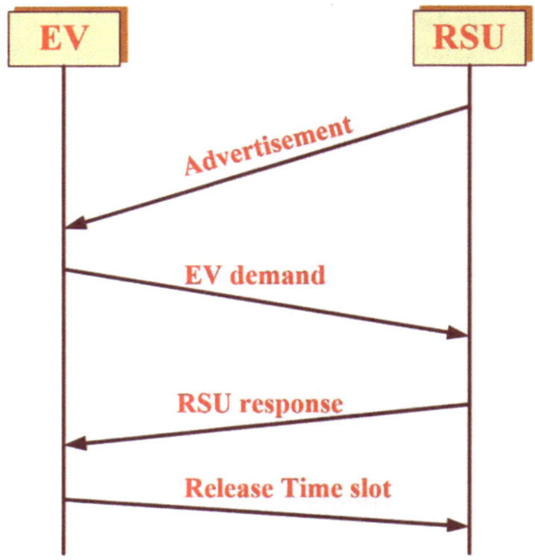

Fig. 5 Exchange of messages between EV and RSU

4 Overview of Solar-Powered EV Charging

Figure 6 shows the schematic overview of solar-powered grid integrated charging in vehicular ad hoc network. The solar panels feed the MPPT module. This is a DC–DC converter with a MPPT algorithm running inside it. The output feeds a common DC bus, from which downstream energy may be provided to the load. Since the system is essentially a solar energy storage setup, it is also possible to use this station to supplement the grid, during peak hours or as a solar farm, using bidirectional grid tied inverters.

5 Conclusion

In this research work, VANET-based charging has been discussed with understanding of communication technology and standards for the V2V, V2I, and V2G, respectively. Communication pattern and interaction model have been discussed for charging application. Finally, an overview of solar-powered grid integrated charging has been discussed to leverage the load from the grid during peak time.

Future work will be concentrated on developing a charging scheme and a more elaborate understating of V2G will be included.

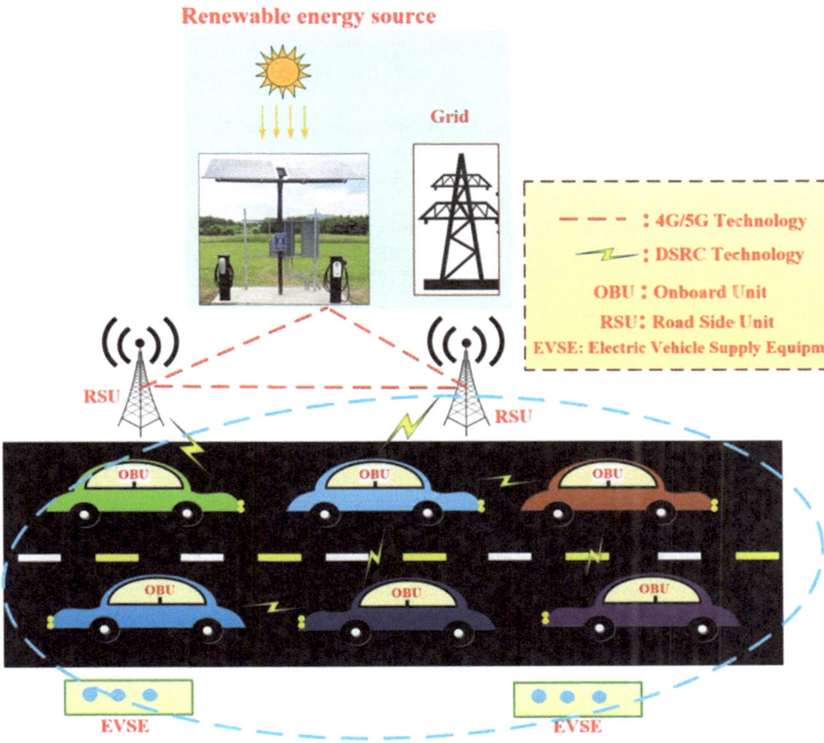

Fig. 6 Schematic setup for solar-powered grid integrated EV charging

References

1. M. Wang, H. Liang, R. Deng, R. Zhang, X.S. Shen, VANET based online charging strategy for electric vehicles, in *Processing of IEEE Global Communications Conference (GLOBECOM)*, (Atlanta, GA, 2013), pp. 4804–4809. https://doi.org/10.1109/glocomw.2013 6855711
2. Hybrid Cars, a comprehensive guide to plug-in hybrid vehicles (2011), Available http://www.hybridcars.com/plug-in-hybrid-cars/#battery
3. Electric Power Resech Institute. Available http://www.epri.com/
4. N. Priyadarshi, A. Anand, A.K. Sharma, F. Azam, V.K. Singh, R.K. Sinha, An experimental implementation and testing of GA based maximum power point tracking for PV system under varying ambient conditions using dSPACE DS 1104 controller. Int. J. Renew. Energy Res. **7**(1), 255–265 (2017)
5. N. Priyadarshi, A.K.. Sharma, F. Azam, A hybrid firefly-asymmetrical fuzzy logic controller based MPPT for PV-wind-fuel grid integration. Int. J. Renew. Energy Res. **7**(4) (2017)
6. N. Priyadarshi, S. Padmanaban, L. Mihet-Popa, F. Blaabjerg, F. Azam, Maximum power point tracking for brushless dc motor-driven photovoltaic pumping systems using a hybrid ANFIS-FLOWER pollination optimization algorithm. MDPI Energies **11**(1), 1–16 (2018)
7. N. Priyadarshi, F. Azam, A.K. Bhoi, S. Alam, An artificial fuzzy logic intelligent controller based MPPT for PV grid utility. Lect. Notes Netw. Syst. **46**. https://doi.org/10.1007/978-981-13-1217-5_88

8. S. Padmanaban, N. Priyadarshi, J.B. Holm-Nielsen, M.S. Bhaskar, F. Azam, A.K. Sharma, A novel modified sine-cosine optimized MPPT algorithm for grid integrated PV system under real operating conditions. IEEE Access **7**, 10467–10477 (2019). https://doi.org/10.1109/ACCESS.2018.2890533
9. S. Padmanaban, N. Priyadarshi, J.B. Holm-Nielsen, M.S. Bhaskar, E. Hossain, F. Azam, A hybrid photovoltaic-fuel cell for grid integration with jaya-based maximum power point tracking: experimental performance evaluation. IEEE Access **7**, 82978–82990 (2019). https://doi.org/10.1109/ACCESS.2019.2924264
10. N. Priyadarshi, N. Padmanaban, J.B. Holm-Nielsen, F. Blaabjerg, M.S. Bhaskar, An experimental estimation of hybrid ANFIS–PSO-based MPPT for PV grid integration under fluctuating sun irradiance. IEEE Syst. J. **14**(1), 1218–1229 (2020). https://doi.org/10.1109/JSYST.2019.2949083
11. N. Priyadarshi, N. Padmanaban, M.S. Bhaskar, F. Blaabjerg, J.B. Holm-Nielsen, F. Azam, A.K. Sharma, A hybrid photovoltaic-fuel cell-based single-stage grid integration with lyapunov control scheme. IEEE Syst. J. https://doi.org/10.1109/jsyst.2019.2948899
12. N. Priyadarshi, M.S. Bhaskar, N. Padmanaban, F. Blaabjerg, F. Azam, New CUK–SEPIC converter based photovoltaic power system with hybrid GSA–PSO algorithm employing MPPT for water pumping applications. IET Power Electron. 1–0 (2020) https://doi.org/10.1049/iet-pel.2019.1154
13. N. Priyadarshi, N. Padmanaban, J.B. Holm-Nielsen, M.S. Bhaskar, F. Azam, Internet of things augmented a novel PSO-employed modified zeta converter-based photovoltaic maximum power tracking system: hardware realisation. IET Power Electron. 1–0 (2020) https://doi.org/10.1049/iet-pel.2019.1121
14. R. Hussain, J. Lee, S. Zeadally, Trust in VANET: a survey of current solutions and future research opportunities. in IEEE Trans. Intel. Transp. Syst. (2020) https://doi.org/10.1109/tits.2020.2973715
15. R. Lu, X. Lin, H. Zhu, X. Shen, SPARK: a new VANET-based smart parking scheme for large parking lots, in *Proceedings IEEE INFOCOM 2009*, (Rio de Janeiro, Brazil, 2009), pp. 1413–1421
16. X. Huang, D. Zhao, H. Peng, Empirical study of DSRC performance based on safety pilot model deployment data. IEEE Trans. Intell. Transp. Syst. **18**(10), 2619–2628 (2017)
17. R.G. Engoulou, M. Bellaïche, S. Pierre, A. Quintero, VANET security surveys. Comput. Commun. **44**, 1–13 (2014)
18. M. Azees, P. Vijayakumar, L.J. Deborah, Comprehensive survey on security services in vehicular ad-hoc networks. IET Intell. Transp. Syst. **10**(6), 379–388 (2016)
19. Federal Communications Commission, Amendment of the commission rules regarding dedicated short-range communication service in the 5.850–5.925 GHz band, (FCC, Washington, DC, USA, Tech. Rep. FCC 02-302, 2002)
20. Z. Lu, G. Qu, Z. Liu, A survey on recent advances in vehicular network security, trust, and privacy. IEEE Trans. Intell. Transp. Syst. **20**(2), 760–776 (2019). https://doi.org/10.1109/TITS.2018.2818888
21. S.M.S. Hussain, T.S. Ustun, P. Nsonga, I. Ali, IEEE 1609WAVE and IEC 61850 standard communication based integrated EV charging management in smart grids. IEEE Trans. Veh. Technol. **67**, 7690–7697 (2018)
22. J.M. de Fuentes, A.I. González-Tablas, A. Ribagorda, *Overview of Security Issues in Vehicular Ad-Hoc Networks* (IGI Global, Hershey, PA, USA, 2010)

Study on Electric Vehicle (EV) and Its Developments Based on Batteries, Drive System and Charging Methodologies in Modern World

V. Dhinakaran, R. Surendran, M. Varsha Shree, and Parul Gupta

1 Introduction

An Electric Vehicle (EV) is a control system that uses one or more electric motors. It can be powered from off-road supply by a collector network or can be fitted into an electricity generator, solar cells or power turbine [1]. In the mid-nineteenth century, electric power first became one of the favorite propulsion devices for vehicles, providing a comfort and operation facility that oil vehicles could not obtain at that time. The main propulsion system of motor vehicles is now the domestic combustion motor for around 100 years, but electric technology has been impacted by certain forms of vehicles, such as trains and cars of every age. The term EV is generally applied to an electric car. In the twenty-first century, EVs saw resurgence because of technological developments and a stronger focus on renewables [2]. Developed a lot of interest in electric cars, and a small group of do-it-yourself (DIY) engineers started to exchange technological information for conversions of electric vehicles. Economic programs, in the United nation, were implemented to improve adoptions. Owing to storage capacity drawbacks, electric cars have not gained much popularity at the moment but electric trains have gained considerable notoriety to their reliability and practical speeds. A hybrid vehicle in the series is basically an electric vehicle to charge the batteries with an onboard power source. In general, a motor is coupled with a generator for generating the power to charge the pump. The system may also be constructed in such a way that the generator can act as a power-propelled load-leveling device. In this sense, the battery size could be reduced but the battery size could be reduced, and the generators and engine needed to be improved [3]. Parallel hybrids can have the cheapest price and versatility to leverage proven technologies

V. Dhinakaran (✉) · R. Surendran · M. Varsha Shree
Centre for Applied Research, Chennai Institute of Technology, Chennai 600069, India
e-mail: dhinakaranv@citchennai.net

P. Gupta
Department of Mechanical Engineering, Moradabad Institute of Technology, Moradabad, India

N. Patel et al. (eds.), *Electric Vehicles*, Green Energy and Technology,
https://doi.org/10.1007/978-981-15-9251-5_6

in motor, battery, and generator output. A parallel hybrid vehicle thus involves a rigorous system of control. The development of metal-oxide-semiconductor (MOS) technology has helped to build new on-road electric vehicles. The MOSFET power and the microcontroller, a single-chip microprocessor form, introduced important developments in the development of electric vehicles. The reverse is intended for parallel hybrid vehicles, which rely on electric motor purposes as well as the generator and engine [4]. Large-scale EV permeation during charging times is expected to increase energy use. Therefore, power flows, losses of grid, and patterns of voltage profiles along the grid will vary radically. In addition, EV"s power supply reliability to the network would also affect grid flows. Combining the two findings which include enhancing the grid elsewhere. Reinforcement postpones will therefore be obtained depending on the technique to be followed for charging the EV [5, 6].

2 Batteries Utilized in Electric Vehicles

The demands for electrical power in the automotive industry while minimizing device weight and volume, impacting vehicle packaging, and optimizing system performance. In addition, these batteries may be recharged (secondary), typically lithium-ion. Such batteries are primarily designed for a high-efficiency ampere-hour (or kilowatt-hour). Electric car batteries vary from SLI batteries because they are equipped to provide electricity for a long time and are deep cycle batteries. These are a battery that is used by electric vehicles. Electric car batteries can be distinguished with a fairly large power to weight ratio, basic capacity, and density; smaller, lighter batteries make the weight of the car attractive and therefore increase its efficiency. New battery kits provide slightly fewer energy space relative to liquid fuels, frequently impacting the entire range of automobiles. The most popular form of batteries in modern electric cars is the lithium-ion and lithium-polymer, due to their high energy efficiency compared with their weight [7].

3 Classification of Batteries Used in Electric Vehicles

3.1 Lead-Acid Batteries

Batteries flooded with acid are the cheapest and most popular car-driven. The two most common types of batteries are electric car batteries and deep cycle batteries. Automotive engine starter batteries are built to provide a low power percentage and fast charging rates for starting machines, while low-cycle electric batteries, such as chariots and golf carts, are used to supply electric vehicles with continuous energy. Deep cycle batteries in recreational vehicles are often used as an auxiliary battery but must be charged in various phases [8]. Flooded batteries must be inspected at electrolyte levels, often drained by water and the gasses are collected during the normal fee. Due to their sophisticated technology, ease of access, and low price, most electric vehicles use lead-acid batteries, except for some early BEVs like Detroit Electric, which used a battery of Nickel-Iron [9]. Because of its advanced technology, low cost, and moderate efficiency, lead-acid batteries are still commonly used in commercially available EVs. Moreover, the recent great development in lead-acid batteries have shown that it is almost unlikely for other modern batteries to replace lead-acid batteries completely in EVs in the near future [10]. Intricate hypotheses occurring within the batteries are still not fully understood. This creates problems for suppliers and consumers because they don't always understand which problems could appear under specific conditions. Such batteries really do have was used without fundamental issues for a long time after there was still no major restriction applied to them [11].

3.2 Nickel Metal Hydride

Intricate hypotheses occurring within the batteries are still not fully understood. This creates problems for suppliers and consumers because they don't always understand which problems would appear under specific conditions. These batteries have actually been used for a long time without fundamental issues after they were still not subject to any major restrictions. While several hybrid electric vehicles (HEVs) rely primarily on batteries built from nickel-metal hydride (NiMH). Lithium-ion (Li-ion) batteries (Li-ion) will therefore soon be usable. The market is expected to dominate, especially with the increase in solely electric Cars plug-in hybrid (PHEV) and battery-powered [12]. Its active ingredients are hydrogen in the form of metal hydride for negative electrodes and positive oxyhydroxide for electrode. When the battery is released, the metal hydride is oxidized into a metal alloy, which decreases nickel oxidation to nickel hydroxide. After charging the pump, the reverse reactions occur. These substances affect the existing profile of discharge and temperature.

3.3 Molten Salt Battery

Molten-salt batteries are a battery type that utilizes molten salts as an electrolyte, giving a good energy capacity as well as high power efficiency. Current non-rechargeable thermal batteries can be retained for long periods of time in their solid state at room temperature before being enabled by heating. The ZEBRA cells are made in discharged condition. The liquid salt $NaAlCl_4$ is impregnated with vacuum in the porous mixture of nickel-salt which forms the cathode [13]. ZEBRA cell's charge capacity is calculated by the amount of salt (NaCl) present in the cathode. They were seen in other EVs, including the industrial vehicle Modec. Zebra batteries are durable for several thousand cycles of cycling and are non-toxic. The drawback to the Zebra battery involves low specific capacity (<300 W/kg) and the necessity that the electrolyte is heated to around 270 °C (518 °F), which consumes some electricity, poses issues with long-term charging storage and is potentially a danger [14].

3.4 Lithium-Ion

Lithium-ion batteries support higher energy costs, exceptional power efficiency, longer service life, and eco-friendliness compared with other commonly used batteries and have proven extensive in electronic goods. But the lithium-ion automobile batteries have a large capacity and a huge number of series parallels, combined with protection, power, accuracy, and the cost constraints of the large lithium-ion car battery and also represented in Fig. 1. The lithium-ion batteries must function securely and consistently over the high temperature- and stress-limited environment. Incapacitating the restrictions of these windows can cause in a rapid battery enactment, amplification, and can even lead to security difficulties. This also diminishes the electrolyte and creates combustible gasses by comparatively lesser voltage or surplus. In Audi A6 (EV) electric applications, voluminous types of lithium-ion batteries are used primarily. The battery power cells most commonly used have the anode of carbon (negative electrodes), and the LTO anode is robust for such anode [15]. Traditional lithium-ion batteries are limited by temperature dependency, low-temperature energy usage, and lack of efficiency induced by ageing. Modern lithium-ion batteries present a fire-safety hazard when punctured, or wrongly charged, due to the volatility of organic electrolytes, the existence of highly oxidized metal oxides, and the thermal weakness of the SEI anodes. Though unbelievably cold, these early cells did not accept energy or supply it with it, heaters might need to warm it up in some climates. This platform is fairly mature [16]. New lithium-ion chemistry variants that lose particular power and strength are used in recent EVs to provide fire safety, eco-friendly loading, and longer service life (as fast as a couple of minutes) [17].

Fig. 1 Lithium-ion battery

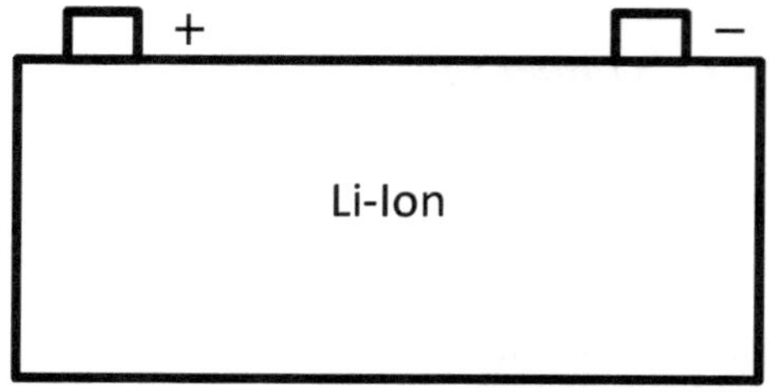

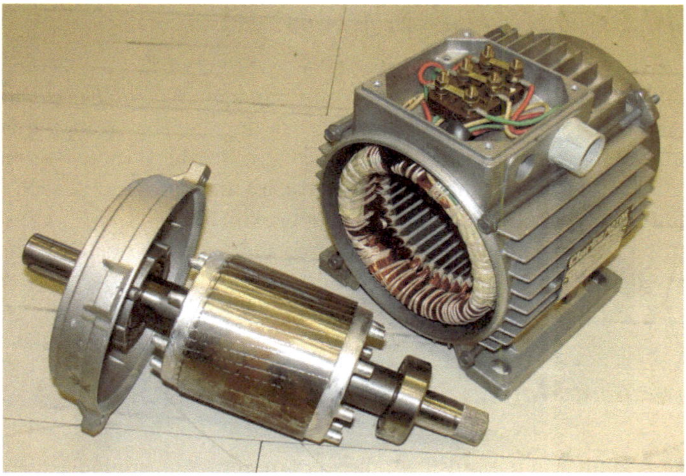

Fig. 2 Representation of electric motor [19]

4 Motors Used in EV

Electric vehicles are not anything fresh for this country, but a tag of future mobility has been granted to it with the technical development and intensified commitment to reduce emissions. Apart from the electric car batteries that substitute the internal combustion engines, the central feature of the EV is an electric motor [18]. With the growing advancement of power electronics and control systems, different types of motors in electric vehicles have been used. The electric motors used for automotive use should have rapid start torque, great electricity density, respectable efficiency, and it is shown in Fig. 2.

5 Classification of Electric Motors Used in EV

5.1 Direct Current Series Motor

The extraordinary starting torque capacity of the Direct Current Series motor brands this as an ideal choice for traction applications. This was the most quite often-used traction engine in the early 1900s. This engine has the advantages of being quick in controlling speed, and can also manage a spontaneous power shift. All of these features make it a perfect traction engine. Since the brushes and shutters, the biggest drawback in the Direct Current series engine is high maintenance. It is found along railroads. This engine is in the Direct Current brushed motor category [20].

5.2 Brushless Direct Current Motors

This is similar to direct current motors with Permanent Magnet. It is called brushless since there is no available switch and brush device. The swapping is achieved remotely in this engine for making maintenance-free Brushless direct current engines. Brushless direct current motors engines have traction properties such as a high starting torque, a decent performance of approximately 95–98%, etc. [21]. The Brushless direct current motors perhaps, the most favored motors for electric vehicle applications, due to their traction capabilities.

5.3 Permanent Magnet Synchronous Motor (PMSM)

This engine is similar to the brushless direct current motors engine, except the rotor has permanent magnets. These engines also have traction properties such as high power density and high speed, comparable to those of brushless direct current motors. The difference is that PMSM has sinusoidal back EMF while brushless direct current motors have trapezoidal back EMF [22]. Higher power ratings are given by permanent Magnet Synchronous motors. PMSM is the perfect option for high demand systems such as cars and buses. Notwithstanding the high expense, owing to better performance than the latter, PMSM is offering stable resistance to induction engines. PMSM is also better than generators from brushless direct current motors [23]. Most car manufacturers use PMSM motors for their hybrid and battery automobiles. For example, it is used as propulsion in PMSM motors for Toyota Prius, Chevrolet Bolt EV, Chevrolet Bolt, Ford Focus Electronic, Nissan Leaf, Hinda Alliance, etc.

5.4 Three Phase AC Induction Motors

The induction engines have no good starting toque, like direct current series engines, under fixed voltage and fixed frequency activity. This attribute can, however, be altered with various control methods like FOC or voltage and frequency. At engine boot time, these control systems make the optimum torque suitable for traction accessible. The squirrel cage induction engines have a long life cycle owing to fewer maintenance. Powerful motors from 92 to 95% can be designed. An induction engine has the downside of involving and being complicated to run a complex inversing circuit. Tesla Model S is an excellent example in contrast with its counterpart for showing strong induction motor control efficiency. Tesla would have sought to reduce its reliance on permanent magnets by using the induction motors. As a result, Indian railways began replacing their direct current engines with AC induction engines [24].

5.5 Switched Reluctance Motors (SRM)

Switched Reluctance Motors is a double-salience variable motor reticence type. Switched Reluctance motors are sturdy and easy to install. The SRM rotor is a piece of laminated steel that is lacking any windings or permanent magnets. This lessens the rotor's inertia and helps with high acceleration. Switched Reluctance Motors is a double-salience variable motor reticence type. Switched Reluctance motors are sturdy and easy to install. The SRM rotor is a piece of laminated steel that is lacking any windings or permanent magnets. This lessens the rotor's inertia and helps with high acceleration. The SRM's greatest downside is the difficulty of the switching circuit to monitor and through. It also has some problems with noise [25]. When SRM joins the commercial industry it should be able to supplement potential PMSM and Induction engines.

6 Charging Methodologies of EVs

There are different approaches for recharging electric car batteries as represented in Fig. 3. Currently, the main issue concerning electric car transportation is the limited usable travel range until recharging is needed. The largest reported range to date was 606.2 miles, accomplished with a Tesla Model 3. This was achieved, though, in very regulated conditions where the car retained a steady speed without the air-conditioning compressor drain being applied. There are three major charging forms of EV-medium, medium, and sluggish. These reflect the power outputs available to charge an EV, and hence the charging speeds [26]. Note that kilowatts (Kilowatt) are used to calculate power.

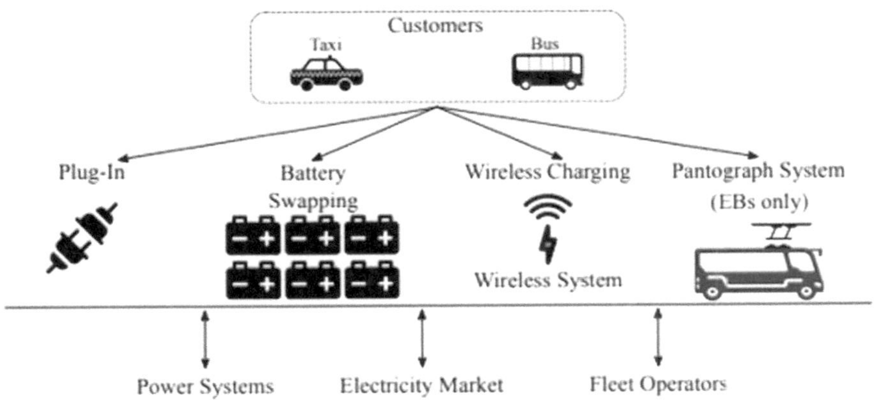

Fig. 3 EV charging methodology [27]

6.1 Rapid Chargers

The easiest way to charge an EV is especially for direct current loading. This is high and fast in two different categories. With the direct current it's only feasible to take out up to 100 KW, sometimes 150 KW and up to 350 KW, extremely quick charging. The bulk of UK's 50 KW direct current fast-charging network, which is really easy to monitor with 43 KW AC, is Conservative Rapid Points. The fast chargers include electricity supplying between 7 and 22 KW and typically charging the EV for up to 3–4 h. The most popular charging point for connectors Type 1 and Type 2 in the UK is a 7 KW unheated inlet level 2. For 3–6 KW slow loaders are used and usually take 8–12 h for pure EV or two-quarters of PHEV during the night service. EVs operate on complicated devices with a 3-pin or type-2 socket link cable for the car [28].

6.2 Constant Current

Constant current charging mechanism regulates the output voltage of charging devices or the sequential resistance with the battery to retain current flow in a steady manner. This uses a constant current value, which changes capacity from start to finish. In both conventional constant voltage and constant current charging methods, the electrolyte uninterruptedly produces hydrogen-oxygen gas, so that the nickel-cadmium batteries are reset to polarize after conservative charging. The oxygen permeates the negative electrode under the impact of internal high pressure and interfaces with the cadmium layer in order to create CdO [29]. As the battery's adequate current capacity decreases gradually as the charge cycle continues, this contributes to the later charging time overloading of the device. Eventually, this would lead to a drastic decrease in battery capacity.

6.3 Constant Voltage

Constant voltage charging is a widely used technique of charging that includes constant tension amongst the battery poles. The starter battery uses an unceasing voltage charge when the engine is in running ailment. If the specified voltage constant value is suitable, it will ensure supreme battery charging with the loss of water and gas. Alternative expansion of the constant current or constant voltage algorithms is the new CC/CV charging algorithm for converters. As a replacement for using constant voltage and current during the complete charge cycle, the charging efficiency is amplified by increasing voltage in the first step, with the battery attaining about 30% of its average charging capacity. After this time the charging algorithm is interchanged to standard CC/CV [30].

6.4 Non-contact Charging Method

Non-contact charging necessitates the usage of magnetic resonance to permit energy in the liquid phase among the device and the battery. This will accomplish an extremely competent energy conversion. It consents EVs to partake a minor battery, as the non-contact charger can endure charging the engine. It's faster, cleaner, and propagates by itself more sustainably. Since the battery is the main supplier to the cost of an EV, the MSRP for an EV is dropped due to the use of non-contact charging. On the other hand, this method of charging needs significant financial resources to build a non-contact charging system. Because of that, many EV producers use conventional methods of charging to keep costs down. Due to the fact that non-contact charging systems be dependent on the electromagnetic field as their mode of process, electronic devices near the charger may be unfavorably affected when charging, their animals are also at risk of being manipulated [31].

7 Short Comings of Fast Charging Points

Fast-charging points are a blessing to eliminate imposed stoppages for the time-crunched EV users. But scientists have found that using high-powered, fast-charging points on an ongoing basis can diminish the life of EV car batteries. The researchers have charged a series of cells utilizing a modern algorithm-based technique that charges the batteries as fast as possible but takes into consideration the internal resistance of the battery. High internal resistance in a battery will trigger issues while charging, according to the engineers. Battery cells may be permanently harmed if these go unaddressed. Commercial fast-charging stations expose EV car batteries to high temperatures and resistance which can trigger cells to break, leak, and loss storage ability [32].

8 Types of Electric Vehicles

The electric vehicles (EVs) are classified into three different categories that utilize electricity as their source of energy. BEVs or electric cars, or conventional hybrid vehicles or hybrid batteries. Power plug-in PHEVs and HEVs. Even BEVs can give direct current at level 3.

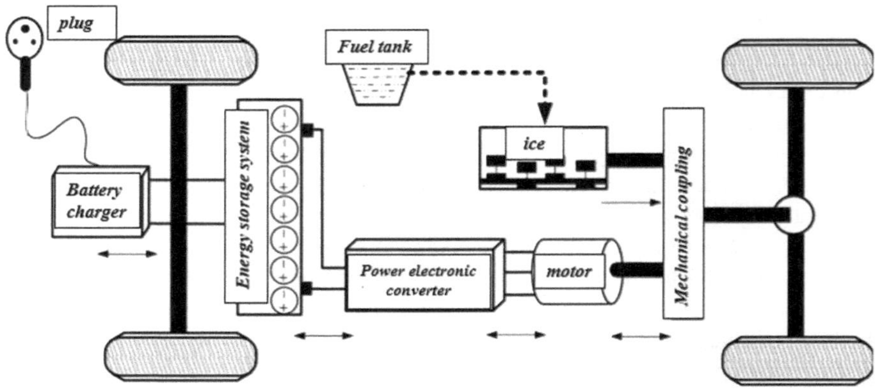

Fig. 4 Representation of plug-in EV [34]

8.1 Plug-in Electric Vehicle

Electric Plug-In Vehicle is any electric vehicle that can be refilled from some external power source, such as electric outlets, and which activates or contributes considerably to wheel activity in rechargeable battery pickings, as seen in Fig. 4. PEV is a subcategory of electric cars that includes hybrid and electric vehicles that are fully or battery driven (BE Automobiles), hybrid and PHEVs and hybrid electric car hybrids. Plug-ins of hybrid electric vehicles (PHEV) are supplied from regenerative and 'plug-in' systems by the battery to external supplies of electrical fuel while "normal" hybrids operate within 1–2 miles (at low speeds) until the petrol engine has turned on, PHEV versions can fly from 10 to 40 miles to the point of delivery of assistance for their gas engines [33].

8.2 Hybrid EVs

A Modern Hybrid Vehicle requires a conventional electric propeller with some sort of control (usually driven by fossil fuel). More than 11 million electric hybrid vehicles have been produced worldwide since its introduction in 1997. Japan is the world's top producer of over five million assembled hybrids, with over four million units in total since 1999 and around 1.5 million hybrids since 2000 supplied by Europe. Japan's hybrid business is the largest percentage in the area. Roughly 30% of the most recent traditional passenger cars were delivered in 2013 and more than 20% of all commercial vehicles such as key vehicles accounted for the hybrid market share. Electric and gas-powered HEVs. The electricity is supplied by the car's own battery charging braking system. Around 30% of the last conventional passenger car shipped in 2013, and over 20% of all industrial car exports, such as main cars,

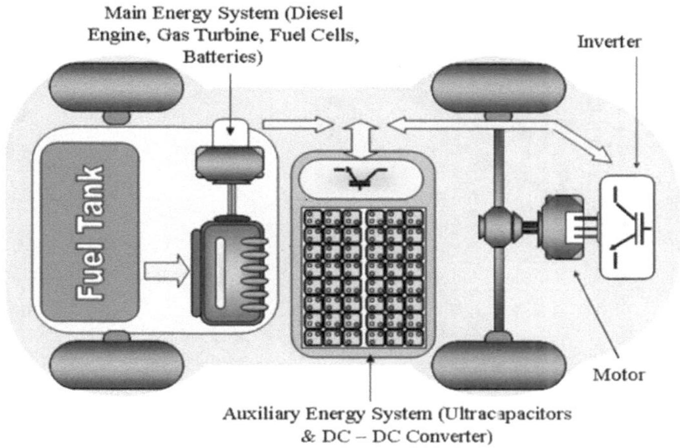

Fig. 5 Hybrid electric vehicle [36]

accounted for the hybrid market share. Electric and gas-powered HEVs. To recharge the battery, the electrical energy is provided by the car's own braking mechanism and it is represented in Fig. 5 [35].

8.3 Railborne EVs

A continuous line configuration allows EVs to be driven fairly straight from permanent overhead lines or electrified three-way lines, eliminating the need for large batteries on board. Electric locomotives, electric trams, roads, trolleys, power lines, and high-speed fast transport are used extensively, particularly in Europe and Asia. Nevertheless, there are no massive internal combustion engines and big batteries in electric cars, which may also follow very small strength to weight ratios. High-speed trains, such as French twin-deck TGVs, are operating at or over 320 km/h (200 miph) speeds, which is far higher in strength than petrol. They also have specialized power for fast acceleration in the short term and the application of reforms would re-engineer and not exacerbate the braking power of the electric grid as it is shown in Fig. 6 [37].

8.4 Electrically Powered Spacecraft

An electrically driven spacecraft propulsion device utilizes electrical, and possibly magnetic fields as well, to adjust a spacecraft's altitude. Most of these forms of propulsion systems in spacecraft operate at high speed by electrically expelling propellant

Fig. 6 Railborne electric vehicle [38]

(reaction mass The thrust is much lower relative to chemical rockets owing to insufficient electrical capacity but electrical propulsion will produce a strong thrust for a long period. For certain deep-space flights, electric propulsion can reach high speeds for long stretches, and can thus perform better than chemical rockets. Electrical propulsion now on spacecraft is a mature and widely used technology. Russian satellites have been utilizing electric propulsion for decades and half of all current satellites are planned to carry full electric propulsion by 2020 as shown in Fig. 7. By 2019, more than 500 spacecraft powered in the Solar System have been utilizing electrical propulsion for station servicing, orbit raising, or primary propulsion. In the future, the most powerful electric thrusters may be able to produce a 100 km/s Delta-v, which is adequate to carry a spaceship to the solar system's outer planets (with nuclear power), but is not appropriate for interstellar flight [39].

Fig. 7 Electric spacecraft [40]

9 Need for Electric Vehicles

A battery-powered electric vehicle (BEV) has much less moving parts than a traditional petrol engine. This reduces much of the maintenance costs associated with an internal combustion engine. The overall carbon content is not only lower but the position and timing of the contaminants are stronger [41]. While a gasoline-powered car produces waste during peak driving hours in the city centre, pollution from charging an EV typically occurs during off-peak driving hours at remote power plants. The total cost of owning EV lies well below the cars powered by gasoline. As with other fuel-powered vehicles, electric cars are subject to the same strength check and test protocols. One would expect airbags to open up in the event of an incident, and battery power to be shut off. This will prevent you and other passengers in the car from getting serious injuries. Electric vehicles rely on electric motors and therefore do not need to lubricate the batteries [42]. As they are louder, electric cars are putting noise pollution at a standstill. Electric motors can provide greater acceleration over longer distances to smooth running. The main mode of transport in the future will most likely be an electric car, considering that the demand for oil would only increase when supplies run out.

10 Cons of Electric Vehicles

Although the promising evidence is becoming very clear, there are many other downsides that need to be considered by each person before deciding to make their next big speculation in an electric car. Development of electric fuelling stations endures. Not many places you will be finding frequently electric charging stations for your vehicle. Electric cars are speed and range-restricted. Most of these vehicles are about 50–100 miles away, and need to be recharged. Electric vehicles are more expensive because the supplier cannot entirely recover the expense of the components that have been scrapped, so modern products and technologies are costly. The distance that electric cars will travel until full battery loss is limited; the maximum range is just around 100 miles. Electric vehicles cannot drive, speed, or ascend quickly enough to contend with automobiles and appliances fuelled by oil, such as air conditioning and radios, which exhaust the battery any further. Electric cars are not fully "emission-free." Whether the energy used is provided by coal or oil-fired engine, the pollutants from the tailpipe are passed to the power plant only [43].

11 Conclusion

There are numerous technologies on the globe to be executed in the forthcoming generations of automobiles. Electric cars offer a means of transportation primarily to minimize overall carbon emissions. This research work offers a technical basis for the effective introduction of electric cars into power systems. The market success of the electric car would entail the construction of an open, simple to use and fairly inexpensive charging system. Unprecedented rates of spending and product production expected by virtually all major OEMs strongly suggest that there is coming a far greater EV sector.

- Automotive firms are designing diesel, hybrid electric, and plug-in hybrid electric cars, with the specifications to diminish pollution and upsurge efficiency of the engine.
- Power electronics is a process that allows the production of environmentally sustainable automobiles and the introduction of modern electrical systems to satisfy the demands for increased electrical charges.

A massive integration of EVs will pose significant challenges to the process and management of future electrical power systems. This set-up will entail the growth of a satisfactory framework for technical and market operations to reduce the adverse effects that EVs could upturn their prospective applications.

References

1. A. Emadi, Y.J. Lee, K. Rajashekara, Power electronics and motor drives in electric, hybrid electric, and plug-in hybrid electric vehicles. IEEE Trans. Ind. Electron. **55**(6), 2237–2245 (2008)
2. J.A. Lopes, F.J. Soares, P.M. Almeida, Integration of electric vehicles in the electric power system. Proc. IEEE. **99**(1), 168–183 (2020)
3. K.T. Chau, C.C. Chan, C. Liu, Overview of permanent-magnet brushless drives for electric and hybrid electric vehicles. IEEE Trans. Ind. Electron. **55**(6), 2246–2257 (2008)
4. A. Sciarretta, L. Guzzella, Control of hybrid electric vehicles. IEEE Control Syst. Mag. **27**(2), 60–70 (2007)
5. K. Clement-Nyns, E. Haesen, J. Driesen, The impact of charging plug-in hybrid electric vehicles on a residential distribution grid. IEEE Trans. Power Syst. **25**(1), 371–380 (2009)
6. B. Nykvist, M. Nilsson, Rapidly falling costs of battery packs for electric vehicles. Nat. Clim. Change **5**(4), 329–332 (2015)
7. L. Lu, X. Han, Jianqiu Li, J. Hua, M. Ouyang, A review on the key issues for lithium-ion battery management in electric vehicles. J. Power Sources **226**, 272–288 (2013)
8. C. Mi, M. Abul Masrur, *Hybrid electric Vehicles: Principles and Applications with PRACTICAL Perspectives* (Wiley, Chichester, 2017)
9. T.R. Hawkins, B. Singh, G. Majeau-Bettez, A.H. Strømman, Comparative environmental life cycle assessment of conventional and electric vehicles. J. Ind. Ecol. **17**(1), 53–64 (2013)
10. O. Egbue, S. Long, Barriers to widespread adoption of electric vehicles: an analysis of consumer attitudes and perceptions. Energy policy **48**, 717–729 (2012)
11. D.B. Richardson, Electric vehicles and the electric grid: a review of modeling approaches, Impacts, and renewable energy integration. Renew. Sustain. Energy Rev. **1**(19), 247–254 (2013)
12. S.F. Tie, C.W. Tan, A review of energy sources and energy management system in electric vehicles. Renew. Sustain. Energy Rev. **1**(20), 82–102 (2013)
13. T.B. Christensen, P. Wells, L. Cipcigan, Can innovative business models overcome resistance to electric vehicles? Better Place and battery electric cars in Denmark. Energy Policy **1**(48), 498–505 (2012)

14. M.U. Cuma, T. Koroglu, A comprehensive review on estimation strategies used in hybrid and battery electric vehicles. Renew. Sustain. Energy Rev. **1**(42), 517–531 (2015)
15. F. He, Y. Yin, S. Lawphongpanich, Network equilibrium models with battery electric vehicles. Transp. Res. Part B Methodological **67**, 306–319 (2014)
16. M. Weiss, M.K. Patel, M. Junginger, A. Perujo, P. Bonnel, G. van Grootveld, On the electrification of road transport-learning rates and price forecasts for hybrid-electric and battery-electric vehicles. Energy Policy **1**(48), 374–393 (2012)
17. M.K. Gray, W.G. Morsi, Power quality assessment in distribution systems embedded with plug-in hybrid and battery electric vehicles. IEEE Trans. Power Syst. **30**(2), 653–671 (2014)
18. M. Yildirim, M. Polat, H. Kürüm, A survey on comparison of electric motor types and drives used for electric vehicles, in *2014 16th International Power Electronics and Motion Control Conference and Exposition*, pp. 218–223 (IEEE, 2014)
19. https://upload.wikimedia.org/wikipedia/commons/5/58/Stator_and_rotor_by_Zureks.JPG
20. R. Cao, C. Mi, M. Cheng, Quantitative comparison of flux-switching permanent-magnet motors with interior permanent magnet motor for EV, HEV, and PHEV applications. IEEE Trans. Magn. **48**(8), 2374–2384 (2012)
21. A.M. Lulhe, T.N. Date, "A technology review paper for drives used in electrical vehicle (EV) & hybrid electrical vehicles (HEV), in *2015 International Conference on Control, Instrumentation, Communication and Computational Technologies (ICCICCT)*, pp. 632–636 (IEEE, 2015)
22. G. Bailey, N. Mancheri, K. Van Acker, Sustainability of permanent rare earth magnet motors in (H) EV industry. J. Sustain. Metall. **3**(3), 611–626 (2017)
23. M. Obata, M. Shigeo, M. Sanada, Y. Inoue, High-performance PMASynRM with ferrite magnet for EV/HEV applications, in *2013 15th European Conference on Power Electronics and Applications (EPE)*, pp. 1–9 (IEEE, 2013)
24. J. Su, R. Gao, I. Husain, Model predictive control based field-weakening strategy for traction EV used induction motor. IEEE Trans. Ind. Appl. **54**(3), 2295–2305 (2017)
25. T. Hadden, J.W. Jiang, B. Bilgin, Y. Yang, A. Sathyan, H. Dadkhah, A. Emadi, A review of shaft voltages and bearing currents in EV and HEV motors, in *IECON 2016–42nd Annual Conference of the IEEE Industrial Electronics Society*, pp. 1578–1583 (IEEE, 2016)
26. A. Ul-Haq, C. Buccella, C. Cecati, H.A. Khalid, Smart charging infrastructure for electric vehicles, in *2013 International Conference on Clean Electrical Power (ICCEP)*, pp. 163–169 (IEEE, 2013)
27. J. Moreno, M.E. Ortúzar, J.W. Dixon, Energy-management system for a hybrid electric vehicle, using ultracapacitors and neural networks. IEEE Trans. Ind. Electron. **53**(2), 614–623 (2006)
28. P. Zhang, K. Qian, C. Zhou, B.G. Stewart, D.M. Hepburn, A methodology for optimization of power systems demand due to electric vehicle charging load. IEEE Trans. Power Syst. **27**(3), 1628–1636 (2012)
29. F.J. Soares, P.R. Almeida, J.P. Lopes, Quasi-real-time management of electric vehicles charging. Electric Power Syst. Res. **1**(108), 293–303 (2014)
30. C.S. Antúnez, J.F. Franco, M.J. Rider, R. Romero, A new methodology for the optimal charging coordination of electric vehicles considering vehicle-to-grid technology. IEEE Trans. Sustain. Energy **7**(2), 596–607 (2016)
31. H. Shareef, M.M. Islam, A. Mohamed, A review of the stage-of-the-art charging technologies, placement methodologies, and impacts of electric vehicles. Renew. Sustain. Energy Rev. **1**(64), 403–420 (2016)
32. G. Li, X.-P. Zhang, Modeling of plug-in hybrid electric vehicle charging demand in probabilistic power flow calculations. IEEE Trans. Smart Grid **3**(1), 492–499 (2012)
33. J. Larminie, J. Lowry, *Electric vehicle technology explained* (Wiley, Chichester, 2012)
34. https://lh3.googleusercontent.com/proxy/k5aXLpFfZ44Zz4bbxi0J26wf55t1DeE403-3_q Rfazm-ruIARmWLe_Foh36BFEFTO5r-Fcj39UNxwVcyPqfLcheh2L47a_UbPNGoT55d-4rYbOg6X9Kz8Q
35. A. Poullikkas, Sustainable options for electric vehicle technologies. Renew. Sustain. Energy Rev. **41**, 1277–1287 (2015)

36. S. Ji, C.R. Cherry, M.J. Bechle, W. Ye, J.D. Marshall, Electric vehicles in China: emissions and health impacts. Environ. Sci. Technol. **46**(4), 2018–2024 (2012)
37. R. Faria, P. Moura, J. Delgado, A.T. De Almeida, A sustainability assessment of electric vehicles as a personal mobility system. Energy Convers. Manage. **1**(61), 19–30 (2012)
38. https://p0.pikrepo.com/preview/788/577/empty-white-and-orange-electric-train.jpg
39. U. Eberle, B. Müller, R. Von Helmolt, Fuel cell electric vehicles and hydrogen infrastructure: status 2012. Energy Environ. Sci. **5**(10), 8780–8798 (2012)
40. https://www.syfy.com/sites/syfy/files/wire/legacy/edu_juno_mission_jupiter.png
41. S. Bubeck, J. Tomaschek, U. Fahl, Perspectives of electric mobility: total cost of ownership of electric vehicles in Germany. Transp. Policy **50**, 63–77 (2016)
42. D. Zhang, Y. Guo, Y. Shen, *A frequency-separated strategy for the integration of different types of electric vehicles into grid frequency regulation*, pp. 167–6 (2019)
43. J.-M. Clairand, P. Guerra-T, X. Serrano-Guerrero, M. González-Rodríguez, G. Escrivá-Escrivá, Electric vehicles for public transportation in power systems: a review of methodologies. Energies **12**(16), 3114 (2019)

An Overview on the Prominence of Phase Change Material Based Battery Cooling and Role of Novel Composite Phase Change Material in Future Battery Thermal Management System

Jay Patel and Rajesh Patel

1 Introduction

In this chapter, a brief overview of the importance of BTMS and features of Novel Phase Change Material is presented. Section 1.1 Background explains the importance of the integration of BTMS with battery cells. Section 1.2 Objectives states the significance of temperature control of battery cells. Section 1.3 Working principle of PCM explains the basic thermal operation (charging and discharging) of PCM. Section 1.4 Selection criteria of various PCM explores the use of specific PCM in certain applications.

1.1 Background

Nowadays, there is enormous usage of Electric Vehicles in the Automobile Sector. Battery cells are energy sources for EVs which generate the electricity by performing reversible electrochemical reactions during discharging and reverse the process in charging. Various types, size, and capacity batteries are available for battery packs in EVs. Battery being an important part of Electric Vehicles (Evs), it is very much essential to give attention. Battery performance and life are vulnerable to the environment conditions, battery conditions, and operating conditions. Battery must operate in a strict environment for proper functioning. Since batteries are very sensitive to

J. Patel (✉)
Mechanical Engineering Department, Ganpat University, Mehsana, Gujarat, India
e-mail: jaydpatel1992@gmail.com

R. Patel
Mechanical Engineering Department, School of Technology, Pandit Deendayal Petroleum University, Gandhinagar, Gujarat, India
e-mail: rajesh.patel@sot.pdpu.ac.in

N. Patel et al. (eds.), *Electric Vehicles*, Green Energy and Technology,
https://doi.org/10.1007/978-981-15-9251-5_7

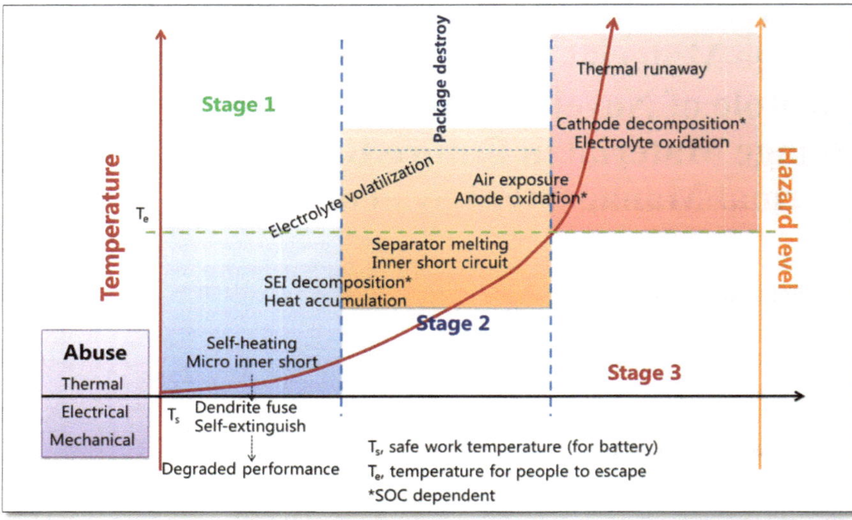

Fig. 1 Stages of heat generation and temperature rise in the battery cell [1]

temperature, it is very valuable to operate a battery within the operating temperature range. For that, BTMS is employed to dissipate the generated heat. It is useful to investigate proper BTMS to reduce the power consumption and increase the range and efficiency of EVs.

The main intention of BTMS is to evacuate the heat from battery cells for the safe functioning of it. When operating conditions become severe, batteries generate large amounts of heat. If some system is not employed to evacuate that heat effectively, it will result in a temperature rise of battery cells. Temperature rise will accelerate the chemical reaction which would further increase the heat generation. This will happen continuously and the battery would reach thermal runaway. When the temperature is beyond 300 °C, there are chances of fire or explosion (Fig. 1).

So, BTMS is not only worthwhile to extend the battery performance and life but also from a safety point of view. It is essential to limit the cell temperature below 40–45 °C and the temperature uniformity inside the battery module must be controlled and should not be increased over 5 °C.

1.2 Objectives

The main purpose of BTMS should not be only to dissipate the heat but also to have a smaller amount of power consumption and less space occupancy. The BTMS have been commonly categorized into two groups from the available research: Active Cooling and Passive Cooling. Both the BTMS will disperse the heat from the battery surface through different mechanisms but dissipate the same via air convection at the end. In active cooling, air or water is forcedly passed over battery surfaces and mechanisms utilize fans or pumps for circulation [2]. Even air cooling will satisfy vehicles' requirements under standard operating time, but with heavy charging and discharge temperatures, the battery temperature would be heavy. The new adapted battery design such as shifting the battery situation, addition of guide plate [3], and adjusting the entrance angle in the battery module, but even this design of air cooling does not fulfill the electric vehicle requirements [4]. The liquid cooling performs well compared to the air cooling but requires an intricate circulation structure, which can increase the vehicle's size and weight [5–7]. PCM based cooling arrangement is the perfect replacement for the mentioned conventional thermal control approaches [8–10]. The PCMs next to the battery surface will absorb the excess heat produced by the electric vehicle's battery. This method of cooling the PCMs is known as Passive Thermal Control. In this PCM based cooling system, the pump or blower is not needed, hence it is price-effective and simple in construction and operation.

1.3 Working Principle of PCMs

PCM is a substance which converts its phase by absorbing or releasing the heat. PCM absorbs the heat and converts into a liquid phase during charging as shown in Fig. 2. During discharging, the absorbed heat is removed and PCM is transformed into a solid phase. Heat absorption or removal changes the internal structure or energy of a substance. Heat which PCM absorbs or releases is transferred at a constant temperature. Since the process is isothermal, it would give good command over temperature control.

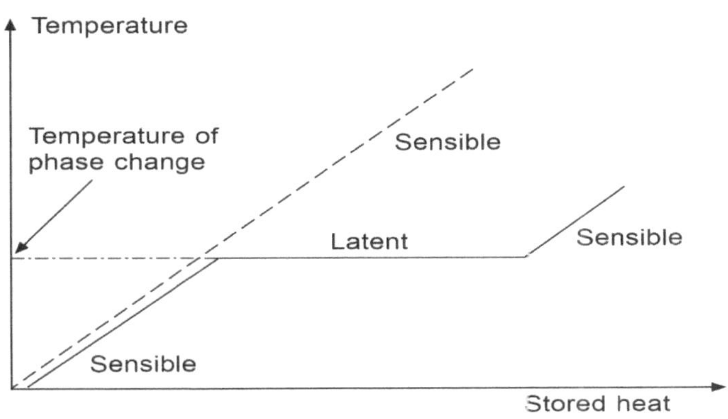

Fig. 2 Working principle of PCM [11]

1.4 Selection of PCMs

PCMs are broadly used in two applications:

1. A Thermal Energy Storage: To accumulate the excess heat when heat is available from a source and release the same to increase the productivity i.e. solar systems, power plants, etc.
2. A Thermal management: To absorb the heat from a source and dissipate the same to the environment to retain the application temperature within a control range i.e. civil construction, textile industry, automobile industry, etc.

The selection of PCM is based on requirements of thermal control, space limitation, weight criteria, cost, environmental impact, etc. One can narrow down his search of selecting the best PCM for his application by answering the following questions:

1. What is the permissible temperature range which needs protection for the payload?
2. What are the required periods? Like how long will the product be in transit or storage?
3. What are the economic consequences of your decision? Thinking of price, usability, and recycling.
4. Which should be considered of the environmental impacts? Trying to concentrate on corrosiveness, toxicity, flammability, and explosive nature of PCMs.

The selection of PCM mainly depends on various parameters which are mentioned in Table 1:

PCMs are classified into three groups:

1. Organic PCMs

Table 1 Selective properties of PCM

Thermal properties	Good thermal conductance Great latent heat of fusion Suitable melting/freezing point Large heat storage density
Physical properties	High density Less change in volume during phase transition No phase separation during fusion Low degree of subcooling
Chemical Properties	Non-corrosive High chemical stability Well-suited with container Non-hazardous Non-flammable
Economic criteria	Inexpensive Easily available

Table 2 Relative assessment between the types of PCM

Criteria	Organic PCMs	Inorganic PCMs
Melting temperature	Low (15–80 °C) and variable melting point	High (50–500 °C) and sharp melting point
Latent heat of fusion	Low	High
Density	Low	High
Change in volume during phase transition	Moderate	High
Thermal conductance	Low	High
Melting behavior	Melt congruently	Incongruous melting and subcooling in some PCMs
Flammability	Flammable	Non-flammable
Availability and cost	Moderate	Easily available and low cost
Chemical nature	Inert to most of the materials	Corrosive to other metals
Application	Thermal management	Thermal energy storage

- Paraffins and Non-Paraffins (fatty acids).

2. Inorganic PCMs
    - Salts, Salt hydrates, and Metallic Alloys.
3. Eutectic PCMs
    - Mixtures of Organic and Inorganic PCMs.

There is a considerable difference between Organic PCMs and Inorganic PCMs. Table 2 shows the relative comparison between them which would help to select PCM for particular applications.

Since organic PCM have a low melting temperature, it is used in the Automobile Industry, Electronic industry, Textile industry, and Building materials. Inorganic PCM shows its usability in solar applications and power plants due to high phase transition temperature.

Among the many PCM, Paraffin (PA) and Polyethylene Glycol (PEG) show excellent thermal behavior for BTMS. Some of the Organic PCM which show their utility for BTMS is listed with their thermal properties in Table 3.

2 Composite PCMs

PCMs have key demerits of low thermal conductivity, regeneration issues, and leakage. Among three disadvantages, thermal conductivity is the major concern when heat is being removed from the battery to the atmosphere in Thermal Management. Lower thermal conductivity of PCM results in less heat transfer rate and

Table 3 PCMs frequently used in BTMS

Materials	Latent Heat of fusion (kJ/kg)	Melting point (°C)	Density (solid/liquid) (kg/m^3)	Thermal conductivity (W/m K)	Specific heat (solid/liquid) (kJ/kg K)
Paraffin					
5913 (N-paraffin)	189	23	900/760	0.21	2.1
6106 (N-paraffin)	189	43	910/765	0.21	2.1
P116 (N-paraffin)	210	41	817/786	0.21	2.1
5838 (N-paraffin)	189	49	912/760	0.21	2.1
6035 (Iso-paraffin)	189	59	920/795	0.21	2.1
6403 (Iso-paraffin)	189	63	915/790	0.21	2.1
6499 (Iso-paraffin)	189	67	930/830	0.21	2.1
n-Alkane					
Hexadecane (C-16)	236	18	770(l)	0.21	2.2
Heptadecane (C-17)	213	22	778(l)	0.21	2.2
Octadecane (C-18)	244	28	814/774	0.21	1.9
Nonadecane (C-19)	222	32	782(s)	0.21	1.91
Eicosane (C-20)	246	37	778(s)	0.21	1.92
Docosane (C-22)	249	44	791(s)	0.21	1.93
Tricosane (C-23)	232	48	764(s)	0.21	1.93
Tetracosane (C-24)	255	51	796(s)	0.21	1.93
Parrafin wax	146	55	880(s)	0.21	1.93
Hexacosane (C-26)	256	57	770(s)	0.21	1.85
Heptacosane (C-27)	235	59	802(s)	0.21	1.91
Fatty acids					
Lauric acid	211.6	41–43	–	1.6	1.16/2.27

(continued)

Table 3 (continued)

Materials	Latent Heat of fusion (kJ/kg)	Melting point (°C)	Density (solid/liquid) (kg/m^3)	Thermal conductivity (W/m K)	Specific heat (solid/liquid) (kJ/kg K)
Capric acid	152.7	32	1004/878	0.153	–
Stearic acid	211.6	41–43	1007/862	1.6	1.16/2.27
Palmitic acid	185.4	58–62	989/850	0.162	–
Erythritol	339.8	117.7	1480/1300	0.733/0.326	2.25/2.61

that would raise the battery temperature. To enhance the response of PCM, some additional material in various forms is introduced into PCM which would constitute a Composite Phase Change Material (CPCM). Temperature uniformity can be increased by enhancing the thermal conductance of CPCM. Uniform temperature distribution in the battery module increases the performance of the battery pack. There are various ways to make CPCM which are explained in Sects. 2.1 and 2.2.

2.1 Addition of High Thermal Conductive Materials

As per the characteristics of PCM, PCM struggles with poor thermal conductivity which makes the non- uniform temperature field. So, the use of PCM alone deteriorates the performance and life of the battery. The addition of thermal conductivity enhancer increases the thermal response of BTMS since thermal conductivity enhancers have higher thermal conductivity (100–300 W/m K) than PCM (0.2–10 W/m K). Thermal conductivity enhancing material is made of either Carbon additive or Metal additive which is listed and compared below:

2.1.1 Carbon-Based Material

Carbon is extensively studied and used for CPCM because of high thermal conductivity, chemically inert, and low density. Since carbon is available in a variety of structures, thermal conductivity enhancer in CPCM is made of carbon particles, expanded graphite (EG), carbon fiber (CF), graphene (GR), and carbon nanotube (CNT). Thermal conductivity enhancement depends on mass fractions and packing density of CPCM.

PCM with Expanded Graphite

Zhang et al. [12] considered the effect of mass fraction of EG in CPCM on cooling performance. For the 2C discharge rate, the maximum temperature of the battery cell is 60.6 °C, 52.7 °C, and 48.8 °C and the maximum temperature difference is 7.8 °C, 5.8 °C and 4.7 °C for 0, 10, and 20% mass ratio of EG in CPCM respectively. They reported that the peak temperature of the battery cell is 60 °C, 56.3 °C, and 52.4 °C, and the peak temperature difference is 8.2 °C, 7 °C and 4.3 °C for 0, 20, and 30% mass ratio of EG in CPCM respectively for 3C discharge rate. Ling et al. [13] used RT44HC as PCM and EG as a conductivity enhancer composite for enhancing the thermal response of the system. They found that 35% EG based CPCM has 30% better thermal conductivity compared with 25% EG based CPCM for the same density. They also stated that by increasing the density, thermal conductivity could be augmented 20–60 times than PCM for the same mass ratio.

Tang et al. [14] explored the use of EG in Eutectic PCM made of a mixture of palmitic and capric acid. They noted that the eutectic mixture with EG enhances the thermal conductance which is of 0.190 W/m K. They investigated that increase in thermal conductivity is by 26.3% and 53.7% with 3 weight% and 5 weight% EG based CPCM, respectively. Yang et al. [15] prepared the CPCM made of polyethylene (PCM) and hybrid graphene aerogels (thermal conductivity enhancer) for increasing the thermal performance. They reported that the enhancement of thermal performance is about 361%. The proposed CPCM improves the thermal performance by increasing the thermal conductance from 0.30 (PCM) to 1.42 W/m K (CPCM).

PCM with Carbon Fibers/Particles

Nomura et al. [16] used CF/Erythritol as CPCM in the experiment. They stated that thermal conductivity is drastically enhanced by using CF from 0.73 (PCM) to 30 W/m K (CPCM). Zhang et al. [17] examined the influence of CF size on the thermal response of CPCM. They used erythritol as PCM and two CF having Aspect Ratio (AR) of 25 and 5 as thermal conductivity additive. They observed that the thermal conductivity of CPCM is improved from 2.46 W/m K (CF of 5 AR) to 3.91 W/m K (CF of 25 AR) of 10% CF-based CPCM. Singh et al. [18] examined the influence of carbon particles in CPCM for better heat transfer. In this study, they used carbon powder of 0–50 μm size as a thermal conductivity enhancer, PEG1000 as a PCM and CPCM contains 8–2.5% carbon by volume. Proposed CPCM improves the thermal performance by increasing the thermal conductance from 0.22 (PCM) to 0.29 W/m K (2.5 volume% carbon powder).

PCM with Carbon Fins/Films

Singh et al. [18] experimentally explored the scope of carbon fins to augment the thermal performance in CPCM. In this revision, carbon fins of 33.96% by weight and 24.68% by volume and polyethylene glycol (PEG1000) are used in CPCM. Carbon fins improve the thermal performance of CPCM from 0.22 (PCM) to 7.3 W/m K (CPCM). Luo et al. [19] inspected the opportunity of graphite film on the cooling of the battery. Constituted CPCM has Paraffin, EG, and Epoxy Resin with 5:2:3 mass ratio and high conductive (1400 W/m K) graphite film. CPCM succeeded to sustain the maximum battery cell temperature under 27 °C and the maximum temperature difference in the battery module under 1.6 °C for 1C discharge. For 3C discharge, the peak battery temperature is 34 °C and the temperature difference in the battery module is 2.8 °C.

PCM with Carbon Nano-Tubes

Carbon nanotubes are two kinds: SWCNTs (single-walled carbon nanotubes) and MWCNTs (multi-walled carbon nanotubes). In recent research, MWCNTs are widely used because of small density, large area/volume ratio, and high thermal conductivity.

Xu and Li [20] used MWCNT with paraffin/diatomite mixtures based on CPCM which intensify the thermal conductivity by 42.45% associated with that of PCM. Zou et al. [21] considered the enhancement of thermal response of CPCM made of paraffin and the addition of MWCNT and graphene. They reported that there is a small decrease in melt temperature from 40.9 °C (Paraffin) to 40.8 °C (CPCM). CPCM with the mass fraction of 3/7 MWCNT/graphene holds the utmost thermal conductivity among mass ratios of 1/1, 2/3, 3/7, and 1/4. Thermal conductivity was increased by 32%, 55%, and 124% associated with graphene-based CPCM, MWCNT based CPCM, and pure PCM respectively.

Zou et al. [22] inspected the influence of embedded carbon materials on the thermal performance of CPCM. Paraffin, EG/paraffin with 15% mass ratio of EG, CPCM (Paraffin-46/EG/carbon nanotubes (CNT)/graphene, 3/7 mass ratio of CNT/graphene) was taken for investigation. They found that the thermal conductivity of Paraffin, EG/Paraffin, and CPCM is 0.4, 4, and 5 W/m K respectively at 25 °C ambient temperature. It was reported that there is a minor decrement in latent heat from 224.8 kJ/kg (Paraffin) to 178.5 kJ/kg (CPCM) as well as in melting temperature from 46.6 (Paraffin) to 46.1 °C (CPCM).

2.1.2 Metal-Based Material

The use of high thermal conductivity of metal enhances the thermal conductance of PCM. Metals like nickel, aluminum, and copper are unified with PCM in various forms like metal particles and metal foams. The addition of metal foams proliferates the effective thermal conductivity according to their loading in metal particles and porosity with pore size in metal foams.

PCM with Metal Particles

Oya et al. [23] used nickel particles in erythritol to constitute CPCM and investigated the result of nickel particles loading on the cooling performance of CPCM. They optimized the thermal conductivity which is 17 volume% nickel particles loading. CPCM's thermal conductivity is 290% greater than PCM alone. Sahan et al. [24] explored the use of Fe_3O_4 in paraffin to constitute CPCM. Outcomes expressed that thermal conductivity with 10 weight% and 20 weight% Fe_3O_4 is 48% and 60% greater than that of PCM alone respectively.

Sahan et al. [24] further explored the other metal oxide ($SiO_2, Al_2O_3, Fe_2O_3, ZnO$) incorporated into paraffin to make CPCM. Mixtures of metal additives based CPCM have the peak thermal conductance of 0.72 W/m K for the charging of 2 weight%. CPCM with Al_2O_3 revealed the highest thermal conductivity of 0.919 W/m K for the charging of 4 weight%. CPCM with Fe_2O_3 showed slightly greater thermal conductivity than that of Al_2O_3 for the charging of 8 weight%. It can be stated that Al_2O_3 is the optimal particle type to increase thermal conductivity. Deng et al. [25] used the polyethylene glycol (PEG) and silver nanowire (Ag NW) based CPCM. CPCMs show decent marks of 0.35, 0.52, and 0.70 W/m K consisting of Ag NW of 7 weight%, 13.7 weight%, and 19 weight% which is far better than that of PCM alone (0.06 W/m K) respectively.

PCM with Metal Foams

Xiao et al. [26] considered the impact of porosity and pore size on the thermal response of CPCM. They carried out experiments with paraffin, nickel foam, and copper foam with different porosity and pore size to form CPCM. Since copper has better thermal conductivity than nickel, copper shows decent results compared to nickel. Copper foam-based CPCM having porosities of 97%, 92%, and 89% and pore size of 25 PPI improves the thermal conductivity by about 14, 32, and 44 times respectively compared to paraffin. Nickel foam-based CPCM increases the thermal conductivity by 3.2, 4.1, and 5.0 times, respectively compared to paraffin with the same porosity and pore size. It may be concluded that thermal conductivity is inversely varied with porosity and merely shows any change with change in pore size.

Pan et al. [27] inspected the effect of cutting copper fiber sintered skeleton (CCFSS) and copper foam in paraffin for BTMS. CCFSS based CPCM has a 28.1% mass ratio of copper in CPCM and copper foam-based CPCM has a 28.6% mass ratio of copper in paraffin. The average maximum temperature increment of the paraffin, the copper foam/paraffin, and the CCFSS based CPCM are about 25, 25, and 22.5 °C respectively. Thapa et al. [28] examined the performance of icosane wax/copper foam-based CPCM. They found that the optimized thermal conductance of CPCM is 3.8 W/m K which would be far better than PCM alone.

2.2 Use of Fin-PCM Structure

When the available area is inadequate to transfer the heat for a given temperature difference, fins are employed. Fins increase the total area to dissipate the heat quickly without any auxiliary devices like fan, pump, heat pipe, or materials like PCM. Combining the fins with PCM, high heat transfer is ensured along with uniform temperature distribution in battery cells.

Ping et al. [29] explored the usage of fins in PCM for enhancement of cooling performance of fin/PCM based BTMS. Optimized fin/PCM structure can retain battery cell surface temperature below 51 under 40 °C ambient temperature due to better heat removal efficiency of the fin structure. Lv et al. [30] examined the effect of fins in CPCM consisting of paraffin, EG, and Low-density polyethylene (LDPE). Due to the high evacuation of heat, CPCM with fins able to keep the battery surface temperature under 50 °C, and the peak temperature difference in the module is decreased to 3.2 °C at a 3.5C discharge rate. Zhong et al. [31] considered the effect of fins set on both sides of the module to investigate the temperature field in the battery module. They found that fins prevent thermal saturation and keep battery surface temperature under 45 °C and peak temperature difference in the module is declined below 5 °C under 5C discharging rate and ambient temperature of 40 °C (Table 4).

3 Prospective CPCM Based BTMS

The addition of carbon material or metal material in CPCM improves the thermal conductivity but lowers down the latent heat and melt temperature. Thermal conductivity enhancer does not contribute to accumulate the heat and elongate the PCM melting time. The novel concept of the addition of low melting alloys in PCM not only increases the thermal conductivity but also increases the melting time. Thus, low melting alloys work as a phase change material and thermal conductance enhancer at the same time. Low melting alloys based CPCM has excellent replacement for conventional CPCM.

Pure gallium (Ga) melts at a temperature of 29.8 °C and has 80 kJ/kg latent heat and thermal conductance of 26 W/m K, whereas gallium eutectics have an operating temperature range of 25–45 °C. Gallium is commonly alloyed with other post-transition metals, including Indium, Zinc, Tin, and Cadmium. Cesium, Rubidium, and Potassium are the other promising aspirants for the Novel low melting alloys based CPCM. Improvement in thermal properties compared to paraffin (latent heat of novel additives) and expanded graphite (thermal conductivity of novel additives) is tabulated in Table 5.

Table 4 Improvement in thermal conductivity of CPCM

PCM/thermal conductivity (W/m K)	Additives/thermal conductivity (W/m K)	Composite fraction (%weight)	Composite thermal conductivity (W/m K)	Authors
RT42/0.20	EG/4-100	26.6	17	Mills et al. [32]
Paraffin/0.20	Graphite	35	70.0	Alva et al. [33]
Paraffin/0.31	Graphite powder/2-90	12	0.460	Cheng et al. [34]
Paraffin/0.2697	EG/4-100	6.25	4.676	Yin et al. [35]
RT44HC	EG	60	9.57	Ling et al. [36]
Paraffin/0.2	Graphite sheet	–	3.95	Lin et al. [37]
Paraffin	Graphene	20	45	Goli et al. [38]
Paraffin	EG	15–25	5.3–11.2	Ling et al. [39]
Paraffin/0.18	EG/LDPE/Double copper mesh (DCM)	–	8.3277	Situ et al. [40]
Paraffin/0.193	Aluminium honeycomb panel/160	–	–	Hasse et al. [41]
Paraffin	Copper foam	–	2.879–3.112	Wang et al. [42]
Paraffin	Nickel foam	–	1.2	Ahmadi et al. [43]
Paraffin	Graphene coated nickel foam	–	4.6	Ahmadi et al. [43]
Paraffin	aluminium	5-30	13.2–78.2	Coleman et al. [44]
Hexadecane/0.15	Aluminium particles	–	1.25	Darkwa and Zhou [45]
N-Tetradecanol/0.32	Silver nanowire	62.73	1.46	Zeng et al. [46]
N-octadecane/0.27	EG/4-100	16.7	1.07	Li et al. [47]
N-docosane/0.20	Graphite powder/2-90	10	0.82	Sari and Karaipekli [48]
N-octadecane/0.15	Inorganic silica shell/1.3	50	0.62	Zhang et al. [49]
Eicosane/0.42	Copper porous foam/4-100	95	3.06	Siahpush et al. [50]
Polyethylene glycol/0.3	EG/4-100	10	1.3	Wang et al. [51]
Stearic acid/0.28	CF/190, EG/4-100	8	0.62/0.8	Karaipekli et al. [52]

Table 5 Comparison of thermal properties of PCM/additives and novel aspirating additives of CPCM

Material as PCM	Melting temperature (°C)	Latent heat of fusion (kJ/kg)	Thermal conductivity (W/m K)
Paraffin (PA)	42	180	0.2
Expanded graphite (EG)	4000	–	30–400
Gallium (GA)	30	80	26
Cesium (Cs)	29	–	25
Rubidium (Rb)	40	30	30
Potassium (K)	63	60	50

4 Challenges in Prospective CPCM Based BTMS

Novel additives of CPCM show excellent thermal properties for BTMS. Novel CPCM will be capable of keeping the battery temperature within the control range due to additional thermal storage of additives and the unvarying temperature field in the battery module owing to the high thermal conductivity of additives. Despite the mentioned benefits, some research work should be carried out in this field. Some challenges need to be addressed to make it the best alternative of additives in CPCM which are listed below:

- Manufacturing of such additives based CPCM.
- Congruent melting process of CPCM.
- Thermal stability of CPCM because of the difference in the melting temperature of additives and PCM.
- Overall heat transfer phenomenon occurred during the charging/discharging of CPCM.
- Toxicity, Flammability, and chemical nature of additives like Potassium, Rubidium, etc.

5 Conclusion and Summary

PCM proves themselves as one of the important methods of battery thermal management systems. Due to passive nature, it will give better results than air cooling or liquid cooling which requires external power to operate. The only problem with PCM is the low heat removal rate from the battery cell to the environment during operating conditions. The heat transfer rate becomes a vital part of BTMS which is influenced by the thermal conductance of PCM. Because of the poor thermal conductivity of the majority of PCM, it would become very important to increase the thermal response of the battery system. CPCM proliferates the cooling efficiency of BTMS by improving the thermal conductivity of the composite. Henceforth, this review emphasizes on approaches to improving the thermal conductivity. Particles having high thermal conductivity are added in PCM to constitute CPCM.

In summary, carbon-based CPCM has benefits of great thermal conductivity, small density, and chemically inert nature. The addition of carbon-based material has different effects on thermal conductivity according to their mass ratio, aspect ratio, and density. Compared to carbon-based CPCM, metal-based CPCM advances the thermal conductivity considerably owing to the great thermal conductivity property of metals. Metal-based CPCM comes with demerits of high density and dispersion issues. The proposed novel CPCM not only owns great thermal conductivity but also gives contribution to heat storage. It will give excellent temperature control on the battery pack in contexts of peak battery cell temperature and temperature uniformity in the battery module.

References

1. X. Wu et al., Safety issues in lithium ion batteries: Materials and cell design. Front. Energy Res. **7**(JUL), 1–17 (2019). https://doi.org/10.3389/fenrg.2019.00065
2. X. Feng et al., Thermal runaway features of large format prismatic lithium ion battery using extended volume accelerating rate calorimetry. J. Power Sources **255**, 294–301 (2014). https://doi.org/10.1016/j.jpowsour.2014.01.005
3. H. Park, A design of air flow configuration for cooling lithium ion battery in hybrid electric vehicles. J. Power Sources **239**, 30–36 (2013). https://doi.org/10.1016/j.jpowsour.2013.03.102
4. S. Park, D. Jung, Battery cell arrangement and heat transfer fluid effects on the parasitic power consumption and the cell temperature distribution in a hybrid electric vehicle. J. Power Sources **227**, 191–198 (2013). https://doi.org/10.1016/j.jpowsour.2012.11.039
5. G. Fang et al., Thermal management for a tube-shell Li-ion battery pack using water evaporation coupled with forced air cooling. RSC Adv. **9**(18), 9951–9961 (2019). https://doi.org/10.1039/c8ra10433f
6. X.H. Yang, S.C. Tan, J. Liu, Thermal management of Li-ion battery with liquid metal. Energy Convers. Manag. **117**, 577–585 (2016). https://doi.org/10.1016/j.enconman.2016.03.054
7. B. Ye, M.R.H. Rubel, H. Li, Design and optimization of cooling plate for battery module of an electric vehicle. Appl. Sci. **9**(4), 754 (2019). https://doi.org/10.3390/app9040754
8. A. Alrashdan, A.T. Mayyas, S. Al-Hallaj, Thermo-mechanical behaviors of the expanded graphite-phase change material matrix used for thermal management of Li-ion battery packs. J. Mater. Process. Technol. **210**(1), 174–179 (2010). https://doi.org/10.1016/j.jmatprotec.2009.07.011
9. W. Wu, G. Zhang, X. Ke, X. Yang, Z. Wang, C. Liu, Preparation and thermal conductivity enhancement of composite phase change materials for electronic thermal management. Energy Convers. Manag. **101**, 278–284 (2015). https://doi.org/10.1016/j.enconman.2015.05.050
10. T. Nomura, C. Zhu, S. Nan, K. Tabuchi, S. Wang, T. Akiyama, High thermal conductivity phase change composite with a metal-stabilized carbon-fiber network. Appl. Energy **179**, 1–6 (2016). https://doi.org/10.1016/j.apenergy.2016.04.070
11. F. Bruno, M. Belusko, M. Liu, and N. H. S. Tay, *Using solid-liquid phase change materials (PCMs) in thermal energy storage systems* (Woodhead Publishing Limited, 2015)
12. C. Zhang, S. Chen, H. Gao, K. Xu, Z. Xia, and S. Li, Study of thermal management system using composite phase change materials and thermoelectric cooling sheet for power battery pack. Energies **12**(1937) (2019). https://doi.org/10.3390/en12101937.
13. Z. Ling, J. Chen, T. Xu, X. Fang, X. Gao, Z. Zhang, Thermal conductivity of an organic phase change material/expanded graphite composite across the phase change temperature range and

a novel thermal conductivity model. Energy Convers. Manag. **102**, 202–208 (2015). https://doi.org/10.1016/j.enconman.2014.11.040

14. F. Tang, D. Su, Y. Tang, G. Fang, Synthesis and thermal properties of fatty acid eutectics and diatomite composites as shape-stabilized phase change materials with enhanced thermal conductivity. Sol. Energy Mater. Sol. Cells **141**, 218–224 (2015). https://doi.org/10.1016/j.solmat.2015.05.045
15. J. Yang et al., Hybrid graphene aerogels/phase change material composites: thermal conductivity, shape-stabilization and light-to-thermal energy storage. Carbon N. Y. **100**, 693–702 (2016). https://doi.org/10.1016/j.carbon.2016.01.063
16. T. Nomura, K. Tabuchi, C. Zhu, N. Sheng, S. Wang, T. Akiyama, High thermal conductivity phase change composite with percolating carbon fiber network. Appl. Energy **154**, 678–685 (2015). https://doi.org/10.1016/j.apenergy.2015.05.042
17. Q. Zhang, Z. Luo, Q. Guo, G. Wu, Preparation and thermal properties of short carbon fibers/erythritol phase change materials. Energy Convers. Manag. **136**, 220–228 (2017). https://doi.org/10.1016/j.enconman.2017.01.023
18. R. Singh, S. Sadeghi, B. Shabani, Thermal conductivity enhancement of phase change materials for low-temperature thermal energy storage applications. Energies **12**(1), 75 (2019). https://doi.org/10.3390/en12010075
19. X. Luo et al., Experimental investigation on a novel phase change material composites coupled with graphite film used for thermal management of lithium-ion batteries. Renew. Energy **145**, 2046–2055 (2020). https://doi.org/10.1016/j.renene.2019.07.112
20. B. Xu, Z. Li, Paraffin/diatomite/multi-wall carbon nanotubes composite phase change material tailor-made for thermal energy storage cement-based composites. Energy **72**, 371–380 (2014). https://doi.org/10.1016/j.energy.2014.05.049
21. D. Zou, X. Ma, X. Liu, P. Zheng, Y. Hu, Thermal performance enhancement of composite phase change materials (PCM) using graphene and carbon nanotubes as additives for the potential application in lithium-ion power battery. Int. J. Heat Mass Transf. **120**, 33–41 (2018). https://doi.org/10.1016/j.ijheatmasstransfer.2017.12.024
22. D. Zou et al., Preparation of a novel composite phase change material (PCM) and its locally enhanced heat transfer for power battery module. Energy Convers. Manag. **180**, 1196–1202 (2019). https://doi.org/10.1016/j.enconman.2018.11.064
23. T. Oya, T. Nomura, M. Tsubota, N. Okinaka, T. Akiyama, Thermal conductivity enhancement of erythritol as PCM by using graphite and nickel particles. Appl. Therm. Eng. **61**(2), 825–828 (2013). https://doi.org/10.1016/j.applthermaleng.2012.05.033
24. N. Şahan, M. Fois, H. Paksoy, Improving thermal conductivity phase change materials—a study of paraffin nanomagnetite composites. Sol. Energy Mater. Sol. Cells **137**, 61–67 (2015). https://doi.org/10.1016/j.solmat.2015.01.027
25. Y. Deng, J. Li, T. Qian, W. Guan, Y. Li, X. Yin, Thermal conductivity enhancement of polyethylene glycol/expanded vermiculite shape-stabilized composite phase change materials with silver nanowire for thermal energy storage. Chem. Eng. J. **295**, 427–435 (2016). https://doi.org/10.1016/j.cej.2016.03.068
26. X. Xiao, P. Zhang, M. Li, Effective thermal conductivity of open-cell metal foams impregnated with pure paraffin for latent heat storage. Int. J. Therm. Sci. **81**(1), 94–105 (2014). https://doi.org/10.1016/j.ijthermalsci.2014.03.006
27. M. Pan, Y. Zhong, Experimental and numerical investigation of a thermal management system for a Li-ion battery pack using cutting copper fiber sintered skeleton/paraffin composite phase change materials. Int. J. Heat Mass Transf. **126**, 531–543 (2018). https://doi.org/10.1016/j.ijheatmasstransfer.2018.06.014
28. S. Thapa, S. Chukwu, A. Khaliq, L. Weiss, Fabrication and analysis of small-scale thermal energy storage with conductivity enhancement. Energy Convers. Manag. **79**, 161–170 (2014). https://doi.org/10.1016/j.enconman.2013.12.019
29. P. Ping, R. Peng, D. Kong, G. Chen, J. Wen, Investigation on thermal management performance of PCM-fin structure for Li-ion battery module in high-temperature environment. Energy Convers. Manag. **176**(September), 131–146 (2018). https://doi.org/10.1016/j.enconman.2018.09.025

30. Y. Lv, X. Yang, X. Li, G. Zhang, Z. Wang, C. Yang, Experimental study on a novel battery thermal management technology based on low density polyethylene-enhanced composite phase change materials coupled with low fins. Appl. Energy **178**, 376–382 (2016). https://doi.org/10.1016/j.apenergy.2016.06.058
31. G. Zhong et al., Researches of composite phase change material cooling/resistance wire preheating coupling system of a designed 18650-type battery module. Appl. Therm. Eng. **127**, 176–183 (2017). https://doi.org/10.1016/j.applthermaleng.2017.08.022
32. A. Mills, M. Farid, J.R. Selman, S. Al-Hallaj, Thermal conductivity enhancement of phase change materials using a graphite matrix. Appl. Therm. Eng. **26**(14–15), 1652–1661 (2006). https://doi.org/10.1016/j.applthermaleng.2005.11.022
33. L. H. Alva S, J. E. González, N. Dukhan, Initial analysis of PCM integrated solar collectors. J. Sol. Energy Eng. Trans. ASME **128**(2), 173–177 (2006). https://doi.org/10.1115/1.2188532
34. W.L. Cheng, R.M. Zhang, K. Xie, N. Liu, J. Wang, Heat conduction enhanced shape-stabilized paraffin/HDPE composite PCMs by graphite addition: Preparation and thermal properties. Sol. Energy Mater. Sol. Cells **94**(10), 1636–1642 (2010). https://doi.org/10.1016/j.solmat.2010.05.020
35. H. Yin, X. Gao, J. Ding, Z. Zhang, Experimental research on heat transfer mechanism of heat sink with composite phase change materials. Energy Convers. Manag. **49**(6), 1740–1746 (2008). https://doi.org/10.1016/j.enconman.2007.10.022
36. Z. Ling, X. Wen, Z. Zhang, X. Fang, X. Gao, Thermal management performance of phase change materials with different thermal conductivities for Li-ion battery packs operated at low temperatures. Energy **144**, 977–983 (2018). https://doi.org/10.1016/j.energy.2017.12.098
37. C. Lin, S. Xu, G. Chang, J. Liu, Experiment and simulation of a $LiFePO_4$ battery pack with a passive thermal management system using composite phase change material and graphite sheets. J. Power Sources **275**, 742–749 (2015). https://doi.org/10.1016/j.jpowsour.2014.11.068
38. P. Goli, S. Legedza, A. Dhar, R. Salgado, J. Renteria, A.A. Balandin, Graphene-enhanced hybrid phase change materials for thermal management of Li-ion batteries. J. Power Sources **248**, 37–43 (2014). https://doi.org/10.1016/j.jpowsour.2013.08.135
39. Z. Ling et al., Experimental and numerical investigation of the application of phase change materials in a simulative power batteries thermal management system. Appl. Energy **121**, 104–113 (2014). https://doi.org/10.1016/j.apenergy.2014.01.075
40. W. Situ et al., A thermal management system for rectangular $LiFePO_4$ battery module using novel double copper mesh-enhanced phase change material plates. Energy **141**, 613–623 (2017). https://doi.org/10.1016/j.energy.2017.09.083
41. C. Hasse, M. Grenet, A. Bontemps, R. Dendievel, H. Sallée, Realization, test and modelling of honeycomb wallboards containing a phase change material. Energy Build. **43**(1), 232–238 (2011). https://doi.org/10.1016/j.enbuild.2010.09.017
42. C. Wang, T. Lin, N. Li, H. Zheng, Heat transfer enhancement of phase change composite material: copper foam/paraffin. Renew. Energy **96**, 960–965 (2016). https://doi.org/10.1016/j.renene.2016.04.039
43. M.H. Ahmadi, M. Alhuyi Nazari, R. Ghasempour, H. Madah, M.B. Shafii, M.A. Ahmadi, Thermal conductivity ratio prediction of Al_2O_3/water nanofluid by applying connectionist methods. Colloids Surfaces A Physicochem. Eng. Asp. **541**(2010), 154–164 (2018). https://doi.org/10.1016/j.colsurfa.2018.01.030
44. B. Coleman, J. Ostanek, J. Heinzel, Reducing cell-to-cell spacing for large-format lithium ion battery modules with aluminum or PCM heat sinks under failure conditions. Appl. Energy **180**, 14–26 (2016). https://doi.org/10.1016/j.apenergy.2016.07.094
45. J. Darkwa, T. Zhou, Enhanced laminated composite phase change material for energy storage. Energy Convers. Manag. **52**(2), 810–815 (2011). https://doi.org/10.1016/j.enconman.2010.08.006
46. J.L. Zeng, Z. Cao, D.W. Yang, L.X. Sun, L. Zhang, Thermal conductivity enhancement of Ag nanowires on an organic phase change material. J. Therm. Anal. Calorim. **101**(1), 385–389 (2010). https://doi.org/10.1007/s10973-009-0472-y

47. H. Li, X. Liu, G.Y. Fang, Synthesis and characteristics of form-stable n-octadecane/expanded graphite composite phase change materials. Appl. Phys. A Mater. Sci. Process. **100**(4), 1143–1148 (2010). https://doi.org/10.1007/s00339-010-5724-y
48. A. Sari, A. Karaipekli, Thermal conductivity and latent heat thermal energy storage characteristics of paraffin/expanded graphite composite as phase change material. Appl. Therm. Eng. **27**(8–9), 1271–1277 (2007). https://doi.org/10.1016/j.applthermaleng.2006.11.004
49. H. Zhang, X. Wang, D. Wu, Silica encapsulation of n-octadecane via sol-gel process: a novel microencapsulated phase-change material with enhanced thermal conductivity and performance. J. Colloid Interface Sci. **343**(1), 246–255 (2010). https://doi.org/10.1016/j.jcis.2009.11.036
50. A. Siahpush, J. O'Brien, J. Crepeau, Phase change heat transfer enhancement using copper porous foam. J. Heat Transfer **130**(8), 1–11 (2008). https://doi.org/10.1115/1.2928010
51. W. Wang, X. Yang, Y. Fang, J. Ding, J. Yan, Preparation and thermal properties of polyethylene glycol/expanded graphite blends for energy storage. Appl. Energy **86**(9), 1479–1483 (2009). https://doi.org/10.1016/j.apenergy.2008.12.004
52. A. Karaipekli, A. Sari, K. Kaygusuz, Thermal conductivity improvement of stearic acid using expanded graphite and carbon fiber for energy storage applications. Renew. Energy **32**(13), 2201–2210 (2007). https://doi.org/10.1016/j.renene.2006.11.011

Battery Electric Vehicles (BEVs)

Ahmad Faraz, A. Ambikapathy, Saravanan Thangavel, K. Logavani, and G. Arun Prasad

1 Introduction

1.1 A Brief Introduction of Battery Electric Vehicles (BEVs)

A battery electric vehicle (BEV) is basically an electric vehicle (EV) that solitary consumes compound energy to run which is put away in rechargeable battery packs, with no other source (e.g., hydrogen energy component, an internal combustion engine, and so forth). BEVs utilize electric engine and system instead of an internal combustion engine (ICEs) for propulsion. They get all force from battery packs and use it to run the engine which further aides in driving the wheels. They are otherwise called unadulterated electric vehicles or just electric vehicles or all-electric vehicle. Battery electric vehicles (BEVs) incorporate different sorts of vehicles, for example, cruisers, bikes, skateboards, railcars, watercraft, forklifts, transports, trucks, and other vehicles.

Battery electric vehicles (BEVs) are basic and simple to work when contrasted with an exemplary internal combustion engine (ICE) vehicles. The engineering of

A. Faraz · A. Ambikapathy (✉) · S. Thangavel · G. Arun Prasad
Galgotias College of Engineering and Technology, Greater Noida, UP, India
e-mail: ami_ty21@yahoo.com

A. Faraz
e-mail: theahmadfaraz@gmail.com

S. Thangavel
e-mail: tsaravcse@gmail.com

G. Arun Prasad
e-mail: arun.prasad@galgotiacollege.edu

K. Logavani
Government College of Engineering—Salem, Salem, Tamil Nadu, India
e-mail: Vani.tulips@gmail.com

N. Patel et al. (eds.), *Electric Vehicles*, Green Energy and Technology,
https://doi.org/10.1007/978-981-15-9251-5_8

Fig. 1 Nissan Leaf and Tesla Model S [1]. *Source* Wikipedia

the battery electric vehicle comprises of high-voltage battery, an electric engine with the power electronics controller and a solitary speed gearbox. Battery electric vehicles are expanding their piece of the overall industry since they are the most practical route towards a perfect and effective vehicle framework. Contrasted to ICE vehicles, the most significant preferences of a BEV are the general high effectiveness, dependability, and the moderately minimal effort of the electric engine.

In 2016, there were 210 million electric bicycles overall utilized day by day. The general deals of battery electric vehicles that are skilled to run on parkway were seen as one million unit till September 2016. Before the finish of 2019, world's top selling makers Nissan and Tesla have done the worldwide offer of 450,000 units of Nissan Leaf and 448,634 units of Tesla Model S separately [1] (Fig. 1).

1.2 Features

The idea of battery electric vehicles was to utilize charged batteries on board vehicles for propulsion. Battery electric vehicles are turning out to be increasingly more alluring with the expansion on fuel costs and headway of new battery innovation (lithium-ion). They have a higher force and energy capacity which gives huge range and more prominent conceivable speeding up when contrasted with more established battery types, for example, lead–acid batteries. For instance, lithium-ion batteries have an energy capacity of 0.8–2.5 MJ/L while lead–acid batteries had an energy capacity of 0.4 MJ/L, which is higher. There is as yet far to go if contrasting it with oil-based energizers and biofuels.

BEVs limit the energy wastage by halting the vehicle when it is stopped and by charging the battery when the brakes are applied ("regenerative braking") like other electric vehicle and hybrid electric vehicles. Electric motors are in like manner more energy compelling than gas or diesel engines.

Battery electric vehicles have the extra favorable position of home stimulating. A 240-V supply is adequate for the vehicle to charge the whole battery in a period of time. Completely energized battery electric vehicles have a driving constraint of 80–100 miles or to a most outrageous extent of 270 miles. Electric motors produce high torque or turning-power, while the torque of ICE vehicle increases with the engine's speed (RPM). This suggests that BEVs have amazingly snappy speeding up appeared differently in relation to standard vehicles [1].

2 History and Evolution of BEV

2.1 Idea of the Electric Car

In the early eighteenth century, the prerequisite needed, apart from the discovery of an electricity, to manufacture electric-driven cars was a battery which can be recharged. In 1828, a Hungarian inventor, Anyos Jedlik made an electric driven motor which consumes electricity as a means of a transportation with the help of a small car model which could be run using his motor.

Later, Thomas Davenport, a Vermont Blacksmith developed an electric vehicle model in 1834 which used to move on a path which is circular. But all those models manufactured were lacking self-mounted power source. The solution to this major problem came out in the year 1859, when French Physicist Guston Plante made the lead–acid battery. In 1881, Camille Alphonse Faure expanded the limit of the battery which prompted the creation of batteries on a large scale. With this, battery-powered energy source creators began exploring different avenues regarding power and velocity [2] (Fig. 2).

Fig. 2 Thomas Parker's EV [3]. *Source* Wikimedia Commons

Fig. 3 GM's The Electrovair [4]. *Source* My Car Quest

2.2 Invention of an EV

In the year 1884, a British Inventor, Thomas Parker delivered the main electric car. It was controlled by his uniquely structured high-limit battery. Be that as it may, Henry G. Morris, a Mechanical Engineer and a Chemist Pedra G. Salom in 1894 in Philadelphia, Pennsylvania, created the principal effective electric vehicle, The Electrobat. It had steel tires to help its substantial edge and huge lead battery.

Later on, in the US William Morrison of Des Moines developed a six-passenger electric car (wagon) which was capable of reaching a speed of 23 km/h. By the time, different creators and architects created numerous models [2].

2.3 First BEV by General Motors

In the mid-1960s, GM's first concept electric vehicle was the Electrovair which was controlled by utilizing a silver-zinc battery pack that could convey upto 532 V. In any case, this idea neglected to make large-scale manufacturing. Their first present-day age electric vehicle to be mass delivered was created in the mid-1990s, named General Motors EV1. California enlivened its planning and fabrication which stated a command that it is necessary for producers to create zero-emanation vehicles [2] (Fig. 3).

2.4 First BEV by Tesla

In 2008, Tesla Motors made their outright first battery electric vehicle (BEV), called as The Roadster. The car was a revolution in the propelled age electric vehicle as it

Fig. 4 Tesla's The Roadster [5]. *Source* Getty Images

used some pattern setting developments, for instance, cutting edge battery advancement and electrically operated powertrain. The main Roadster is a battery electric vehicle, and it was the primary lawful sequential creation of all-electric vehicle which had ever used a lithium-ion battery as a force source. It is likewise first BEV which is fit for traveling in excess of 320 km for every charge. It can likewise arrive at a mind blowing top speed of 200 km/h. During its creation years (2008–2012), about 2500 Roadsters were bought in more than 25 nations around the globe [2] (Fig. 4).

3 Comparison Between BEVs and Other EVs

3.1 BEV

B = Battery, E = Electric, V = Vehicle.

Battery electric vehicles are the vehicles which totally runs on the power. They do not utilize some other fuel (petroleum or diesel) to run. They are 100% all-electric vehicle. The power is put away in the batteries introduced in vehicles (Fig. 5).

They are condition neighborly as they do not discharge any sort of unsafe outflows in nature or we can say that they are zero-emanation vehicle. The charging of the battery electric vehicle is simple as it is very well-finished with the typical home unit supply, or at the charging outlet if accessible. A portion of the instances of battery electric vehicles are Tesla Model 3, Peugeot e-208, MG ZS EV, Nissan Leaf, Renault Zoe, Hyundai Kona Electric, Kia e-Niro, Jaguar I-Pace, and so on [6].

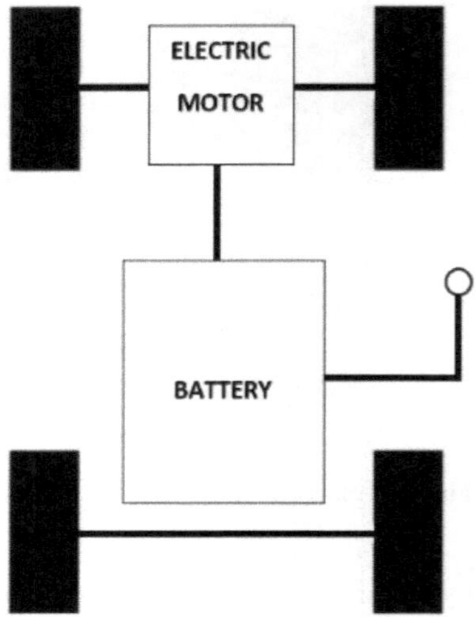

Fig. 5 BEV

3.2 HEV

H = Hybrid, E = Electric, V = Vehicle.

These vehicles are the vehicles which are controlled through the internal combustion engine with the assistance of fuel, for example, petroleum or diesel. They additionally have a battery which supplies the ability to run the electric engine. Be that as it may, the downside is, this battery cannot be legitimately charged by the external charger or supply. The battery is charged uniquely with the regenerative slowing down which implies when we apply the brakes to stop the vehicle then the engine turn over filling in as a generator and consequently helps in charging the battery (Fig. 6).

Such kind of vehicle frequently runs on a battery at low speeds, and when the vehicle needs to go quicker, they switch on to the internal combustion engine and turn over running on fuel. A portion of the instances of hybrid electric vehicles are Toyota Corolla Hybrid, Toyota Yaris Hybrid, Lexus RX450h, Ford Mondeo Hybrid, Honda NSX, and so on [6].

3.3 PHEV

P = Plug-in, H = Hybrid, E = Electric, V = Vehicle.

These electric vehicles are the vehicles like the half breed vehicles, yet the distinction here is we can charge our vehicle by connecting the charger remotely dissimilar to the hybrid electric vehicles where the battery is charged just regenerative slowing

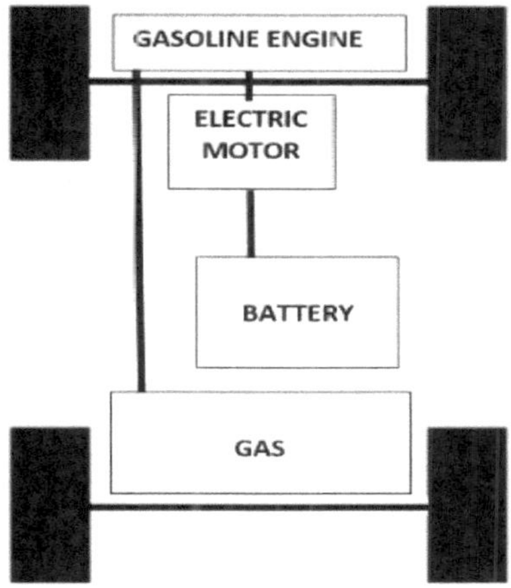

Fig. 6 HEV

down. As such type of vehicles has both internal combustion engines just as batteries along these lines so that they can be run on both the modes (Fig. 7).

The plug-in hybrid electric vehicles can run upto 30 miles on battery on a solitary charge, after that the vehicle begins running on a fuel. A portion of the instances of the plug-in hybrid electric vehicles is Mitsubishi Outlander, Volvo XC60 Twin Engine, BMW 225xe, Volkswagen Golf GTE, Toyota Prius Plug-in, Mercedes-Benz E350 e SE, and so forth [6].

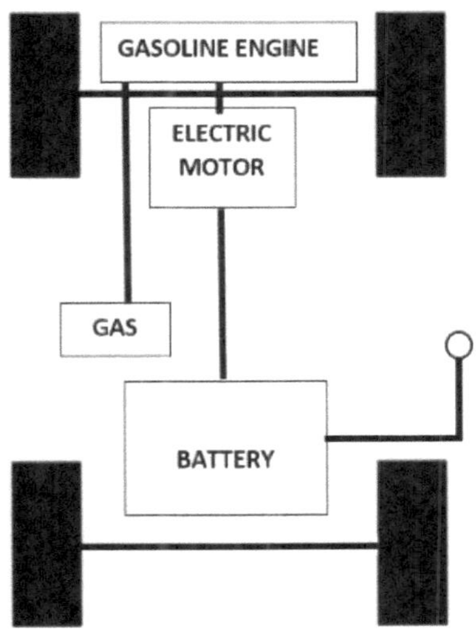

Fig. 7 PHEV

4 Construction of BEV

4.1 Key Components

The parts labeled in Fig. 8 are the major key components of any battery electric vehicles.

The auxiliary battery plays a major role in an electrically driven vehicle which offers capacity to control vehicle decorations. There is a charging port which allows the vehicle to communicate with an external power supply so that the battery of the vehicle can be charged. These types of vehicles contain a type of converter which are capable of converting DC supply to DC supply. They basically convert high DC voltage into low DC voltage because low DC voltage helps in operating the parts of the vehicle and also to provide energy to the battery. Another major component of such vehicle is the traction motor which uses power from the battery and helps in moving the wheels so that the vehicle can drive. Such type of motors can be used as both to drive and to recover. For the charging of the batteries, we need a charger which is known as the on board charger. The role of this charger is to extract the AC power supply from the household or any other power source which is generally 240 V. Now this AC power supply is then converted into a DC supply to charge the battery. It also checks the voltage, current, temperature, and various other factors of the battery. There is another device called as the power electronics controller which helps in the conveying of the power supply from the battery and the speed and torque control of the vehicle. The system which handles or manages the temperature of certain parts of the vehicle is also installed along with the other components, and this system is known as the thermal system. This system maintains the temperature of

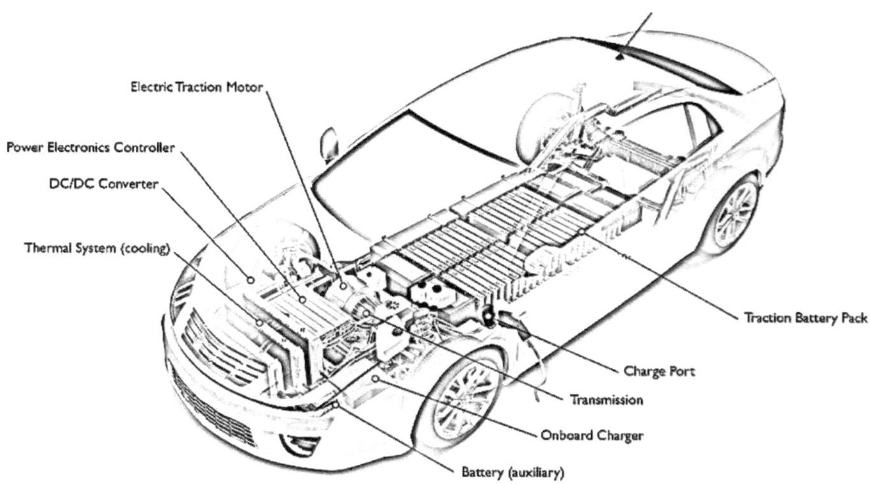

Fig. 8 Construction of BEV

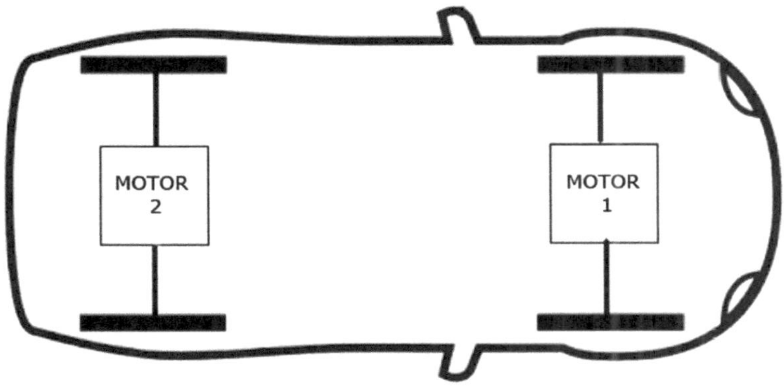

Fig. 9 Anatomy of Tesla Model S

motor, engines, etc. There is a traction battery pack which is another key components of the vehicle. It is used to store power which is further used in the engine of the electric vehicle. And lastly, there is a component of transmission. The role of this component is to transmit the mechanical force from the motor to the wheels so as to move the vehicle [7, 8].

4.2 Anatomy of Tesla Model S

The Tesla Model S is one of the superior battery electric vehicles. It has two electric engine for traction, one for the front pivot and the other one on the back hub independently. Additionally, both the motors have their own individual power electronics controller. This double engine design for both the axles is the bit of leeway for the vehicle as it gives all-wheel drive (AWD) abilities and furthermore a decent exhibition as far as increasing speed and driving elements [9] (Fig. 9).

4.3 Anatomy of Rimac Concept One

Another exceptionally elite BEV is Rimac Concept One. This is the principal battery electric hypercar. It has a significant level of execution and the driving elements. The powertrain of this vehicle comprises of four separate motors, one for each wheel. Every one of these motors has their own apparatus box, the single speed gear box in the front and the two-speed gearboxes with carbon fiber grips in the back. The high-voltage battery is uprooted in a "T" shape, between the front and the back axles as appeared in Fig. 10 [9].

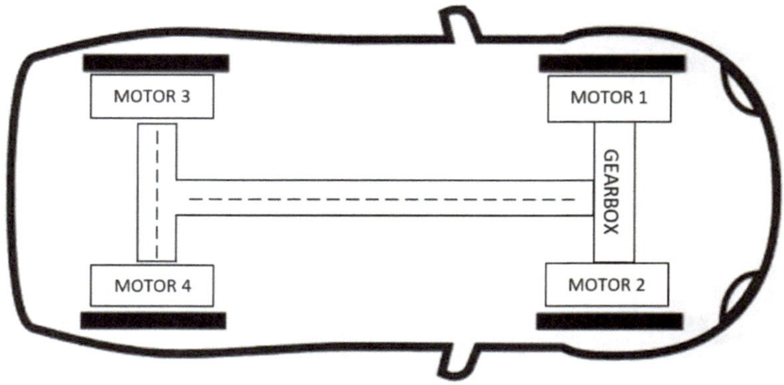

Fig. 10 Anatomy of Rimac Concept One

5 Working of BEVs

5.1 Working Principle

Since we realize that an electric vehicle utilizes electrically driven engine unlike an internal combustion engine. The electric vehicle uses the force put away in the huge battery pack which supplies energy to the electric engine. Each electric vehicle has own port of charging, which allows the vehicle to communicate to an external power supply which helps in the charging of the battery. Charging station or the charging outlet gives charging of the large battery packs. The principle capacity of this battery is to give power to the embellishments in an electric vehicle (Fig. 11).

All the vehicle need a low voltage DC power for working, with the goal that a DC/DC converter is introduced in it. The electric motor which is associated with the

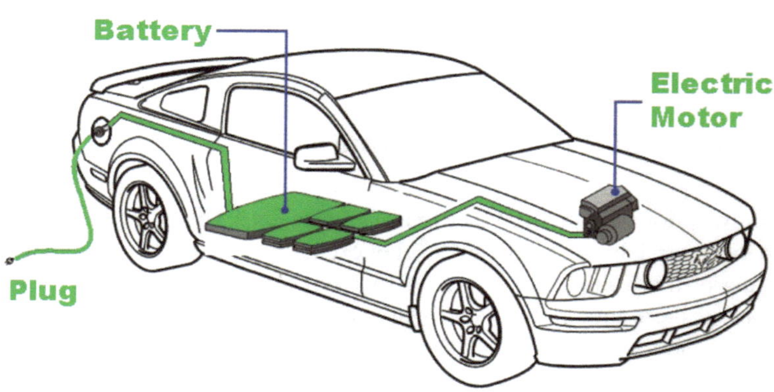

Fig. 11 Internal architecture of BEV [10]. *Source* Circuit Digest

wheels of the vehicle utilizes the energy of the batteries which helps to run the motor or engine so as to move the vehicle. The speed and the torque of the motor created in it are constrained through the power electronics controller, and it moreover manages the movement of electrical energy passed on by the battery. This is the fundamental working guideline of the battery electric vehicles (BEVs).

It has anon board charger, which changes over AC power to DC power to charge the batteries of the vehicle. The characteristics of the standard battery, for instance, voltage, temperature, current, and state of charge are obliged through it while charging the battery pack. The working temperature extent of the electric motor, power contraptions and various parts are kept up with the help of the cooling system [11].

5.2 Modes of Operation

5.2.1 Electrical Driving

This method of driving is the fundamental driving mode, wherein the vehicle is in the running condition. The power electronics of the vehicle are given power with the help of the high-voltage battery. The converter changes over the direct voltage into an alternating voltage which helps in driving the electric engine and subsequently the vehicle moves. Then again when the motor runs as an alternator the rotating voltage is changed over into the immediate voltage and afterward took care of to the battery which prompted the charging of the battery [12].

5.2.2 Regenerative Braking

In the event that the electric vehicle is moving with no assistance of the driving torque from the electric engine then a piece of a dynamic energy is taken care of into the high-voltage battery with the help of the electric engine which works as an alternator. Fundamentally, regenerative braking is the change of the vehicle's kinetic energy into energy which is put away in the battery, where it is very well-utilized to drive the vehicle later. It is regenerative in light of the fact that the energy is recovered to the battery where it tends to be utilized once more.

The kinetic energy put away in a moving vehicle is identified with the mass and speed of the vehicle through the condition $E = 1/2mv^2$. All else being equivalent, if the vehicle is twice as heavy it has double the kinetic energy and in the event that it is moving twice as quick it has multiple times the kinetic energy. Whenever the vehicle hinders the active energy put away in the vehicle must be recoupled with the assistance of an engine as engine changes over the formed torque into the fitting voltage and current waveforms to deliver the instructed torque in the engine in the most effective way. The torque order can be negative or positive. At the point when the torque serves to slow the vehicle, then energy is stored back to the battery, and henceforth, this is called regenerative braking [12].

5.2.3 Climate Control While Vehicle Is Stationary

In the event that the electric vehicle is remaining in a car influx, no yield is required from the electric engine or generator. The solace necessities of the inhabitants are met with a high-voltage warming framework and climate control system blower. There is no utilization of energy when the vehicle isn't moving or fixed. An atmosphere control gadget for a fixed control of the atmosphere in an engine vehicle has a heatpump circuit. There are two heat exchanger, one heat exchanger for taking up heat, and another heat exchanger for releasing heat into the vehicle interior. There is an electrically or precisely drivable blower which is situated in the stream heading of the heat siphon circuit between the first and the second heat exchanger, and a supporter set for delivering mechanical and electrical force. The electrical and mechanical force is being utilized in any event to some degree to drive the blower. Also, there is a third heat exchanger by which the fume heat of the sponsor set is moved to the heatpump circuit. Along these lines, great proficiency is guaranteed for the atmosphere control gadget in the warming activity while the vehicle is not moving [12].

6 About Batteries in BEVs

The fundamental and the most significant piece of any battery electric vehicle is the high-voltage battery. Different parts and the components of the vehicle are altogether affected through the parameters of the battery, for example, greater traction motor torque, most extreme recovery brake torque, vehicle run, weight of vehicle, vehicle cost, and so on.

6.1 Types of Batteries

A battery comprises of at least two electrochemical cells combined. The battery changes over the chemical energy to electric energy. A solitary battery cell is made of a positive cathode and a negative anode, both are associated with an electrolyte. The compound response between the electrolyte and terminals produces power. Battery-powered batteries can turn around the substance response by switching the current. Along these lines, the battery can be energized. The material and their types utilized for the cathodes and electrolyte decides the battery particulars. Various batteries accessible are appropriate for EVs. These batteries are depicted here.

6.1.1 Lead/Acid

These batteries are one of the most established class of a battery utilized in battery electric vehicles. The positive terminal of the battery is lead oxide, and the negative terminal of the battery is lead. Both the terminals of the battery are drenched in an electrolyte of sulfuric acid. During the release procedure, the lead on the negative terminal and the lead oxide on the positive terminal respond to sulfuric acid. Because of the response lead sulfate is shaped and the electrolyte transforms into water, energy is discharged during the chemical reaction and when energy is involved the reaction will be reversed.

The cost of the particular energy of the lead/acid batteries is low when contrasted with the other battery innovations. A common lead/acid battery has a particular energy of about 40 Wh/kg and a particular intensity of about 240 W/kg [13].

6.1.2 Nickel Metal Hydride

These types of batteries, on the positive terminal of the battery, use nickel. These are of four sorts: nickel zinc (Ni–Zn), nickel iron (Ni–Fe), nickel metal hydride (Ni–MH), and nickel cadmium (Ni–Ca). Due to their minimum life and less explicit force, the Ni–Zn and Ni–Fe types of batteries are not viewed as a possibility for EVs. The best invention was the cadmium-based Ni-Ca batteries. These batteries are in contrast with the Ni-MH batteries. The Ni-MH battery reaction is such that the metal hydride reacts with the nickel oxyhydroxide and gives nickel hydroxide and metal separately.

They have high explicit force upto 2000 W/kg and have explicit energy between the scope of 45 and 70 Wh/kg [13].

6.1.3 Lithium-Ion

These type of batteries are the most dependable and prevailing innovation in the field of batteries for BEVs just as different EVs. The development of the lithium-ion batteries has been expanding from the last few years. Maximum BEVs in the future are going to be outfitted with lithium-ion batteries. Examples of vehicles which use such type of batteries are the Tesla Roadster and the Mitsubishi i-Miev. These batteries are really reasonable for superior EV batteries as a result of the principle characteristics of this metal which is high expected potential. This means they have high potential energy (in Wh/kg), all things considered. Lithium is additionally light in weight among all the accessible metallic materials.

The particular energy of the lithium-ion battery ranges from 60 to 250 Wh/kg while the force capacity can be as high as 2000 W/kg [13].

6.1.4 Metal Air

The metal air batteries are not as easy to work as different batteries referenced on the grounds that the greater part of them cannot be revived by turning around the current rather we need to supplant the anodes with different materials. These batteries are precisely rechargeable and are practically identical in energy units. The significant favorable position of such batteries is that it has only one reactant and other reactant is supposed to be due to which the weight of the battery is reduced. Among all the metal air batteries, the zinc air battery is the most popular while other batteries are in progress to develop. The enormous energy capacity of the battery is valuable in little gadgets requiring a dependable battery.

They have a high explicit energy of about 220 Wh/kg, yet they have a generally low explicit intensity of about 110 W/kg [13].

6.1.5 Ultra Capacitors

Ultra capacitors are totally not quite the same as batteries, as they store their energy genuinely though the batteries store their energy artificially. The two plates of the capacitors are known as collectors which are separated in an electrolyte. The role of capacitors is to store the energy on high surface materials. Ultra capacitors can be charged and released a lot quicker than batteries and are truly reasonable for putting away the energy from regenerative slowing down. The ultra capacitor has high explicit intensity to 5000 W/kg yet a low energy capacity of 5 Wh/kg [13] (Fig. 12).

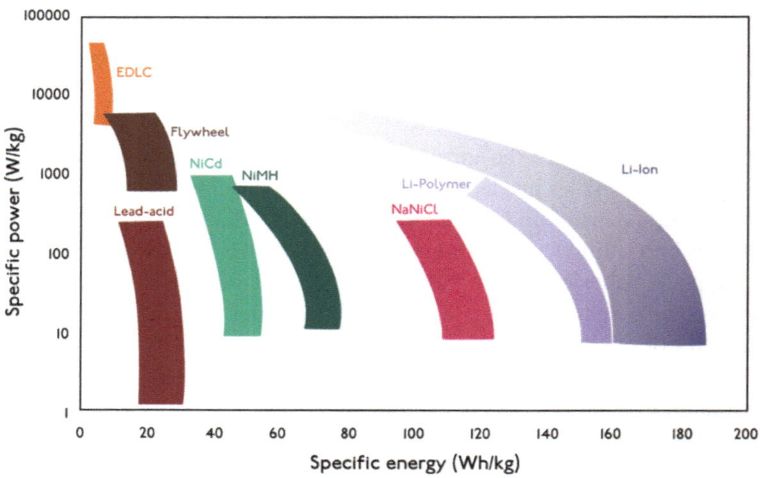

Fig. 12 Power and energy of different batteries [9]. *Source* x-engineer.org

6.2 Battery Pack Architectures

A battery pack of high-voltage battery is the mix of a number of battery cells organized in strings and modules. The battery cell is considered as the smallest unit of the battery pack. When the individual battery cells are gathered together or arranged together, they are called as modules. Further, when these modules are joined together and orchestrated in the arrangement or pattern, they structure battery packs. Depending upon the battery parameters, there might be a few degrees of measured quality.

The quantity of cells orchestrated in arrangement decide the all out battery pack voltage. Consider the model demonstrated as follows shows the series connection of the cells.

So as to build the present limit of the battery, the blend of strings has to be associated equally. Think about the model as demonstrated as follows: The limit and the current ability of the battery will be multiple times when the series of cells are associated equally. From Figs. 13 and 14, the series and the parallel combinations of batteries and their representation can be understood [9].

6.2.1 Mitsubishi i-MIEV Battery Pack

The high-voltage battery pack of Mitsubishi i-MIEV has 22 modules which contains 88 cells which are associated in arrangement. These modules contain 4 cells each. The voltage of each cell is 3.7 V, and the absolute voltage of the battery pack is 330 V which can be seen from Fig. 15 [9].

Fig. 13 Battery cells in series

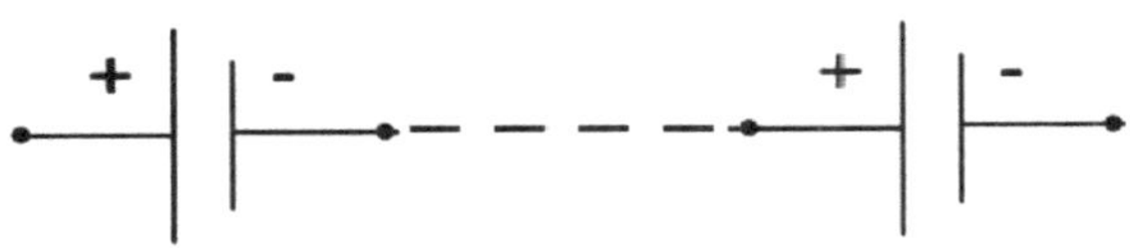

Fig. 14 Battery cells in parallel

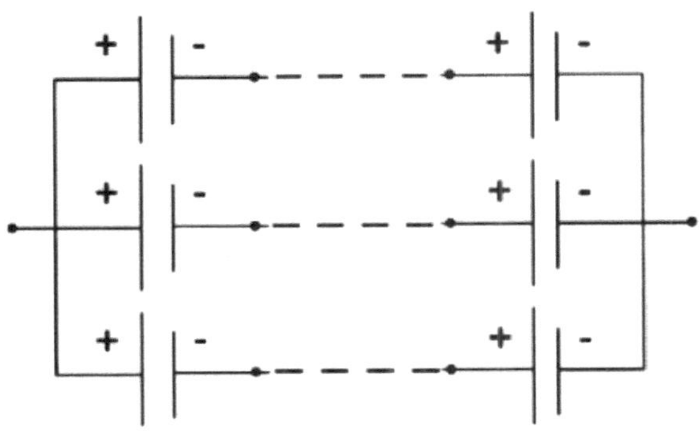

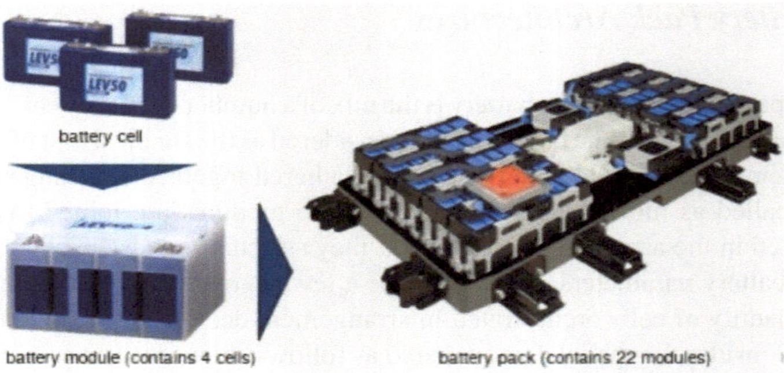

Fig. 15 Battery pack of Mitsubishi [14]. *Source* Mitsubishi

Fig. 16 Battery pack of Tesla Model S [14]. *Source* Tesla

6.2.2 Tesla Model S Battery Pack

The high-voltage battery pack of Tesla Model S has 16 modules involved 6 gatherings each. Each gathering present in every module is a mix of 74 cells; in this manner, all out number of cells present in the battery pack are 7104. This can be seen in Fig. 16 [9].

6.3 Battery Pricing

Battery estimating is the most significant detail of the battery electric vehicle (BEVs). Today the NiMH battery is as yet less expensive than lithium-ion batteries; however, lithium-ion batteries will in the long run be more affordable than the NiMH battery when being mass delivered.

The Pb/A battery won't decline in value in the future as the innovation is practically developed. NiMH battery costs are additionally cannot be predicted significantly may be as more as the cost of nickel which make up the vast majority of the battery cost. The first delivered lithium-ion pack is about 1600 €/kWh and at a creation stock of one lakh packs per year the manufacturing cost will be as minimum as 150 €/kWh.

The advancement proportion determined can give a sign of how the innovation will create in coming years when one lakh battery packs are delivered. An advancement proportion for lithium-ion batteries for purchaser items was determined and evaluated as of 83% somewhere in the range of 1993 and 2003. An other report assessed an advancement report of 92.5% somewhere in the range of 2010 and 2030. Similarly, as with different advancements the cost will diminish more in the beginning time of improvement [9].

7 Classification of Charging

The charging of the battery electric vehicle is arranged into three levels: level one, level two and level three or DC fast charging.

7.1 Level One Charging

From Fig. 17, this level utilizes a similar 120-V supply found in home unit outlets, and this charging can be done by using the normal cable of charging which other

Fig. 17 Level one charging [11]. *Source* New Car Kids

electrically driven vehicle use. Making this sort of charging accessible is as straightforward as introducing devoted 120 V outlets in the parking area, home or work environment. The charging of about 10 h is sufficient to travel roughly 70–80 miles [15].

Advantages

No establishment cost if an outlet is accessible close to where the vehicle is left.

This has a low effect on electric utility.

Disadvantages

Slow charging, commonly in one-hour of charge it can travel 5–6 miles.

7.2 Level Two Charging

From Fig. 18, this type of charging requires a specific station which gives power at 240 V. This sort of charging needs installation of charging station unit and electrical wiring fit for dealing with higher voltage power. Such type of charging is beneficial for daily work commute. It takes nearly 4–5 h to completely charge a battery so that the vehicle can travel around 70–80 mile [15].

Advantages

Faster charge time normally 10–20 miles per every hour of charge.

Energy effective, about 3% gain in proficiency.

Assortment of EV charging makers gives separated items which can plan charging, track use, and gather expenses.

Fig. 18 Level two charging [16]. *Source* Arthur D. Little

Disadvantages

More costly than level one.

It has possibly higher effect on top kilowatt request charges for organizations.

7.3 Level Three Charging or DC Fast Charging

From Fig. 19, DC quick charging of 480 V gives good vehicles a 80% charge in 30 min to 1 h by changing over AC capacity to DC power capacity in EV batteries. For DC quick charging, there are three attachment types utilized by various automakers: the CHAdeMO, Society of Automotive Engineers (SAE) consolidated charging framework (Combo/CSS). Tesla's Supercharger hardware is just good with Tesla vehicles, inspite of the fact that they offer a connector which permits Tesla proprietors to utilize CHAdeMO gear [15].

Advantages

Charge time is decreased radically, and it is almost as quick as refueling a fuel vehicle.

Disadvantages

Altogether costlier than the other two levels of charging.

Conceivably expanded charges for business areas.

Diverse fitting sorts are confounding to potential EV purchasers and charging station administrators.

The contingent upon the vehicle and charging gear, quick charging can be eased back during chilly climate.

Fig. 19 Level three charging [16]. *Source* Arthur D. Little

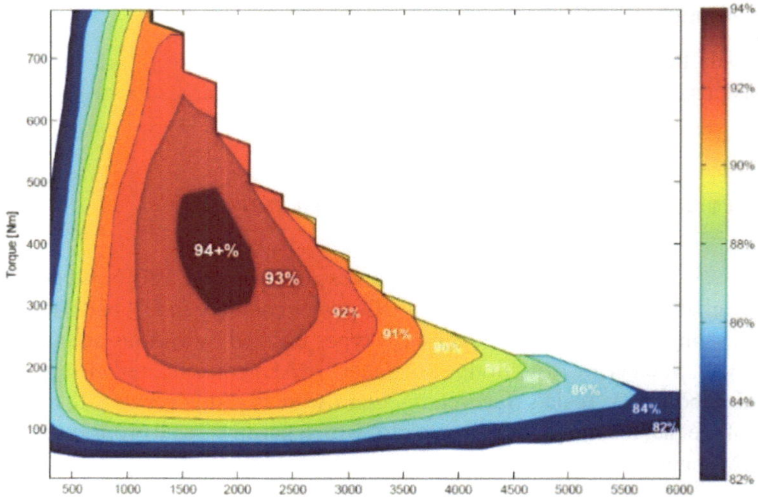

Fig. 20 Efficiency map [9]. *Source* x-engineer.org

8 Advantages of BEVs Over Other EVs and Ice Vehicles

8.1 Efficiency

In battery electric vehicles, the substituting flow (AC) electric engine which has permanent magnet or induction rotors offer high generally efficiency. The engine utilized for the vehicle is permanent magnet electric engine which has a base productivity of around 65–80% and a limit of 90–97%, though the efficiency of an ICE is a lot of lower around 10–40%.

An electric engine can be handily worked at a normal productivity of 80–90% on a driving cycle; however, the normal efficiency of an ICE vehicle is 20–30% which implies that the effectiveness of BEVs is multiple times higher than the internal combustion engines [9] (Fig. 20).

8.2 Dynamic Performance

Figure 21 gives information regarding torque and power characteristics. The engine of the battery electric vehicle has a perfect traction trademark. The electric engine can convey the most extreme torque at a zero speed, which gives an excellent beginning to the vehicle. An electric engine is equipped for giving a peak torque to a constrained measure of time in the beginning and a ceaseless torque as long as there is an inventory given to the vehicle. For a top torque in the beginning, the engine requires a very

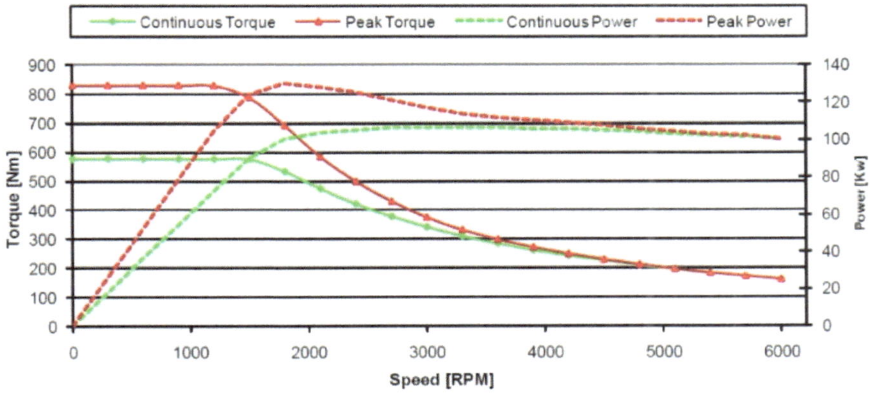

Fig. 21 Torque and power characteristics [9]. *Source* x-engineer.crg

measure of current from the battery which creates a high temperature in the force hardware and engine. To shield the parts from higher temperature, the peak torque is constrained in time yet the torque is accessible enough to help the speeding up execution of the vehicle [9].

8.3 *Reliability*

The battery electric vehicles have not many prospects of disappointments when contrasted with the internal combustion engine vehicles in light of the fact that BEVs have less moving parts. Likewise because of high torque and fast qualities of the electric engine, there is no requirement for a multistage gearbox which implies a solitary advance mechanical gearbox is sufficient for traction prerequisites. Though ICE vehicles have many moving parts and furthermore extra framework, for example, a fuel framework, a weariness framework, and so on which can prompt disappointments and an extra wellspring of disappointments is a multi-stage gearbox. In this way, BEVs are more dependable than other vehicles [9].

8.4 *Torque Vectoring*

For an all-wheel drive (AWD) battery electric vehicle, the soundness during the cornering is significant. This stability during cornering can be improved with the assistance of controlling the torque of the wheels. An electric engine of a battery electric vehicle has quicker torque reaction and is equipped for creating negative torque additionally, which then again, internal combustion engine vehicle cannot do.

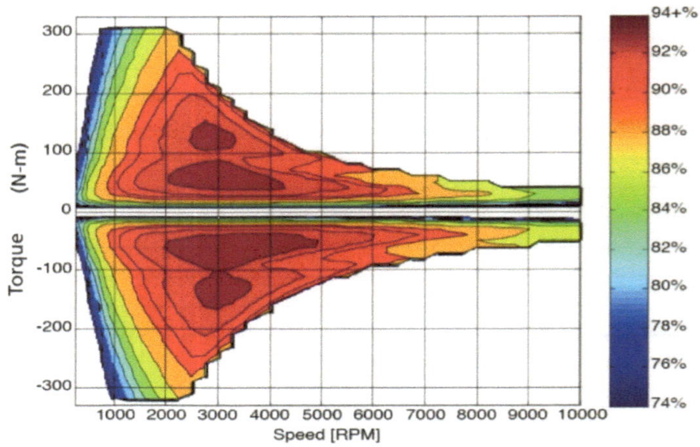

Fig. 22 Positive and negative torque map [9]. *Source* x-engineer.org

Rather than ICE vehicles, they have constrained commitment to the vehicle solidness and they lessen the measure of torque conveyed [9].

8.5 Energy Recuperation and Brake Regeneration

The utilization of an electric machine in the battery electric vehicles is one of the unmistakable favorable position of BEVs over different vehicles. The electric machines are reversible which implies that they can deliver torque in the event that they are provided through the electrical energy, and they can likewise create electrical energy when they have an input torque because of the latency of the vehicle. At the point when the electric machine is creating torque, then it is in engine mode and when the electric machine is delivering electrical energy, then it is in generator mode. Figure 22 shows that the negative torque can be as high as the positive torque which implies the electric machine can create a lot of the slowing down torque [9].

8.6 Cost and Maintenance

The battery electric vehicles have a bit of leeway of much lower running expense when contrasted with different EVs just as the regular internal combustion engine vehicles. The expense of power required to charge a battery electric vehicle is 33% as much per kilometer as the expense of purchasing petroleum for other customary vehicles. Additionally having less moving parts and segments, a battery electric vehicle is simpler and less expensive to keep up.

8.7 *Better for an Environment*

Less Pollution: The battery electric vehicles help in lessening the destructive air contamination from exhaust emanations as the BEVs have zero fume discharges [17].

Renewable Energy: By utilizing the sustainable power source to energize the battery electric vehicle, the green house gas outflows can likewise be decreased. Utilizing solar PV to revive the vehicle is a bit of leeway as it requires typical household supply to energize its battery which should be possible by solar powered PV framework or some other sustainable power source. Also, renewable energy such as wind energy can be used to produce power which can be further utilized to charge the vehicle [18–20].

Eco-friendly Materials: In the cutting edge age, there are manufacturers who are using eco-friendly materials for the development of vehicles, for instance, recycled plastic materials, bio-based materials, or recycled household waste materials. In this way, the manufacturers are concerned to build the vehicles with an eco-friendly materials.

9 Conclusion

From the invention of the electric vehicles in the early eighteenth century to the modern age world of today, there have been many evolutions and developments in the electric vehicles. After years of growth and development in this field, battery electric vehicles demand is finally increasing. A number of factors such as better performance, cutting edge technologies, connectivity, environment friendly, quieter ride, and low running cost are combining to focus the attention on the adoption of battery electric vehicles.

Despite having so many features, there are a few concerns as well before adoption such as driving range, initial cost, lack of charging infrastructure, and time required to charge the vehicle. The average driving range of the battery electric vehicles is 150–200 miles which are quite less than that of ICE vehicles. But it is better for the people who have a daily commute of 70–100 km s. Similarly, the initial cost of BEVs is more in contrast with the ICE vehicles, whereas the maintenance cost and the running cost of BEV are much less than that of ICE vehicles, and therefore, BEVs are better for long run. One of the major concerns about BEVs is the lack of charging infrastructure. For the vehicles used for the daily commute, the home or workplace charging is more than sufficient. But for the long-range drive of BEVs, there is a need of charging stations at various places just like the fuel stations for the ICE vehicles. With the help of charging stations, batteries of the vehicles can be recharged and they can travel over the longer distance. Addition to this, time taken to charge the vehicle is another concern about BEVs. On an average, a typical battery electric vehicle takes near about 8 h to charge a 60 kWh battery from empty-to-full which can cover up to 200 miles of distance. This charging time can be reduced to minimum of 30 min with the help of fast-charging stations. Also, benefiting from the battery technology improvements, the time required to charge at fast-charging stations can be reduced further.

Considering all these factors and the technological advancements, the market of the battery electric vehicles will be at the tipping point in the coming few years.

References

1. Z. Shahan, *1 Million Pure EVs Worldwide: EV Revolution Begins* (Clean Technica, 2016)
2. J. Fuller, What is the history of electric cars? How Stuff Works, 14 July 2009
3. *World's First Electric Car Built by Victorian Inventor in 1884* (The Daily Telegraph, London, 2009)
4. E. Rishavy, W. W. Bond, T. Zechin, Electrovair—A Battery Electric Car, SAE Technical Paper 670175, 1967
5. G. Berdichevsky, K. Kelty, J.B. Straubel, E. Toomre, *The Tesla Roadster Battery System* (Tesla Motors, 2006)
6. D.B. Sandalow, *Plug-in Electric Vehicles: What Role for Washington?* (The Brookings Institute, 2009). ISBN 978-0-8157-0305-1
7. A. Anand et al., Optimal selection of electric motor for E-rickshaw application using MCDM tools, *Cognitive Informatics and Soft Computing* (Springer, Singapore, 2020), pp. 501–509
8. K.C. Manjunatha, A.K. Bhoi, K.S. Sherpa, Design and development of buck-boost regulator for DC motor used in electric vehicle for the application of renewable energy, *Advances in Smart Grid and Renewable Energy* (Springer, Singapore, 2018), pp. 33–37
9. T.R. Crompton, *Battery Reference Books*, 3rd edn. (Newnes, 2000). ISBN 978-0080499956. Retrieved 18 Mar 2016
10. How Exactly Do Electric Cars Work? Green Car Future, 11 November 2018. Retrieved 22 Nov 2018
11. Components and Systems for Electric Vehicles (HEVs/EVs), Hitachi Review. Retrieved 18 Mar 2016
12. *Seminar on Electric Vehicles* (Automative Research Association of India, 2017)
13. D. Bakker, *Battery Electric Vehicles Performance, CO_2Emissions, Lifecycle Costs and Advanced Battery Technology Development Emissions, Lifecycle Costs and Advanced Battery Technology Development* (Utrecht University, Netherland, 2010)
14. The Battery Pack of Mitsubishi i-MIEV, Green Car Congress Energy, Technologies, Issues and Policies for Sustainable Mobility, 14 May 2008
15. *Electric Vehicle Charging Types, Time, Cost and Savings* (Union of Concerned Scientists, 2018)
16. J.W. Brennan, T.E. Barder, Battery electric vehicles vs internal combustion engine vehicles, *A United States-Based Comprehensive Assessment* (Arthur D Little, Strategy and Organization, Boston)
17. N. Priyadarshi et al., A closed-loop control of fixed pattern rectifier for renewable energy applications, *Advances in Greener Energy Technologies* (Springer, Singapore, 2020), pp. 451–461
18. N. Priyadarshi, F. Azam, A.K. Bhoi, A.K. Sharma, A multilevel inverter-controlled photovoltaic generation, in *Advances in Greener Energy Technologies* (Springer, Singapore, 2020), pp. 149–155
19. A. Sahu et al., Design of permanent magnet synchronous generator for wind energy conversion system, *Advances in Smart Grid and Renewable Energy* (Springer, Singapore, 2018), pp. 23–32
20. S. SenGupta, A.F. Zobaa, K.S. Sherpa, A.K. Bhoi, *Advances in Smart Grid and Renewable Energy* (2018)

Communication Standards for Interconnections of Smart Grid Infrastructure and Intelligent Electric Transportation System

Lipi Chhaya

1 Introduction

Smart grid is an evolutionary technology as it will completely transform traditional power grid. Some of the significant features of smart grid technology are bidirectional communication, active consumer participation, dynamic billing phases, integration of renewable energy resources, supervisory control and data acquisition (SCADA), sensing and automation, optimization of energy and advance metering infrastructure [1–3]. Communication architecture of smart grid consists of hierarchical and diverse network layers such as home, neighborhood and wide area networks [2–5]. This technology is proficient to address the problems such as scarcity of primary energy resources, GHG emissions, blackouts, limitations and challenges of renewable resource integration, reduced efficiency of coal-based power plants, lack of consumer participation, lack of automation and control, etc. [2–6]. Electric vehicles are the advanced form of vehicles for eco-friendly transportation [7]. Smart grid and EV integration can facilitate various ancillary services such as regulation of voltage and frequency, peak power shaving, as well as spinning reserve. Moreover, electric vehicles can solve many issues such as pollution, scarcity of fossil fuels and geopolitical issues related to fossil fuels. According to the study conducted in [8], by the year 2040, approximately 50 million grid allied electric vehicles are expected in the market. Smart grid technology is equipped with advanced automation, control and communication architecture to facilitate advanced metering infrastructure (AMI) [9–14]. Penetration of communication technologies is extensive in the entire world, and thus, integration of electric vehicles and smart grid can be facilitated with advanced automation and control with real-time communication of charging, distance, traffic, storage and fault diagnosis. It can serve as an independent distributed energy source in microgrid. Microgrid is a miniaturization of smart power grid in order to facilitate

L. Chhaya (✉)
Ahmedabad, India
e-mail: lipi.chhaya@gmail.com

N. Patel et al. (eds.), *Electric Vehicles*, Green Energy and Technology,
https://doi.org/10.1007/978-981-15-9251-5_9

an integration of distributed energy resources [13]. Distributed renewable energy sources provide variable response according to weather. Assimilation of renewable and alternative energy resources with smart power grid is the most crucial and challenging aspect due to their intermittent, unpredictable and unstable supply of energy. Moreover, integration of electric vehicles and smart grid will result in complex architecture and will require real-time communication of charging, controlling, traffic, scheduling, etc. Smart grid communication infrastructure can facilitate various requirements for charging, scheduling, optimization and aggregation of electric vehicles in accordance with renewable energy resources and grid infrastructure. Communication of real-time information will be facilitated through heterogeneous communication network layers of smart grid architecture [14–17]. Assimilation of electrically powered vehicle with smart power grid infrastructure can compensate for an intermittency of renewable energy sources and provide stability and reliability to the grid. An assimilation of the smart power grid and plug-in vehicles requires heterogeneous communication networks for communication for charging, controlling and location-based services. The first set of communication technologies will be required for communication between an electric vehicle and the charging facility of smart grid infrastructure. The second set of communication standards is required for communication between vehicles for mobile vehicular communication. The third set of communication technologies is required for control center and communications such as parking, traffic and various transportation facilities. This paper covers vehicle-to-grid (V2G), vehicle-to-vehicle (V2V) and vehicle-to- everything communications (V2X) for deployment of integrated and intelligent transportation systems. The diverse set of communication standards can be used on the basis of various requirements, such as distance, data rates, noise, etc. Figure 1 shows the different constituents of smart grid technology.

2 A Vehicle-to-Smart Grid Infrastructure Communication

An energy exchange between smart grid infrastructure and electric vehicle can be achieved by either taking power from the smart grid or else giving power to the smart grid. But for communication of signals, vehicle-to-smart grid interaction can be realized by bidirectional communication. The wireless sensor network is an inevitable technology for communication between electric vehicle and the smart grid. It consists of sensors, actuators, transmitters, receivers and embedded electronics devices [17]. An integration between smart grid and electric vehicle can be the flow of energy from the grid to vehicle if energy storage in vehicle is less, flow of energy from vehicle to power grid if stored energy in electric vehicle is high and energy flow from vehicle to home for using batteries of electric vehicles for household appliances [15]. Multiple communication scenarios are required for communication between electric vehicles, smart meters, an electric vehicle management system and smart grid data centers of

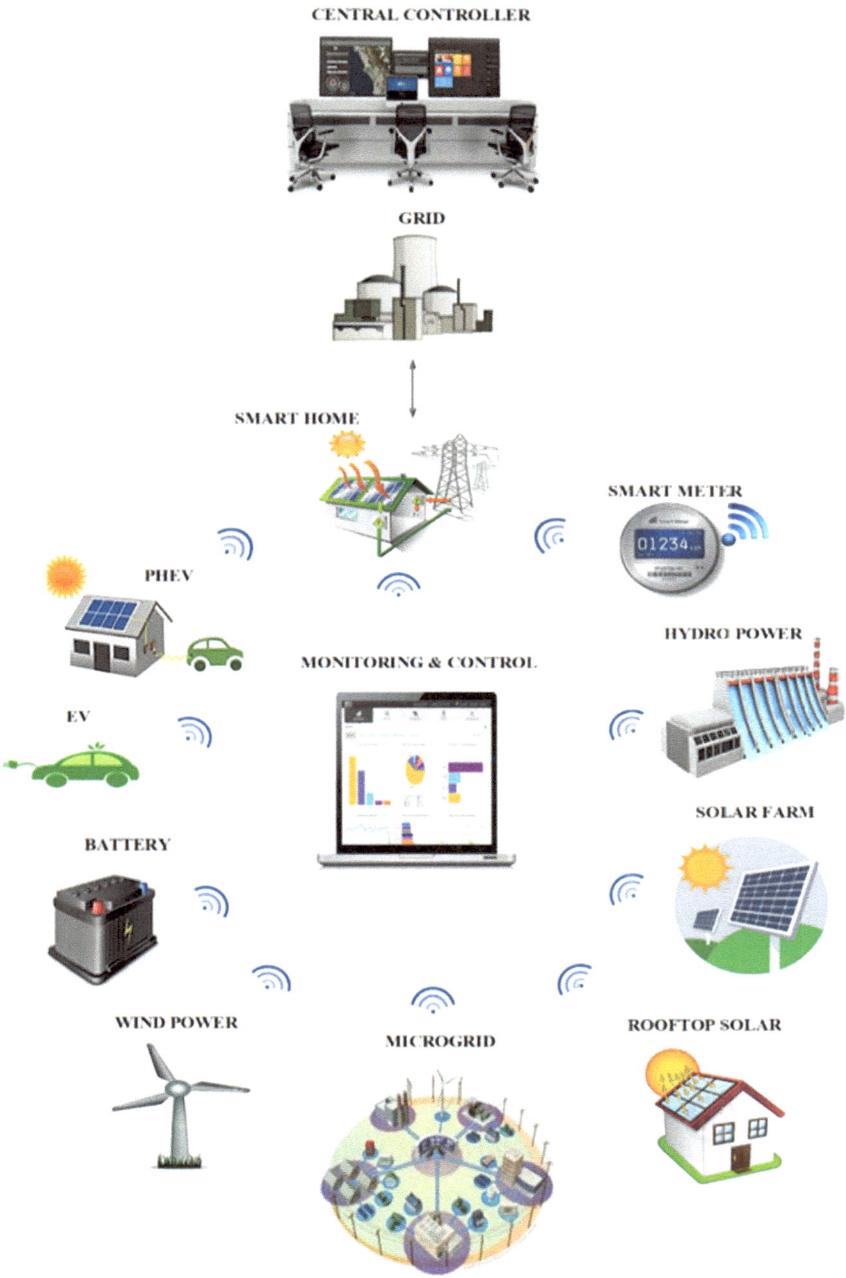

Fig. 1 Constituents of smart grid technology

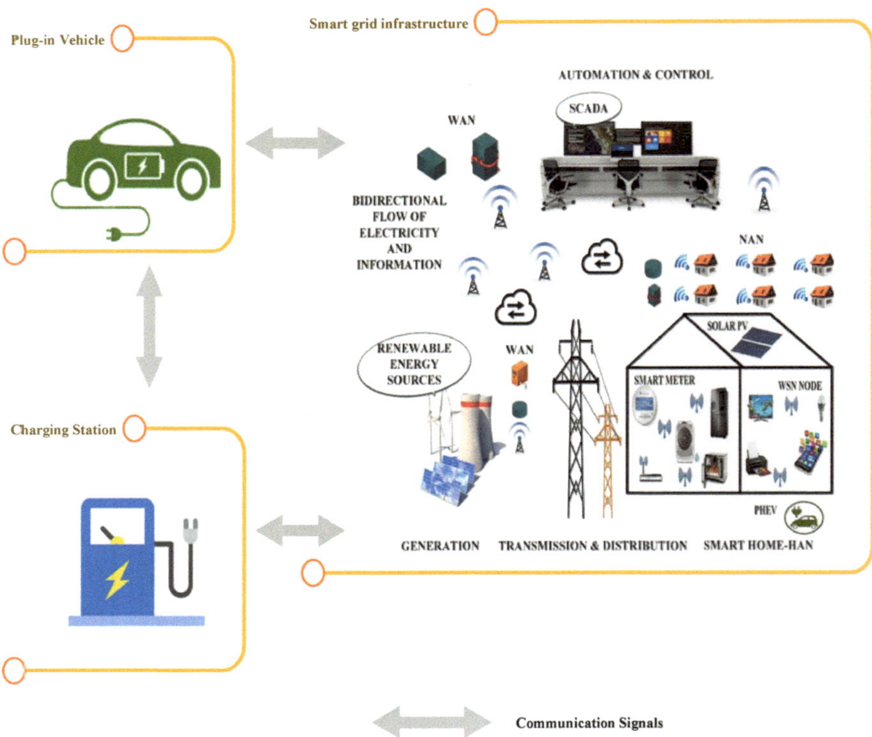

Fig. 2 Communication between electric vehicle and smart grid interconnections

various network layers [14–16]. Diverse set of communication technologies is necessary for transmission as well as reception of information related to charging, intelligent scheduling, metering, dynamic billing, energy forecasting, energy demand, energy consumption, readings of real-time voltage and current, dynamic load adjustment, renewable integration data, peak shaving, voltage regulation, frequency regulation, load statistics, etc. Figure 2 shows the communication between an electric vehicle and smart grid interconnections.

2.1 Communication Standards for Vehicle-to-Smart Grid Interconnections

A number of communication technologies and standards for interconnections between smart power grid infrastructure and plug-in vehicle are described below.

- **Bluetooth**

Table 1 Technical features of Bluetooth

Bluetooth features/characteristics	Technical specifications
Frequency band	2400–2483.5 MHz, Industrial, Scientific and Medical band (ISM)
Frequency Hopping Spread Spectrum (FHSS)	79 channels of 1 MHz each from 2402 to 2480 MHz
Modulation technique	Gaussian frequency shift keying (GFSK)
Channel bandwidth	1 MHz
Hops per second	1600
Frequency spectrum management	Dynamic
Symbol rate/maximum transmission rate	1 Mbps
Wired backbone network	IEEE 802.3-Ethernet or PPP
Authentication and encryption	128 bit key
Maximum data throughput	700 Kbps
Receiver sensitivity	−70 dBm for BER 0.1%
Power classes Class 1 Class 2 Class 3	Three power classes 100 mW to 100 m range 2.5 mW to 10 m range 1 mW to 1 m range

Bluetooth wireless technology is based on IEEE 802.15.1 standard. Bluetooth is a personal area network for short distance wireless communication among various devices [16]. This standard supports point-to-point and point-to-multipoint wireless communications. This technology functions in an ISM band of 2.4 GHz [18]. Table 1 shows different technical features of IEEE 802.15 standard. Progressive versions of Bluetooth such as 2.0 and 2.1 provide data output of 2.1 Mbps.

- **WirelessHART**

WirelessHART standard is developed on the base basis of IEEE 802.15.4 standard issued in 2006. It can be used for mobile devices, sensors, actuators, routers and gateway devices [16]. This standard can be used for control and automation applications. Table 2 shows the different technical characteristics of WirelessHART protocol.

- **Zigbee**

This technology is based on IEEE 802.15.4 standard. This technology forms a wireless personal area network (WPAN) with lesser data rates. It functions in an unlicensed band. The physical and media access layer are implemented from IEEE 802.15.4 standard, and the Zigbee alliance has determined rest of the layers [14]. It is the most extensively used wireless personal area network. Numerous technical features of Zigbee standard are depicted in Table 3.

Table 2 Technical features of WirelessHART

WirelessHART features/characteristics	Technical specifications
Frequency band	2.4 GHz, Industrial, Scientific and Medical band (ISM)
Spread spectrum technology	Direct sequence spread spectrum (DSSS)
Modulation technique	O-QPSK
Transmitter power	10 dBm
Multiple access scheme	TDMA
Wired backbone network	Wired HART, IEEE 802.3-Ethernet
Authentication and encryption	AES 128 bit key
Data throughput	250 kbps
Coverage area	Approximately 200 m

Table 3 Technical features of Zigbee

Zigbee features/characteristics	Technical specifications
Frequency band	2.4, 868, 915 MHz
Spread spectrum technology	DSSS
Modulation technique	BPSK, O-QPSK
Transmitter power	−25 to 0 dBm
Multiple access scheme	DSSS
Wired backbone network	Ethernet
Authentication and encryption	CBC-MAC, AES
Data throughput	20–250 kbps
Coverage area	10–100 m

- **6LOWPAN**

6LOWPAN is an abbreviation of IPv6 over low-power wireless personal area network. It is intended for insertion of small devices over IPv6 network. It is developed on the foundation of IEEE 802.15.4 standard. The Internet of Things is the core feature of smart grid information and communication network. The wireless sensor network (WSN) is the key module of IoT. Personal area network is appropriate for communication between various devices placed in the wireless sensor network. IoT consists of IP-enabled smart devices. IP-based communication is very multifaceted in terms of processing, control and management. It is difficult to integrate low-power miniature devices with restricted processing abilities for IoT network applications. The IETF 6LOWPAN association has established the simple resolution for IoT [14]. Nodes are allotted a shared 64-bit IPv6 prefix so that regardless of the locality of a node, IPv6 prefix of the nodes remains identical. This standard enables an amalgamation of IPv6 with embedded electronics devices having less power, lower data output and restricted processing speed through optimization. Table 4 shows the various technical characteristics of 6LOWPAN standard.

Table 4 Technical features of 6LOWPAN standard

Features/characteristics of 6LOWPAN	Technical specifications
Frequency band	2.4, 915, 868 MHz
Modulation technique	O-QPSK
IP support	IPv6
Coverage area	200 m
Multiple access technique	TDMA
Wired backbone network	Ethernet
Authentication and encryption	AES 128 bit
Data throughput	20–250 kbps 20 kbps for 868 MHz 40 kbps for 915 MHz 250 kbps for 2.4 GHz

- **Wireless Fidelity/WLAN**

Wireless fidelity is designed on the foundation of IEEE 802.11 protocol which was introduced commercially in the year 1997. IEEE 802.11 defines the physical and media access layer. Different versions of this standard such as IEEE 802.11a, b, and g were also developed with diverse technical features. WLAN is appropriate for home automation, advanced metering and IoT applications [14, 16, 19]. Technical features of various versions of this standard are presented in Table 5.

- **WiMAX**

Wireless interoperability for microwave access standard is developed on the basis of IEEE 802.16 protocol. WiMAX is intended for the wireless metropolitan area network (WMAN). It is a scalable broadband wireless standard. It operates in the frequency band of 2–66 GHz. WiMAX has the ability to substitute complete wired network and to deliver last mile connectivity. It can be categorized into fixed WiMAX (IEEE802.16d) and mobile WiMAX (IEEE802.16e). Fixed WiMAX provides point-to-multipoint connectivity up to 50 km [16]. Mobile WiMAX is designed and developed for mobile wireless clients with the coverage range up to 15 km. WiMAX technology uses multiple input multiple output technology as well as smart antennas for higher data speed. Technical features of various versions of WiMAX are shown in Table 6.

- **Cellular communication standards**

Present cellular communication architecture can be optimized for various applications of smart grid such as vehicular communication, AMI, WSN, etc. LTE and 4G wireless communication standards are the appropriate choices for smart grid and plug-in vehicle communication [9, 16]. These technologies can serve high-throughput requirements for various applications such as charging, scheduling, traffic

Table 5 Technical features of WLAN standard

Features of WLAN	Technical specifications			
	IEEE 802.11	IEEE 802.11 a	IEEE 802.11 b	IEEE 802.11 g
Frequency band	2.4 GHz ISM band	5 GHz	2.4 GHz ISM band	2.4 GHz ISM band
Physical layer specifications	Direct sequence spread spectrum, frequency hopping spread spectrum	Orthogonal frequency division multiplexing	Direct sequence spread spectrum	Orthogonal frequency division multiplexing
Maximum rate of transmission	2 Mbps	54 Mbps	11 Mbps	54 Mbps
Maximum data throughput	1.2 Mbps	32 Mbps	5.5 Mbps	24 Mbps
MAC specification	Carrier sense multiple access-collision avoidance	Carrier sense multiple access-collision avoidance	Carrier sense multiple access-collision avoidance	Carrier sense multiple access-collision avoidance
Wired backbone network	Ethernet	Ethernet	Ethernet	Ethernet
Encryption method	40 bit RC4	40 bit RC4	40 bit RC4	40 bit RC4

Table 6 Technical features of WiMAX standard

Features/characteristics of WiMAX	Technical specifications		
	IEEE 802.16	IEEE 802.16a/d	IEEE 802.16e
Frequency spectrum	10–66 GHz	2–11 GHz	< 6 GHz
Transmission	Line of sight (LOS)	Non line of sight (NLOS)	Non line of sight (NLOS)
Modulation technique	16 QAM, 64 QAM, QPSK	16 QAM, 64 QAM, 256 QAM, OFDM-QPSK	16 QAM, 64 QAM, 256 QAM, OFDM-QPSK
Data rates	32–134 Mbps	70–100 Mbps	15 Mbps
Mobile/Fixed	Fixed	Fixed	Mobile
Radius of cell	1–3 miles	3–5 miles	1–3 miles
Channel bandwidth	20, 25, 28 MHz	1.25–20 MHz	5 MHz

congestion, etc. An uplink and downlink data rates of long-term evolution technology are 50 Mbps and 100 Mbps, correspondingly. LTE advanced and 4G technology can provide an uplink data rate of 500 Mbps and downlink data rate of 1 Gbps correspondingly. 4G technology is a combination of PAN, wireless, cellular and satellite network. It attains data speed up to 100 Mbps [16]. Some of the key features of 4G are MIMO, smart antenna, OFDM, software-defined radio, etc. The 5th generation mobile wireless standard is the successor of 4th generation technology. 5G is suitable

Table 7 Technical features of 4G and 5G wireless technologies

Features of cellular technology	Technical specifications	
	4G	5G
Technology	OFDM	BDMA, FBMC
Data throughput	LTE-Downlink-3 Gbps, Uplink-1.5 Gbps	10–50 Gbps
Frequency spectrum occupancy	LTE-1.8, 2.6 GHz	1.8, 2.6, 30–300 GHz
Channel bandwidth	LTE-1.4–20 MHz	60 GHz
Type of switching	Packet switching	Packet switching
Error correction	Turbo codes for forward error correction	Low-density parity check codes for forward error correction
Application support	HDTV	Ultra-high video, virtual applications

for an Internet of Things, machine-to-machine communication, sensor networks, etc. The anticipated data output of 5G is 10–50 Gbps with the bandwidth of 60 GHz. 5G offers World Wide Wireless Web for integrated wireless network connectivity. 5G is developed for heterogeneous, pervasive and universal services. Table 7 shows different technical specifications of 4G and 5G standards.

- **PLC Technologies**

Various powerline technologies can be used for communication between the home area network of smart grid infrastructure and charging facilities [16]. Home plug powerline, PRIME and G3 PLC can be used for home area applications of the Smart grid which also includes advanced metering and plug-in vehicle charging. Home plug AV, Home plug AV2 and Home plug PHY are different versions of Home plug standard [16, 20]. IEEE 1901.2 standard describes narrowband powerline technologies. G3 PLC and PRIME PLC are categorized as narrowband powerline standards. Table 8 shows the technical features of G3 and PRIME PLC technologies.

- **SAE Standards**
- Society of automotive engineers have defined a set of standards for interconnections between electric vehicles during charging [7].

  SAEJ2836/1 and J2847/1: This standard defines requirements for communication between grid and EV during energy transfer.
  SAEJ2836/2 and J2847/2: This standard defines requirements for communication concerning EV and grid as well as charger.
  SAEJ2931: This standard defines requirement for digital communication for off-board charging of electric vehicles.
  SAE J2931/1: This standard depicts power line communication for EVs.
  SAEJ2931/2: This standard describes requirements for physical layer communication between EV as well as electric vehicle supply equipment (EVSE).

Table 8 Technical features of PLC technologies

Features	Technical specifications	
	PRIME	PLC G3
Frequency band	42–89 kHz	35–91 kHz
Value of sampling frequency	250 kHz	400 kHz
Data throughput	128.6 kbps	33.4 kbps
FEC	Convolutional code	Convolutional code, reed Solomon code, repetition code
Interleaving	Per symbol of OFDM	Per packet of data
Modulation technique	Differential binary phase shift keying, differential quadrature phase shift keying, differential eight phase shift keying	Differential binary phase shift keying, differential quadrature phase shift keying

3 Vehicle-to-Vehicle and Vehicle-to-Everything Communication

Vehicle-to-vehicle communication technology is meant for wireless communication between various plug-in vehicles for tracking, traffic congestion, prevention of accidents and congestion information for charging purpose [21]. Traffic, speed and distance are the three important parameters for vehicular communication. The vehicle-to-vehicle communication can be facilitated by various cellular wireless communication technologies such as WiMAX, 4G, 5G, etc. Vehicular communication is required for various features such as parking, emergency breaking, turning, public charging stations, collision warning, etc. WPAN can facilitate vehicular communication. Vehicular communication systems can be used to enhance the security of vehicles by taking decisions in emergency accidental situations. Moreover, parking and traffic problems can be addressed and overcome using vehicular communications.

The vehicle-to- everything (V2X) communication is meant for communication between vehicles and roadside management system, parking system, public charging stations, pedestrians, mobile phone, electronics devices [22], etc. V2X communication system is equipped with sensors, cameras, parking lane markers, street lamps, traffic lights, etc. V2X communication system can be deployed by using various short- as well as long-distance communication standards, such as wireless personal area networks and cellular communication standards. Zigbee, WirelessHART, 6LOWPAN, WLAN, WiMAX and other 4G and 5G technologies can be used to facilitate the communication between the vehicle and various devices (Fig. 3).

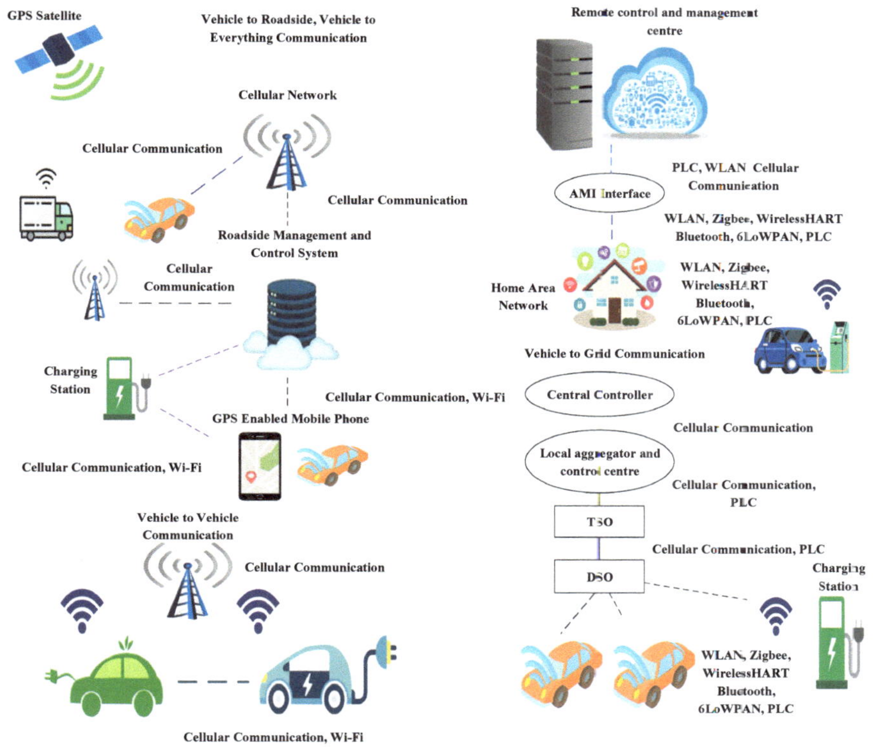

Fig. 3 V2G, V2V and V2X communication

4 Conclusion

Smart grid is an inclusive technology in the context of applications and standards. It includes intelligent generation, transmission, distribution, renewable integration and plug-in vehicles. An intelligent transportation system is an integral part of smart grid technology. It can be categorized as V2G, V2V and V2X interactions. This paper is anticipated to serve as a complete analysis of the wired and wireless communication standards required for vehicular communications. Selection of a particular standard and technology depends upon numerous factors such as data rates, bandwidth requirement, security, noise immunity, distance coverage, infrastructure requirement, scalability, etc. In this paper, vehicular communication technologies are categorized to discuss the suitability of the specific set of standards for particular application. Moreover, this paper focuses on integration of smart grid technology and plug-in vehicles. Various communication standards can be used for charging of plug-in vehicles in the public charging system or in the home area network of smart grid infrastructure. Plug-in vehicles are very important to deliver auxiliary services like peak power shaving, grid reliability, stability, regulation of voltage and frequency, renewable integration, spinning reserve, etc. The vehicle-to-vehicle interactions and vehicle-to-infrastructure communication are also included in the paper. Smart grid and intelligent transportation system technologies will bring a revolution in terms of integrated technologies, Internet of grid and vehicular communication systems. Interconnections between smart grid and intelligent transportation systems can be realized through an advanced metering infrastructure, communication standards, IoT and intelligent transportation system.

References

1. N. Binti Mat Isa, T.C. Wei, A.H.M. Yatim, Smart grid technology: communications, power electronics and control system. in *International Conference on Sustainable Energy Engineering and Application (ICSEEA)*, Bandung (2015), pp. 10–14
2. L. Chhay, P. Sharma, A. Kumar, G. Bhagwatikar, EMI concerns, measurements and standards for smart grid, *Advances in Greener Energy Technologies* (Springer, Singapore, 2020), pp. 123–136
3. M. Erol-Kantarci, H.T. Mouftah, Energy-efficient information and communication infrastructures in the smart grid: a survey on interactions and open issues. IEEE Commun. Surv. Tutor. **17**, 179–197 (2015)
4. J. Huang, H. Wang, Y. Qian, Smart grid communications in challenging environments, in *IEEE Third International Conference on Smart Grid Communications*, Tainan (2012), pp. 552–557
5. L. Chhaya, P. Sharma, A. Kumar, G. Bhagwatikar, Cybersecurity for smart grid: threats, solutions and standardization, *Advances in Greener Energy Technologies* (Springer, Singapore, 2020), pp. 17–29
6. L. Chhaya, P. Sharma, A. Kumar, G. Bhagwatikar, Integration of cognitive radio with heterogeneous smart grid communication architecture, in *Intelligent Communication, Control and Devices*. Advances in Intelligent Systems and Computing, vol. 624 (Springer, Singapore, 2018)
7. I.S. Bayram, I. Papapanagiotou, A survey on communication technologies and requirements for internet of electric vehicles. J Wirel Commun Netw **2014**, 223 (2014)
8. Technology roadmap, Electric and plug-in hybrid electric vehicles. Technical report, International Energy Agency (2011)
9. V.C. Gungor, D. Sahin, T. Kocak, S. Ergut, C. Buccela, C. Cecati, G.P. Hancke, Smart grid technologies: communication technologies and standards. IEEE Trans. Ind. Inform. **7**, 529–539 (2011)
10. Y. Yan, Y. Qian, H. Sharif, D. Tipper, A survey on smart grid communication infrastructures: motivations, requirements and challenges. IEEE Commun. Surv. Tutor. **15**, 5–20 (2013)
11. S. Bera, S. Misra, M.S. Obaidat, Energy-efficient smart metering for green smart grid communication, in *IEEE Global Communications Conference*, Austin (2014), pp. 2466–2471
12. J. Momoh, Pathway for designing smart grid, *Smart Grid: Fundamentals of Design and Analysis* (Wiley, Hoboken, 2012), pp.122–139
13. H. Zhang, S. Li, Research on microgrid, in *Advanced Power System Automation and Protection (APAP)* (2011), pp. 595–598
14. C. Landsteiner, F. Andren, T. Strasser, Evaluation and test environment for automation concepts in Smart Grids applications, in *IEEE First International Workshop on Smart Grid Modeling and Simulation*, Brussels (2011), pp. 67–72
15. B.P. Bhattarai, M. Lévesque, M. Maier, B. Bak-Jensen, J.R. Pillai, Optimizing electric vehicle coordination over a heterogeneous mesh network in a scaled-down smart grid testbed. IEEE Trans. Smart Grid **6**, 784–794 (2015)
16. L. Chhaya, P. Sharma, G. Bhagwatikar, A. Kumar, Communication theories and protocols for smart grid hierarchical networks. J. Electr. Electron. Eng. 43–48 (2017)

17. L. Chhaya, P. Sharma, G. Bhagwatikar, A. Kumar, Wireless sensor network based smart grid communications: cyber-attacks, intrusion detection system, and topology control. Electronics **6**(1), 1–22 (2017)
18. R. Piyare, M. Tazil, Bluetooth based home automation system using cell phone, in *International Symposium on Consumer Electronics*, Singapore (2011), pp. 192–195
19. L. Chhaya, P. Sharma, G. Bhagwatikar, A. Kumar, Development of wireless data acquisition and control system for smart microgrid, in *Advances in Smart Grid and Renewable Energy*. Lecture Notes in Electrical Engineering, vol. 435 (Springer, Singapore, 2018).
20. S. Galli, A. Scaglione, Z. Wang, For the grid and through the grid: the role of power line communications in the smart grid. Proc. IEEE **99**(6), 998–1027 (2011)
21. F. Arena, G. Pau, An overview of vehicular communications. Future Internet **11**, 1–12 (2019)
22. D. Jomaa, S. Yella, M. Dougherty, A comparative study between vehicle activated signs and speed indicator devices. Transp. Res. Procedia **22**, 115–123 (2017)

Smart Grid, V2G and Renewable Integration

K. Logavani, A. Ambikapathy, G. Arun Prasad, Ahmad Faraz, and Himanshu singh

Smart Grid concept is a combination of number of technologies like internet of things, power electronic control strategies, end-user solutions and equips a number of regulatory policies. Technological development in the field of generation, transmission and distribution enables a complete automated and remote operation of power system feasible.Renewable integration is the important characteristics of the Smart Grid infrastructure. Distributed energy sources like solar Photo Voltaic (PV), wind energy and other micro resources like biogas, biomass plays a vital role in the Distributed Generation scenario. In case of solar PV and wind, intermittent power generation could cause fluctuations in the power delivered as to be higher or lower than the demand more often and poses a problem in the load dispatch. However, these sources are feasible if they could store the energy and discharge excess energy when demand arises. Only a Smart Grid setup could provide a practical solution for balancing the renewable generation with the demand through energy storage systems and controllable loads.

K. Logavani
Government College of Engineering, Salem, Tamil Nadu, India
e-mail: vani.tulips@gmail.com

A. Ambikapathy (✉) · G. Arun Prasad · A. Faraz · H. singh
Galgotias College of Engineering College and Technology, Greater Noida, UP, India
e-mail: ami_ty21@yahoo.com

G. Arun Prasad
e-mail: arun.prasad@galgptiacollege.edu

A. Faraz
e-mail: theahmadfaraz@gmail.com

H. singh
e-mail: himanshuk282@gmail.com

N. Patel et al. (eds.), *Electric Vehicles*, Green Energy and Technology,
https://doi.org/10.1007/978-981-15-9251-5_10

Smart Grid helps to improve the reliability, quality of power and grid security through predictive maintenance and self-healing nature of the system. For efficient smart control of the grid, the technologies required are Internet of Things via communication platforms, sensing and measurement, control, technologies, power modulators and energy storage.

High aggregate technical and commercial losses and power theft necessitate the upgradation in the metering front and old meters are replaced by smart meters. With many DERs playing in the market, the smart metering in the power system network can acquire information about the power fluctuations for the change in demand thereby regulating the economic scheduling of renewable generation and distribution to the aggregators. In smart metering, a greater number of smart meters are included in the system, the wireless communication seems to be cost-effective and thus the ICT plays a major part in the coordination of renewable integration in the Smart Grid environment. In the smart grid setup, the owner of the electric vehicle communicates with the utility grid and utilizes the power flow and the economic benefits of electricity.

Smart metering is one of the thrust elements of the smart grid as it provides the status of the power flows in the smart grid network and the loads. Through effective control and monitoring, the states of the power system could easily be estimated and this makes the control and coordination effective. Various renewable sources could be integrated into the system so that the fossil-fired plants could be operated cost-effectively and replaced in near future by the Smart Grid.

Implementation of Smart Grid technologies would improve every sector of power system. It also increases the possibilities of distributed generation. The minimum the distance from the power generation to the distribution side, the more economical and efficient the system would be. Smart Grid is a two-way flow of electricity and information between the grid and the load. It is capable of coordinating all controls from the power plant level to the individual consumer level. Smart Grid aims in real-time information flow and it enables an optimal balancing of generation and demand. Smarter Grid also helps to reduce the cost of meeting the peak demand. Smart grid enables the participation of all like producers, consumers, aggregators and accommodates every part of micro and macro generation, the storage capabilities and provides power quality, utilize the assets optimally, predicts the disturbances, responds and self-heals to any grid disturbances.

Smart Grid encompasses the technology that integrates the solar, wind and many other micro generations and intelligently controls these distributed generation resources. This system allows for the large penetration of PV systems into the consumer side. Overall the Smart Grid system is intelligent, accommodating all resources, efficient in terms of smart metering, motivating for more resource participation, opportunistic in terms of allowing more plug and play devices, quality-focused due to digitalization, resilient in terms of self-healing for sudden disturbances and more organic as it slowly replaces conventional fossil-fired plants through PV and wind resources. Asset optimization is increased through smart grid setup. The control capability to manage all sources of power has paved way for the coordination of more distributed energy sources into the Smart Grid environment [1].

1 Vehicle to Grid (V2G)

Electric Vehicle is basically an electricity dependent load that can be used as a remotestoring device that could involve in the load adjustment of utility grid. This also provides a steady pace for the renewable energy integration, automation and control. This technology also offers benefits form the power grid as well for the EV owners. The renewable integration would prove to be effective if the grid is able to have its own self-healing feature and predictive maintenance along with the storage and smart communication technology embedded to it so that all devices in the smart network could communicate between themselves.

Vehicle to Grid is the complete incorporation of an electric vehicle into the utility grid. The grid stability could be increased by interconnecting many electric vehicles. Payback is involved for the EV owners as some money is given back for pumping the power into the grid. Since many decades, fossil fuels are the prominent energy source for electricity generation and transportation. According to an analysis in the year 2007, it is has been predicted that the availability of coal at global level would sustain for another 120 years and that of oil and natural gas for another 20 and 40 years, respectively. With the depleting scenario of fossil fuels and its increased price, the penetration of renewable energy resource is seen to be cost-effective and climate supportive. Solar Photo Voltaic and Wind energy are rapidly replacing conventional resources.

With the greater number of Distributed Energy Resources(DERs) in the power system network, energy storage becomes vital and this revolutionized the emerge of Electric Vehicle (EV) technology and renewable integration. Electric Vehicle is a carbon-free vehicle as it does not emit pollutants associated with the carbonized vehicles. The advantages of EV are: no carbon emission as the vehicle is operated by the battery, battery-powered motors cost less, safety and efficiency are high due to the absence of fuel combustion engine and the most important feature is that the battery is powered by the naturally available free renewable source. One type of EV is the Battery Electric Vehicle (BEV) that uses chemical energy and the major components of the BEV are the electric motor, controller and rechargeable battery. Important part of the BEV is the inverter that converts the stored DC power into the AC power. An example for BEV is the Nissan Leaf and that its battery is charged from the Grid. Whereas the Plug-in Hybrid Electric Vehicle (PHEV) uses the combination of electric motor and an internal combustion engine for powering the vehicle.

Incorporation of the transportation sector into the utility grid which comprises of various types of Electric Vehicle has been possible with the advancement of EV technology. Vehicle to Grid is the control and management of an electrically operated vehicle between EV load and the power grid using communication protocols. In the present scenario mostly, the vehicles use either the gasoline or the petroleum for the propulsion of the engine and they do not have any interconnection with the electricity grid in any aspect. But with the advancement in the battery technologies and EV market, the concept of plugging the vehicle to the grid for charging during low load period and discharging to the grid during peak demand has become easier and practical. There are many practical benefits like load shaving, load averaging and voltage regulation by plugging the vehicle to the grid. This results in cost benefits for both the grid system and the aggregator. There are many challenges when integrating the EV to the grid.

Faster battery degradation due to battery's increased charging cycles is identified as major issue in which more studies have been conducted on battery life cycles. V2G implementation would result in frequent battery charging and discharging cycles that result in more conversion losses. An intelligent controller will have the option of whether to control the charging or discharging process of EV battery. There are three types of V2G

1. Hybrid vehicle
2. Battery powered vehicle
3. Solar powered vehicle.

In the hybrid vehicle, the power is generated from storable fuel and the generator in it produces power and pumps into the grid during peak electricity demand period. In this case, the vehicle serves as one of the microgeneration systems giving power to the grid. A plug-in hybrid vehicle on the other hand produces power from its excess rechargeable battery to the power grid during a period of peak demand. During light load periods these vehicles are recharged at cheaper prices. In this case, the vehicle serves as a distributed storage system. A solar-powered vehicle uses extra solar energy tapped in the battery to power the grid. This kind of small renewable plants is used to act as a buffer for the grid.

2 Concept of V2G

Variations in the daily load curve, voltage and frequency regulation of the grid are the prime factors that influence the efficiency and operating costs of the grid. The efficiency of the power grid is not stable due to the varying nature of daily load curve that shows the load fluctuations and instability in the voltage and frequency regulation. This increases the cost of the power grid operation and results in energy waste. When the demand is greater than the base load, to meet out the load, peak power plants are operated and when the demand goes lower than the base load capacity the energy is wasted as unused [2].

To overcome the problems and to maintain the grid stability, the concept of V2G is proposed and the fundamental impression is to utilize the stored energy of Electric Vehicles as the supporting power grid and the DERs. So when the demand is at peak level, the energy could be pumped into the grid from the Electric Vehicle and when the expected demand is low the energy could be stored in the EVs thereby avoiding the energy waste. So the benefit is twofold for the EV users, wherein the energy could be bought from the grid during low demand when the price is low and energy could be pumped into the main grid whenever the load demand is high at a higher price [3].

Nowadays more battery operated Electric Vehicle (EV) and Plug-in Hybrid Electric Vehicle (PHEV) emerge into the market. As these vehicles are equipped with more battery storage capacity, they can deliver the energy to the grid as a buffer during their large parking period as an idle asset. The capacity of the battery is high with a greater number of vehicles and this could be treated as a separate DERs. Various literature studies have shown that the V2G principle and the benefits reaped economically are remarkable. At the same time, there were implementation issues like power grid stability and renewable energy integration. There are various implementation methods for V2G.

In Centralized V2G electric cars are seen together in a certain area and it charges the grid according to the demand based on an effective control strategy. An intelligent charger is built on the ground and the charging control is managed by the algorithm for the optimal power use. The charging and discharging process of the system is coordinated by the instantaneous grid status. The charging station and the grid scheduling system in a coordinated way decide the peak load adjusting during the charging of the electric vehicle. The power demand P_D of the grid is the basis for the peak adjusting process and the state-of-charge of each EV battery is used as the evaluation index. P_i is the discharging feed power of the EV.

The objective function is given by the equation

$$P_D = \sum_{i=1}^{N} P_i \min\left[\mathrm{SoC}_i - \left(\sum_{i=1}^{N} \mathrm{SoC}_i\right)/M\right] \tag{1}$$

N represents the number of EVs contributing for grid stability in the centralized V2G.

Similarly, minimum difference between the electric vehicle charging station and the group of EVs is given by the equation

$$\varepsilon = \min\left(P_D - \sum_{i=1}^{N} P_i\right) \tag{2}$$

The intelligent charger in the case of the centralized V2G is installed in the ground and this achieves the overall optimization due to centralized coordination [3].

In Autonomy type V2G the electric cars are parked in a scattered manner and it follows a vehicular intelligent charge controller and they are operated according to the reactive power demand, state of Charging of the car's battery and the electrical characteristics of the grid as well. In network-based V2G, the electric car is integrated into piconets, a set of Bluetooth devices sharing a common channel. In this method, the subject of importance is not the grid but the piconets. It provides support for the micro-level distribution network which comprises of load and micropower supply [3].

With the growing number of DERs and the development in electric vehicle technology, the existing ageing grid infrastructure is lacking smart support. Solar and

Wind energy are harnessed more replacing the conventional energy resources and the Electric Vehicles are becoming more popular and cost-effective due to low oil dependency and low emission. The smart grid system allows a bidirectional distribution and communication between the power distributors and aggregators. The conventional system outline of V2G network is shown [4] (Fig. 1).

The power flow between the EV and the V2G is shown (Fig. 2).

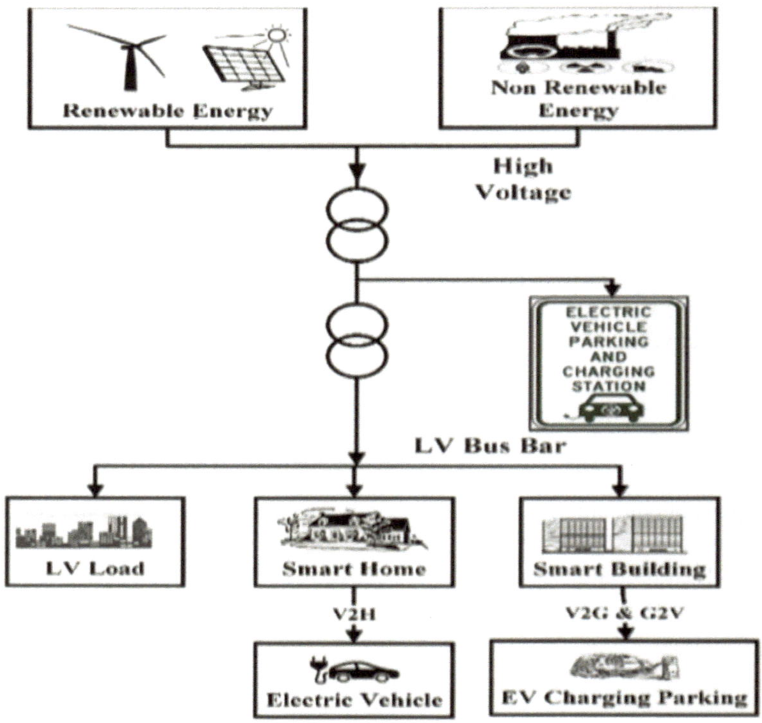

Fig. 1 Conventional V2G network [4]

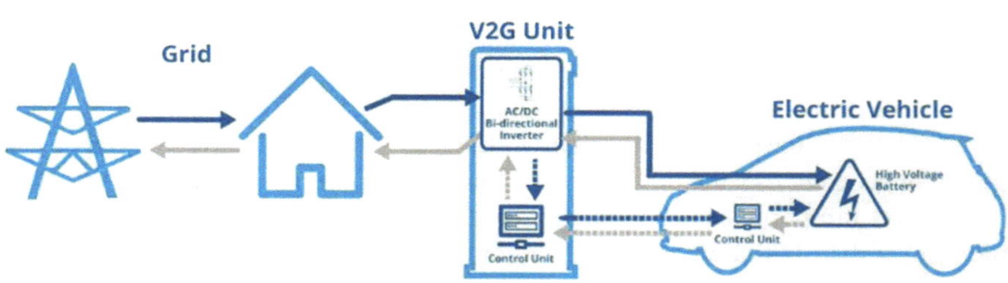

Fig. 2 Power flow between EV and V2G [19]

In the bidirectional V2G the flow of power is in both directions for achieving maximum benefits. The V2G charger consists of bidirectional AC to DC and DC to DC converters respectively. In charging mode, AC to DC converter operates and while discharging the DC current is inverted into an AC before injecting into the grid. Choppers are employed for bidirectional power flow. The advantages of a bidirectional V2G system is that it offers reactive power support, power factor control, active support for the network and allows the renewable integration. The thrust area of research in Vehicle to Grid technology is on lightweight and high-power batteries. Grid synchronization, harmonic distortion, injection and absorption of reactive power are the areas where the V2G is to be concentrated upon to have better grid stability and economic benefits for the owner. When compared to other EV batteries, Lithium-ion polymer batteries have the best efficiency, specific power density, least self-discharge rate and nominal voltage. In Indian scenario, Ni-cadmium battery is preferred due to its average discharge rate and large charging cycles [4].

The main objective of renewable energy integration is to enable the grid to reduce the carbon emissions through incorporating many renewable energy sources, enhancement of reliability, security and resiliency in microgrid operations, supporting the Plug-in electric vehicle operations with the grid and to increase the asset usage through integration of distributed system and the consumer loads in order to reduce the peak demand and its associated cost of electricity.

Independent operation of distributed energy sources and systems poses high threats related to electrical parameters like voltage and power fluctuations, frequency instability, harmonic distortion and other critical problems related to protection of devices. With the evolution of smart grid technology, greater number of distributed energy sources could be integrated into the grid thereby providing effective instantaneous load demand balance and optimal and anticipated control [5]. Smart Grid represents the coordinated and integrated control of the conventional grid with greater number of renewable energy sources penetrating and this ensures the power quality. The association of all three are given in (Fig. 3).

There are many complications associated with the integration of DERs into the Smart Grid environment. As many DERs participate in the generation, balancing of power and demand is critical as these resources are intermittent in nature. Reserve powers could be created or the V2G implementation helps in maintaining Grid stability. In Indian context, the penetration of Renewable Energy sources to smart grid is allowed by the groups working on distributed generation and Indian Smart Grid Task Force. (ISGTF) [6].

Almost many renewable energy resources contribute to electricity generation from micro to megawatt level and in particular, the wind power contributes to about two-third of the grid-connected renewable resource [7]. To reap the benefits of almost all renewables, integration is important. Almost many of the connected DERs are intermittent in nature, storage system has to be adopted. This necessitates the development of Smart Grid technology. Smart Grid enables the customers to actively participate through information and communication bidirectional flow mechanisms [7].

With the integration of greater number of Distributed Energy Resources into the Smart Grid [2, 8–12], many challenges occur with respect to the protection of devices.

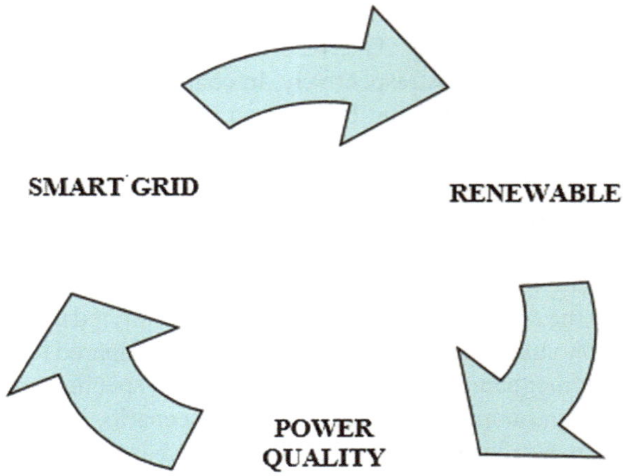

Fig. 3 Association of three concepts [6]

When a fault occurs in the power system, there is a higher chance that the distributed energy resources in the microgrid set up ay get disconnected from the utility grid. The situation of isolation of microgrid from the utility grid is called the Islanding scheme. The detection of islanding is very essential for the following reasons:

1. Frequency and Voltage stability
2. During islanding the control strategies of the inverter is changed as it receives data from the Point of Common Coupling (PCC)
3. Protection of like fuses, relays, reclosure switches as these may fail during islanding.
4. If synchronous DG is connected this would result in the equipment damage.
5. Threat to operating personnel during the islanding mode of operation.

Many types of islanding techniques exist in the literature.

1. Active islanding techniques
2. Passive islanding techniques
3. Hybrid islanding techniques
4. Communication based techniques.

Passive islanding techniques do not require any additional circuitry for the detection of islanding. Parameters like voltage, current, harmonics, phase jump of the PCC are measured. In the active method of detection, small disturbance is injected into the PCC and when the islanding occurs the disturbance is amplified and the distributed energy resource stops delivering power to the load. Hybrid islanding techniques use both active and passive islanding techniques and the communication-based techniques use the WAMS, SCADA signals for islanding detection. As more and more DERs penetrate into the system [2, 8–12], the security and reliability of the system are monitored using these islanding detection techniques in the smart grid [13].

A central controller in the Smart grid network will determine the mode of operation of grid as Grid-connected or Islanded.

$$\Delta P = P_L - P_{\mathrm{DG}} = P_{\mathrm{GRID}} \tag{3}$$

$$\Delta Q = Q_L - Q_{\mathrm{DG}} = Q_{\mathrm{GRID}} \tag{4}$$

The above two equations determine the power mismatch that exists in the grid. If ΔP is large P_L is more and the voltage is reduced. So the under frequency relay experiences a trip signal. As per the IEEE1547 standards, Under frequency/Over frequency relays, Under voltage/Overvoltage relays operate based on their limits to determine whether the distributed generation is active or not. As per the standards, the islanding detection should persist not more than 2 s. Various Passive islanding techniques stated in the literature include power quality based, impedance-based, Rate of Change of Frequency (ROCOF), Rate of Change of Phase Angle, Wavelets and Artificial Intelligence based techniques, respectively.

In PQ based detection technique, the ripple in voltage is measured. In impedance method, at the point of common coupling, the voltage, current and impedance are measured. The variation in impedance states that the microgrid is in Islanded mode. ROCOF method of islanding detection method fails for 0% power mismatch identification. Moreover, it is also influenced by the feeder length and X/R ratio. ROCPA methods detect the islanding through energy content variation during phase angle change [14].

ROCO superimposed negative sequence impedance method collects the three-phase voltage and currents. The fundamental component is calculated using the least square error method. The negative sequence voltage and current are obtained using this method. Generally unbalancing occurs in power system due to single-phase loads. So negative sequence components are present. The negative sequence voltage and current before disturbance and during islanding are calculated through the islanding relays at the terminals of DERs. The superimposed negative sequence impedance determines the islanding condition in the system. This technique works for the 0% power mismatch. The passive islanding detection technique results in Non-Detection Zone (NDZ) or so-called the blind spot [15].

This is overcome in the active islanding technique. Low frequency, high frequency, negative sequence signals and pulse signals are applied through inverters and the voltage and current are measured at the PCC. In positive feedback techniques, voltage phase angle is given as feedback to the controller of the inverter. Various smart solutions to manage power balancing in the power system network are listed below.

1. By connecting smart devices for connection and advanced power electronic control devices.
2. Unified storage system
3. By using real-time management system
4. Intelligent algorithms for effective forecasting
5. Smart meters for demand response.

By proper voltage and power flow using phasor measurement units(PMUs), the measurement of parameters is very crucial in the renewable sources integration to avoid the islanding issues. For integrating the distributed energy resources into the smart grid system, the communication between all electronic and protective devices is very essential as this ensures the effective functioning of the grid infrastructure in a restructured network [16].

The conventional and the smart grid attributes are shown in Fig. 4. As shown in the above figure, smart metering from the utility side offers effective peak demand shaving and islanding issues. From the customer side, it provides an economic benefit in terms of real-time pricing for managing their energy consumption. A large amount of data is obtained from metering technology that helps to identify and reduce the Aggregate Technical and Commercial Losses (ATC) and theft, monitoring and controls from WAMS and SCADA platform and this enable two-way communication between the utility and the consumer/operator [17]. There exist many optimization techniques in literature for providing Vehicle to Grid services for renewable energy integration. Mathematical modelling, classical optimization algorithms, metaheuristic optimization techniques and hybrid optimization techniques. The very basic objective of the optimization technique is to minimize the operating cost of the grid. All these algorithms help to obtain better economic scheduling of the available resources for the V2G charging and discharging process during peak power demand and low load period.

In restructured power systems like where a large number of DERs play a prominent role, the commitment of generating units in the grid for scheduling is of prime importance. The participation of units determines the economic load dispatch. Unit commitment is a complex decision-making procedure in which the status of participating units for generation is identified as the optimal solution for the minimum cost objective function. A mathematical modelling method called modified dynamic programming approach is applied to a 10 unit IEEE system and 7 unit NTPS system

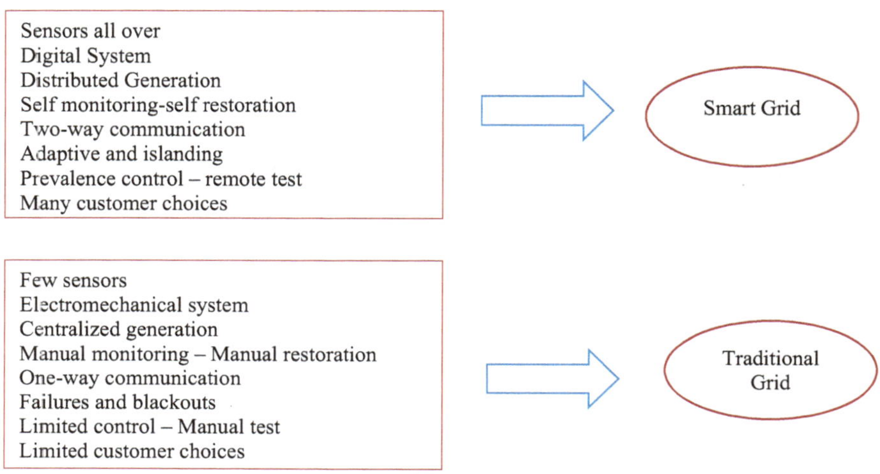

Fig. 4 Conventional grid and smart grid attributes [15]

for checking the validity of the objective function with respect to various constraints. Unit Commitment constraints like minimum up time, minimum down time, start-up cost, shut-down costs are considered for the coal-fired thermal plant. The same algorithm could be effectively used to solve the objective function for many DERs connected in the microgrid setup prior to the implementation in Smart Grid environment using various other intelligent algorithms like Genetic Algorithm, Particle Swarm Optimization, Ant Colony Optimization, Bat algorithm and many more. It has been found that the profit obtained is cost-effective compared to conventional dynamic programming approach [18].

3 Conclusion

The smart grid network serves as an effective platform in terms of integrating various distributed energy resources and the Vehicle to Grid concept. There are numerous techniques to detect the stability of the grid from the PCC terminal, load, generation, transmission, distribution, inverter control and charge controller terminals respectively. The objective of the Smart Grid network, its challenges, concept of V2G, renewable integration, islanding detection techniques in renewable integration has been discussed. The cost benefits from customer and utility side in terms of operating 2G are also discussed. Optimization of the peak demand and economic scheduling is achieved through many intelligent algorithms and a modified dynamic programming type of mathematical modelling was also discussed in this paper.

Effective functioning of Smart Grid is possible only when all sources, data, communication, parameters, control strategies are coordinated by a smart system of network.

References

1. J. Ekanayake, N. Jenkins, K. Liyanage, J. Wu, A. Yokoyama, Smart grid technology and applications, Wiley Publications
2. N. Priyadarshi, F. Azam, A.K. Bhoi, A.K. Sharma, Dynamic operation of grid-connected photovoltaic power system, in *Advances in Greener Energy Technologies*, (Springer, Singapore, 2020), 211–218
3. I. Sami, Z. Ullah, K. Salman, Hussain, S.M. Ali, B. Khan, C.A. Mehmood, U. Farid, A bidirectional interactive electric vehicles operation modes: Vehicle-to-grid (V2G) and grid to vehicle (G2V) variations within smart grid
4. M. Singh, P. Kumar, I. Kar, Analysis of vehicle to grid concept in Indian scenario, in *14th International Power Electronics and Motion Control Conference*, EPE-PEMC 2010
5. C.H. Lo, N. Ansari, The progressive smart grid system from both power and communications aspects. Commun. Surv. Tutorials **14**(3), 799–521 (2012). IEEE
6. S. Gheorghe, N. Golovanov, G.-C. Lazaroiu, R. Porumb, Smart grid, integration of renewable sources and improvement of power quality, in *21st International Conference on Control Systems and Computer Science,* 2017

7. S. Mukhopadhyay, S.K. Soonee, R. Joshi, A.K. Rajput, On the progress of renewable energy integration into smart grids in India, in *IEEE Power and Energy Society Conference*, 2012
8. N. Priyadarshi, F. Azam, A.K. Bhoi, A.K. Sharma, A multilevel inverter-controlled photovoltaic generation, in *Advances in Greener Energy Technologies*, (Springer, Singapore, 2020), pp. 149–155
9. N. Patel, D. Porwal, A.K. Bhoi, D.P. Kothari, A. Kalam, An overview on structural advancements in conventional power system with renewable energy integration and role of smart grids in future power corridors, in *Advances in Greener Energy Technologies,* (Springer, Singapore, 2020), pp 1–15
10. T. Choudhary, N. Priyadharshini, P. Kumar, F. Azam, A.K. Bhoi, A fuzzy logic control based vibration control system for renewable application, in *Advances in Greener Energy Technologies*, (Springer, Singapore, 2020), pp 651–663
11. A. Sahu, S. Gupta, V.K Singh, A.K. Bhoi, A. Garg, K.S. Sherpa, Design of permanent magnet synchronous generator for wind energy conversion system, in *Advances in Smart Grid and Renewable Energy,* (Springer, Singapore, 2018), pp 23–32
12. K.C. Manjunatha, A.K. Bhoi, K.S. Sherpa. Design and development of buck-boost regulator for DC motor used in electric vehicle for the application of renewable energy, in *Advances in Smart Grid and Renewable Energy*, (Springer, Singapore, 2018), pp. 33–37
13. C.H. Rami Reddy, K. Harinadha Reddy, Islanding detection techniques for grid integrated distributed generation–a review. Int. J. Renew. Energy Res. **9**(2), 2019
14. H. Samet, F. Hashemi, T. Ghanbari, Islanding detection method for inverter based distributed generation with negligible NDZ using energy of rate of change of voltage phase angle. IET Gener. Transm. Distrib. (2015). ISSN 1751-8687
15. H. Muda, P. Jena, Rate of change of superimposed negative sequence impedance based islanding detection technique for distributed generations. IET Gener. Transm. Distrib. 3170–3182 (2016). ISSN 1751-8687
16. A. Gaviano, K. Weber, C. Dirmeier, Challenges and integration of PV and wind Energy facilities from a smart grid point of view, Energy Procedia **25**, 118–125 (2012)
17. Y. Yan, R.Q. Hu, S.K. Das, H. Sharif, Y. Qian, An efficient security protocol for advanced metering infrastructure in smart grid. Network **27**(4), (2013). IEEE
18. L. Kandasamy, S.K. Selvaraj, Lambda optimization of constraint violating units in short-term thermal unitcommitment using modified dynamic programming. Turkish J. Electr. Eng. Comput. Sci. **25**,1311–1325, (2017)
19. www.forbes.com—An article dated Dec 18 2017, “Electric vehicle-to-grid services can feed, stabilize power supply” by Constance Douris

Blockchain-Based Smart Contract Design for Crowdfunding of Electrical Vehicle Charging Station Setup

Manish Kumar Thukral

1 Introduction

Electrical vehicles are emerging out to be prominent mode of transportation in order to counter the growing environmental concerns because of fossil fuel pollution [1, 2]. Indian government is promoting in big way by supporting development of ecofriendly technologies like EV. In this direction, Indian government through ministry of Heavy Industries and Public Enterprises launched FAME scheme in year 2015. Prime aim of this scheme is to promote electric mobility and provide incentives in financial terms to companies involved in EVs production and designing charging infrastructure [1–3]. In India, there are 150 electric vehicle charging stations only. This number is too less to have electric vehicle as major source of transportation in India. Considering this shortfall, government of India is planning to set up one charging station every three Km in cities and every 25 km on both sides of highways [2–4]. Government of India has planned to switch over to 100% electric mobility by 2030 [3–5]. In this direction, most of the automobile industries have been complaining of acute shortage of electrical vehicle charging stations [2–4]. Looking at the prospective aim of government of India to transit to 100% electric mobility in future and the charging stations and piles required, there is clearly a huge gap. In an attempt to fill this gap, government of India is planning to set up 2636 charging stations across India [4–6]. Not only Indian government but many private companies like Tata Power is planning to set up 700 electrical vehicle charging stations in India by 2021 [5–7]. Few more private companies like ABB, Acme industries, Fortum India and state-run companies like NTPC, GAIL INDIA, Indian oil corporation and power grid have shown their interest in setting up charging stations in India. Besides this, government of India is giving subsidy up to Rs. 10 million for encouraging the entrance of private investment [6–8]. For a country like India aiming to significantly transit toward electric

M. K. Thukral (✉)
Electrical Engineering Department, Manipal University Jaipur, Jaipur, India
e-mail: manish.thukral@jaipur.manipal.edu

N. Patel et al. (eds.), *Electric Vehicles*, Green Energy and Technology,
https://doi.org/10.1007/978-981-15-9251-5_11

mobility by 2030, mobilizing financial investment in setting up charging station is critical [7–9]. It can be observed that prime investors in setting up electric vehicle infrastructure involve government bodies, big private enterprises and grid companies but a common a low budget investor is not having platform to participate. Advent of crowdfunding concept can play a pivotal role in this direction. In crowdfunding, a good number of people fund a project or one can say money is raised through people for a venture mainly from Internet [8–10]. With time, this source of funding has become extremely popular because of its success [9–11]. Since crowdfunding involves gathering funds from Internet and since small clients can also participate, large amount of capital can be gained. Crowdfunding can play a vital role in expansion of electric vehicle charging stations. Since crowdfunding is Internet-based financing method, it demands complete safety and transparency. In crowdfunding project, the one who starts the crowdfunding campaign called idea person or manager collects the money from contributor through Internet. Ideally, the manager diverts the money to the vendors who supply the basic parts of the project. But in real world, cases have been reported where the manager malfunctions and divert the fund to private bank accounts and does illegal transactions. In the presented research work, a blockchain-based crowdfunding platform has been designed for raising funds to set up EV charging stations infrastructure. For this, a smart contract is designed on Ethereum blockchain platform. To best of the author knowledge, this is first of its kind of smart contract designed for EV charging station infrastructure fund raising through crowdfunding platform. The proposed smart contract provides secure way of crowdfunding. The presented paper is divided into different sections. In first section, basic concept of blockchain is explained followed by insight into EV charging station crowdfunding smart contract. In Sect. 4, proposed smart contract is tested.

2 Blockchain Basics

As the name suggests, blockchain is nothing but chain of blocks which are connected to each other by a means of cryptographic technique called hash pointers. Each block of blockchain has record of transactions; hence, it can be termed as ledger as well. The main aim of blockchain technology is to move from centralized system to decentralized system. In traditional centralized system, there is complete reliance for data security on single node which is not a safe practice. The advent of this technology was with aim of decentralize transactions between peer to peer in a secure manner [12]. Each participating node contains copy of the blockchain so that all the transactions are visible to all the participating nodes. Through synchronization process, any change in the ledger record across any node, the same is reflected at all the nodes. In one way, blockchain is a distributed ledger. One of application of blockchain technology is Bitcoin which is peer-to-peer payment network that involves no central authority [13].

Major characteristic of Blockchain can be stated as follows

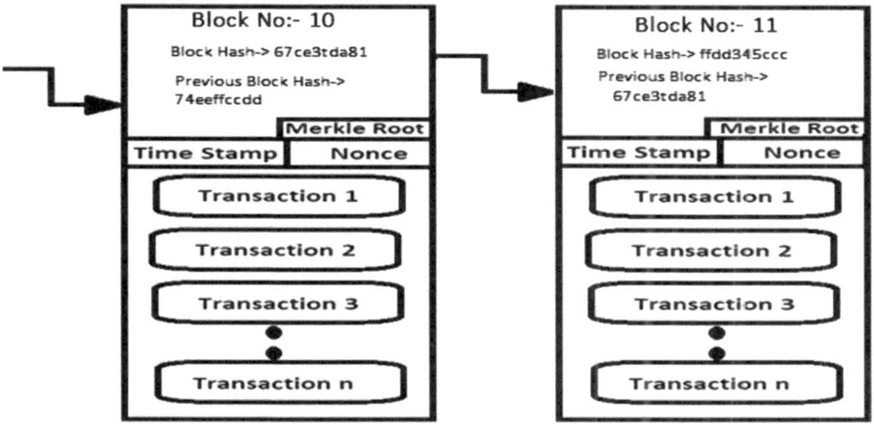

Fig. 1 Basic blockchain

- It is accessible to all therefore open
- There is no central authority to maintain the ledger that is why it is distributed among the participating parties with identical copies.
- It is scalable and quite fast in terms of transactions execution and recording so can be termed as efficient.
- The transactions recorded in block are immutable which means that they cannot be change.

Every of block of a blockchain has two main parts, i.e., header and body [14]. The block header of a block contains information like block number, current block hash, previous block hash, time stamp, nonce and address of the creator. Similarly, block body contains information of transactions. A basic blockchain diagram is shown in Fig. 1.

Each block of a blockchain has two main components, i.e., header and body [14]. Block header comprises of block number, hash value of previous block, hash value of current block, time stamp, nonce, address of the creator. The block body contain transactions. Figure 1 shows a basic design of a block.

Block hash is nothing but cryptographically generated number for a block using hash functions. Hash functions are the functions that used for cryptography of the data [15]. In blockchain, block hash is generated by hashing three parameters, i.e., previous block hash (H_{k-1}), Merkle root (T) of the current block and nonce which is nothing but a random value. This is done by what we call as miner for which reward is given. In other sense, a miner has to solve a mathematical puzzle $H_K = \text{Hash}(H_{k-1} || T || \text{Nonce})$ such that the resulting block hash has certain number of zeros prefixed which is called difficulty level. The miner has to find the nonce for that. Data of any arbitrary length is mapped to a fixed length data using hash functions. Hash functions have one-way property in the sense that given a data X if its computed hash value $H(X)$, then it is not possible to compute X given its hash value $H(X)$. In

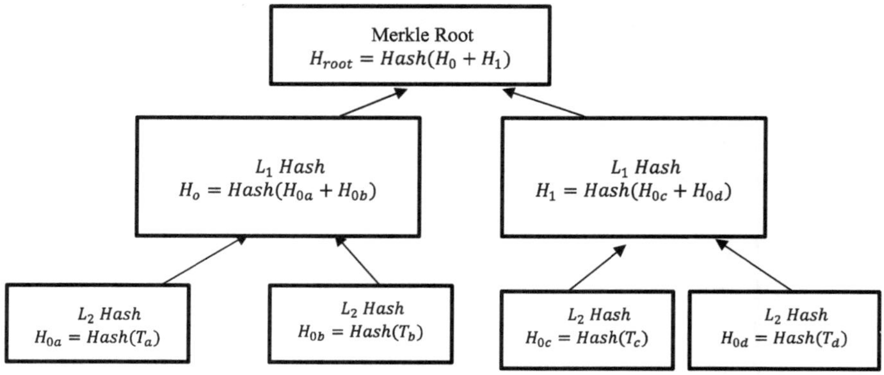

Fig. 2 Merkle tree

most of the literatures, X is stated as message and $H(X)$ as message digest. There are many hashing functions but SHA256 is mostly used in blockchain technology. SHA256 hash function generates 256-bit length encrypted data.

One very important aspect of blockchain in Merkle root which is obtained from Merkle tree [16] which is shown in Fig. 2. This tree-like structure is formed by hashing the transactions which is shown as an example in Fig. 2. In Merkle tree if any intruder tries to change any transaction data, then Merkle root will change which will eventually change all the block hash values which do not obey difficulty level. This will immediately report tampering. Because of this technique blockchain is called as immutable record of transaction and hence secure.

3 Smart Contract Design

Smart contracts are nothing but the agreement between the participating parties which are implemented by digitally through coding and then embedded into the blockchain [17]. In order to code smart contracts, programming language called solidity is used. Once smart contract is embedded in blockchain as a transaction, then it cannot be tampered by any participating parties. Blockchain technology saw its first application in Bitcoin form but smart contracts further extend this technology in terms of its application in variety of domains. Smart contracts are used to implement business logics and contracts digitally. The smart contracts are executed automatically once certain conditions are met.

3.1 Proposed Smart Contract Algorithm

- Smart contract for electrical vehicle charging station is deployed in blockchain by idea person by setting minimum donation amount.
- The donators contribute to the smart contract
- If the donator contributed amount greater than minimum donation amount, then he is considered as valid donator and has voting right.
- After good amount of donation is done, the idea person initiates spending request by entering the details of spending request, i.e., reason for spending request, amount required to be paid to vendor, address of the vendor.
- Legitimate donators vote for the spending request.
- A spending request gets approval only if more than have legitimate donators have voted in favor of spending request.
- Once the spending request gets approval through majority voting, the amount is transferred from smart contract to the vendor account.

The smart contract code written in solidity language and implemented on REMIX IDE Ethereum platform is given in Appendix 1. The design of functions and variables used in smart contract is given in Table 1.

Table 1 Functions and variables defined for proposed smart contract

Variables	Variable type	Role of variable
Ideaperson	Address	Person who has ideate campaign
Minimaldonation	Uint	Least amount of donation to qualify to be an approver
Consenters	Address[]	A list containing address of every donator
Requests	Requests[]	Struct type which contains details of each request made by manager
Functions		Role of Function
Chargingstationcampaign		Constructor Function
Donate		Called when a contributor wants to be become part of the chargingstation campaign and become consentor
Spendingrequest		Called when ideaperson wants to create spending request
Requestconsent		Called by every donator to give consent for spending request by ideaperson

4 Smart Contract Testing

The smart contract is executed and tested in Remix IDE platform provided by Ethereum. The smart contract is tested in four phases as per following explanation.

4.1 Deployment Phase

First, we will deploy "chargingstationcampaign" with setting minimum contribution of 100 Wei. The idea person who launches or deploys the contract has account number 0xca35b7d915458ef540ade6068dfe2f44e8fa733c. The smart contract deployment execution on REMIX IDE is shown in Fig. 3. When we test the minimal contribution after deployment, it shows 100 Wei as evident in Fig. 3.

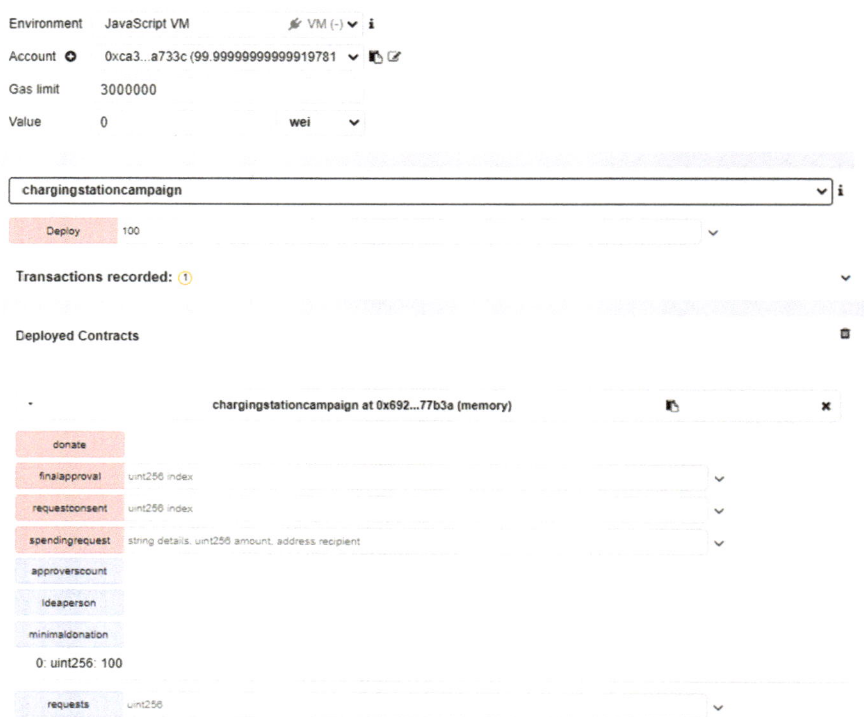

Fig. 3 Testing of "chargingstationcampaign" smart contract deployment on remix IDE

4.2 Donation to "Chargingstationcampaing" Phase

In next testing phase, the idea person itself contributes 1 ether to the campaign. His account has balance of 99.99 ether before donation as shown in Fig. 4a. After donation, his account balance reduces to 98.99 ether as shown in Fig. 4b.

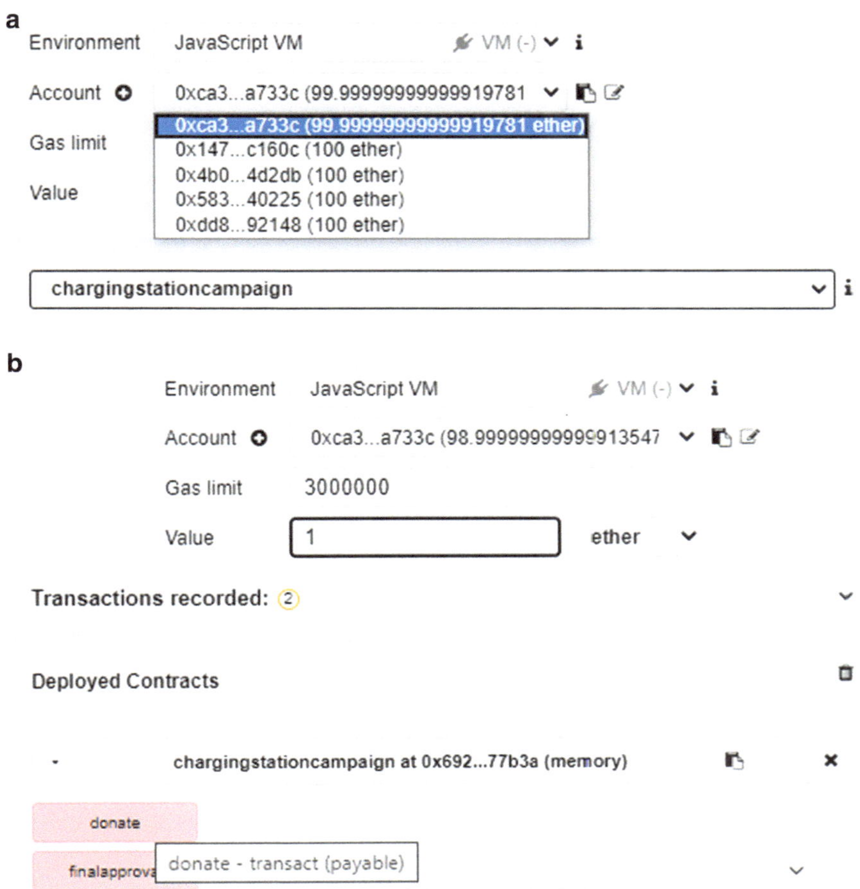

Fig. 4 **a** Idea person account balance before donation. **b** Donation of 1 ether by idea person to charging station campaign

4.3 Spending Request Phase

In this phase, the idea person places spending request to buy accessories for EV charging station setup from a vendor. It is to be mentioned here that following accessories are required to set up a EV charging station

Charging station accessories [18]

- As per society of automotive engineers the J1772 connector.
- Charging chord of minimum length of 20 feet.
- Power availability at 120 V single phase 12–16 A for level-1 with maximum 1.44 KW power, 240 V single phase up to 80 A for level-2 with maximum 1.92 KW for AC charging.
- Power availability at 50–1000 V up to 80 A with maximum power 80KW for level-1, 50–1000 V, up to 400 A with maximum power 400 KW for level-2 DC charging.
- Charge stands.
- Attachment plugs.
- Vehicle connector.
- On board or off board charger.

Various vendors like ABB, Siemens, Bosch, Nissan, etc. are providing the charging station setup accessories [18]. The proposed smart contract facilitates secure method of spending the funds collected from crowdfunding for buying the accessories.

Here, we select account 0x14723a09acff6d2a60dcdf7aa4aff308fddc160c as recipient address of the vendor from whom cord and J1772 connector for charging station will be ordered by the idea person. It is to be noted that before approval of the spending request, the balance of the idea person is 100 ether as shown in Fig. 5a. In Fig. 5b a, spending request is executed by idea person. The details entered in spending request field are to order cord and J1772 connector for charging station with payment of 9999999 Wei to the vendor having account 0x14723a09acff6d2a60dcdf7aa4aff308fddc160c. After pressing "requests" button, the details of spending request executed can be seen as in Fig. 5b.

4.4 Spending Request Approval Phase

For testing purpose, the idea person gives consent to spending request and index 0 through button "requestconsent." After this, final approval is executed by idea person through button "finalapproval" for index 0 as shown in Fig. 6. It can be observed in Fig. 6 that the balance of the vendor has increased from 100 ether. Also, it can be seen from Fig. 6 that the "consentorscount" has increased to 1 as compared to 0 in Fig. 5b.

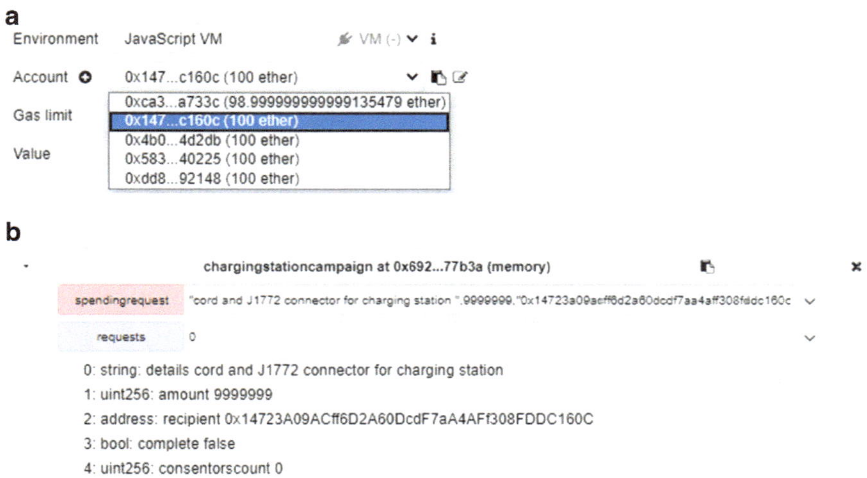

Fig. 5 **a** Vendor address and balance before spending request approval, **b** spending request executed by the idea person

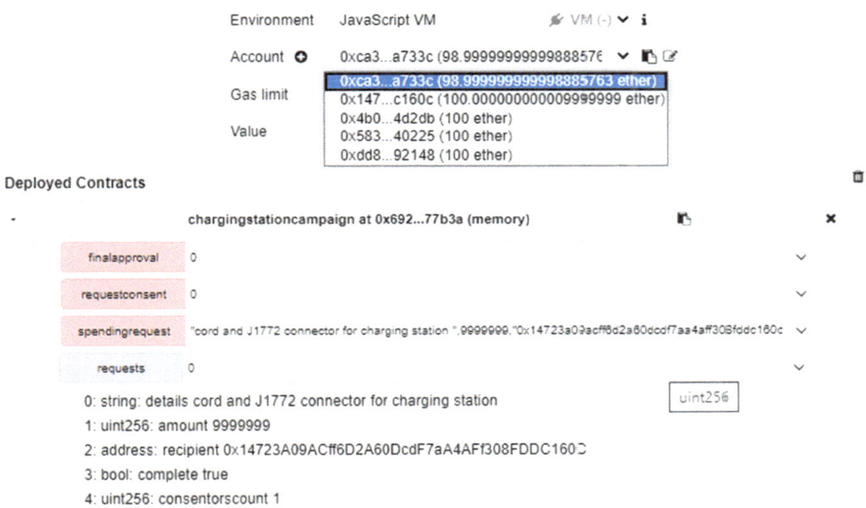

Fig. 6 Smart contract approval

It can be seen from the testing phase that the smart contract works as desired and hence ready to be deployed to blockchain.

5 Conclusion

In the presented work, a smart contract has been designed for secure crowdfunding in order to set up electrical vehicle charging station. The smart contract is designed on Ethereum blockchain platform using solidity language. The proposed smart contract is tested in four phases. It is observed that it passes all the test phases as it works as desired. It is expected that the proposed smart contract would be a guiding platform for the those desirous of designing decentralized application for electrical vehicle charging station crowdfunding on blockchain.

Appendix 1

```
pragma solidity ^0.4.17;
contract chargingstationcampaign {
    struct Request {
     string details;
      uint amount;
      address recipient;
       bool complete;
       uint consentorscount;
    mapping(address => bool) approvals;
  }
    Request[] public requests;
  address public Ideaperson;
    uint public minimaldonation;
     mapping(address => bool) consentors;
    uint public approverscount;
      modifier restricted() {
    require(msg.sender==Ideaperson);
    _;
  }
    function chargingstationcampaign(uint minimum) public {
```

```
        Ideaperson=msg.sender;
        minimaldonation=minimum;
    }
    function donate() public payable {
        require(msg.value>minimaldonation);
        consentors[msg.sender] = true;
    approverscount++;
}
function spendingrequest(string details, uint amount, address recipient) public
restricted {
  Request memory newRequest = Request({
    details:details,
    amount:amount,
    recipient:recipient,
    complete:false,
    consentorscount: 0
    requests.push(newRequest);
}
function requestconsent(uint index) public {
  Request storage request = requests[index];
    require(consentors[msg.sender]);
    require(!request.approvals[msg.sender]);
    request.approvals[msg.sender]=true;
  request.consentorscount++;
}
function finalapproval(uint index) public restricted {
  Request storage request = requests[index];
  require(request.consentorscount>(approverscount/2));
    require(!request.complete);
    request.recipient.transfer(request.amount);
```

```
    request.complete=true;
  }
}
```

References

1. R.S. Tripathi, M.K. Thukral, Switching angles computation of multi-level inverter for electrical vehicle application, in *2019 Global Conference for Advancement in Technology (GCAT),* (IEEE, 2019), pp. 1–5
2. M.K. Thukral, Design and simulink implementation of electrical vehicle charging using wireless power transfer technology, in *Optical and Wireless Technologies,* (Springer, Singapore, 2020), pp. 631–640
3. https://www.fame-india.gov.in/
4. https://www.prnewswire.com/news-releases/setting-up-ev-charging-stations-in-india-2019-exploration-report-300979844.html
5. https://economictimes.indiatimes.com/india-will-stick-to-plan-of-having-100-electric-mobility-by-2030-nitin-gadkari/articleshow/60772337.cms?from=mdr
6. https://entrackr.com/2020/01/govt-to-establish-2636-evs-in-india/
7. https://economictimes.indiatimes.com/industry/auto/auto-news/subsidy-proposals-this-week-to-set-up-5000-e-charging-stations/articleshow/69749146.cms?from=mdr
8. https://powerline.net.in/2020/03/07/scaling-up-2/
9. https://www.nrdc.org/experts/sameer-kwatra/mobilizing-finance-electric-vehicle-charging-india
10. S. Mamonov, R. Malaga, Success factors in Title III equity crowdfunding in the United States. Electron. Commer. Res. Appl. **27**, 65–73 (2018)
11. L. Zhu, Q. Zhang, H. Lu, H. Li, Y. Li, B. McLellan, X. Pan, Study on crowdfunding's promoting effect on the expansion of electric vehicle charging piles based on game theory analysis. Appl. Energy **196**, 238–248 (2017)
12. U. Cali, A. Fifield, Towards the decentralized revolution in energy systems using blockchain technology. Int. J. Smart Grid and Clean Energy **8**(3), 245–256 (2019)
13. S. Nakamoto, Bitcoin: A peer-to-peer electronic cash system. (2008)
14. O. Pal, B. Alam, V. Thakur, S. Singh, Key management for blockchain technology. ICT Express (2019)
15. A.J. Menezes, J. Katz, P.C. Van Oorschot, S.A. Vanstone, Handbook of applied cryptography. CRC press (1996)
16. J. Xu, L. Wei, Y. Zhang, A. Wang, F. Zhou, C.Z. Gao, Dynamic fully homomorphic encryption-based merkle tree for lightweight streaming authenticated data structures. J Netw Comput Appl **107**, 113–124 (2018)
17. L. Yu, W.T. Tsai, G. Li, Y. Yao, C. Hu, E. Deng, Smart-contract execution with concurrent block building, in *2017 IEEE Symposium on Service-Oriented System Engineering (SOSE),* (IEEE, 2017), pp. 160–167
18. https://www.driveelectricvt.com/Media/Default/docs/electric-vehicle-charging-station-guidebook.pdf

Modeling and Designing of E-bike for Local Use

Rohit Tripathi, Adwait Parth, Manish, and Manoj K. Shukla

1 Introduction

Due to the fast depleting natural resources such as petrol, diesel and natural gas, the energy crisis is one of the major concerns in today's world. Environmental decay is an added factor which is donating to the depletion of the natural resources which is an alarming notification. Present study suggests the resolution for this above hazardous glitches. The model for the magnetic field study was completed by spreading extents decided by the repetition of the model design. The magnetic flux density was taken low since the increase in iron losses in the rotor that used a solid core and the competence and the torque modification were less than 0.5%. Therefore, this very design model was proposed [1]. The operator and the citizens that live and use the city are the key focuses that must be addressed when evolving a bike sharing system. The reduction of the cost, i.e., focusing on cost-effective fabrication process and also needs added value design features which denotes extra service and saving costs in maintenance for the operator [2].

According to the proposed design and model, if the proposed E-bike has been compared with the commercial one, then it leads to approx. same cost and complexity. The observation over the generally used lead-acid battery was considered less effective than the Li-ion cells as it has large number of life cycles [3]. The BLDC motor is used as the main driving unit as of its long life span in comparison to others, i.e., brushed DC motors as it is very fruitful to spend over 7–8 K at one time instead of spending the more in installments as they (the 2nd one) wear out early after continuous use [4]. A wireless charging is proposed in this model for which an innovative recharge system for E-bike batteries has been predicted for which through a magnetic

R. Tripathi (✉) · A. Parth · Manish
Department of Electronics & Communication Engineering, Galgotias University, G. Noida, India
e-mail: rohittripathi30.iitd@gmail.com

M. K. Shukla
Department of Electronics Engineering, Harcourt Butler Technical University, Kanpur, India

N. Patel et al. (eds.), *Electric Vehicles*, Green Energy and Technology,
https://doi.org/10.1007/978-981-15-9251-5_12

coupling structure, the power from the grid is managed to the load wirelessly. The anticipated wireless solution permits an efficient recharge [5–7].

This model depicts the unique features of dual modes of transmission drive system, i.e., pedal drive and other is motor drive system which would be very beneficial; if somehow the motor would be incapable of carrying the load, there would be another option of the pedal drive which would be an effective method in cost reduction as well as the efficiency remains constant. The significant electrical and magnetic properties such as speed characteristics and torque of motors are equated according to their limitations. The permanent magnet motor with outer and inner rotor for the drive system of the E-bike is verified theoretically and experimentally [8]. The main constituents of the electric bike such as the motor, battery and controller, material for chassis, brakes and suspension system are the design standards for the hardware design of an electric bike [9–11].

In present study, the electric bike has been designed, fabricated and developed which is able to run on outside roads with maximum 40 km/h speed. It contains one battery pack which delivers 48 V and 25 Ah, voltages and current, respectively. It contains one BLDC motor with motor controller also. This electric bike runs through motor that is powered by the battery, for a local trip. This bike has various benefits both to the members of the team and also external benefits. thereby making awareness of using alternative modes of transport. The main purpose of using this E-bike is that it is user-friendly, economical and relatively cheap. In another approach, this battery can be charged through solar panels also which can be alternative source.

2 System Description and Principal

In present study, E-bike having major parts like; chassis, chain drive and electric system as shown in block diagram, Fig. 1. The electric system contains motor, controller and battery pack. For E-bike, this electric system is backbone. In this section, all individual sections have been explained separately in details.

Here, one new design has been attempted, where the overall weight of chassis is 8.5 kg and cost of fabrication with material is INR 6400. One gear has been placed in rear wheel of the bike, which is matched to motor gear shaft. The chain has been connected to this gear to motor gear and this complete system is also called as chain drive system. One BLDC motor which has the ratings of 900 W and 3400 rpm. It is the same, which is used in e-rickshaw, which is able to run the bike with maximum load, 150 kg at maximum speed 40 km/h with 3.5–4 hr and 70–80 km distance.

The same parameters have been also observed at half load also. The lithium ion battery has been used which is rechargeable. The rating of this battery pack is 48 V and 50 Ah and it takes 8 h to charge completely. The overall cost of E-bike is found as INR 26,500, including all components as Tables 2 and 3.

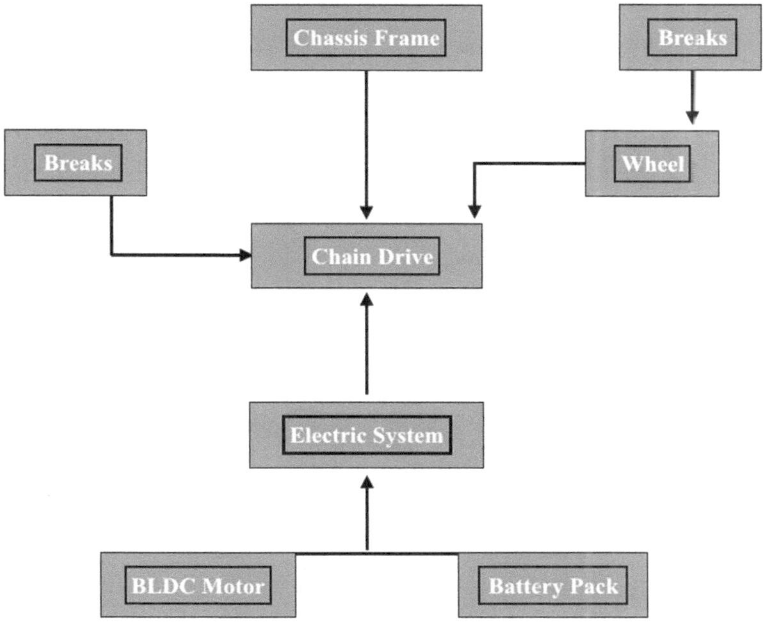

Fig. 1 Block diagram of proposed E-bike

This bike has been designed and developed for two loads, **half load**: where only one person is ridding this bike and overall weight over motor is 70–80 kg and **full load**: where two persons are riding on bike and overall weight over system is 140–150 kg.

The designing picture and actual fabricated pictures of present bike have been shown in Figs. 2 and 3.

According to both loads, the detailed study has been explained and technical point have been focused in next section.

3 System Modeling

3.1 Chassis or Body of Bike

The selection of the material in design depends on various factors such as load, function, climatic condition, lifetime and overall expenditure. Taking the above factors into consideration, material selection has been performed in order to design an efficient and economical type of frame. Steel alloys, aluminum and its alloys, titanium,

Fig. 2 Proposed design of E-bike

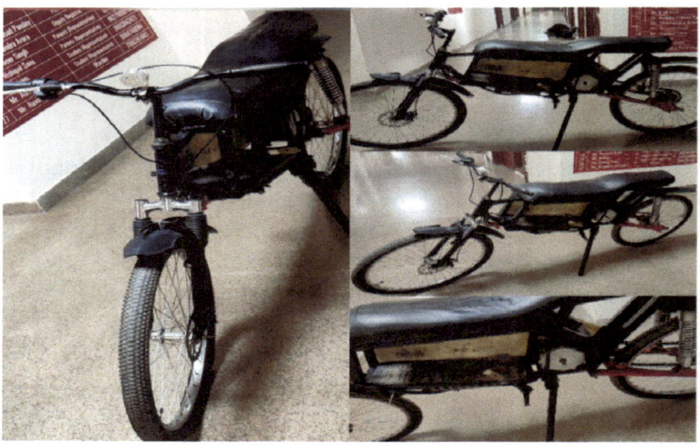

Fig. 3 Actual picture of different angle of fabricated model

carbon fiber have been preferred type of materials during selection [12]. Comparatively, AISI 4130 Alloy Steel has been used in the present study as it is easily available, cost effective and has improved mechanical properties as shown (Table 1).

The cost of chassis, with all minor and major components, has been described in Table 2.

Fig. 4 Single module of Li-ion cell

Table 1 Material composition

Element	Content (%)
Iron (Fe)	97.03–98.22
Chromium (Cr)	080–1.10
Manganese (Mn)	0.40–1.10
Carbon (C)	0.280–0.330
Silicon (Si)	0.15–0.30
Molybdenum (Mo)	0.15–0.25
Sulfur (S)	0.040
Phosphorous (P)	0.035

Table 2 Chassis component details

S. No.	Components	Rating	Costing (Approx.)	Weight (Approx.) (kg)
1.	Throttle	12–72 V	INR 200	0.115
2.	LED	12 V	INR 100	0.02
3.	Sprockets	High quality C45 steel	INR 900	0.4
4.	Gears	1:4 Mild Steel	INR 800	0.2
5.	Disk brake	Aluminum alloy standard, 6 bolt	INR 400	0.1
6.	Glass fiber seat (Self mold)	Glass fiber	–	0.700
7.	Chassis (Self analyzed and designed on SolidWorks)	AISI 4130 alloy steel 20–20 mm^2 hollow pipes	INR 4000	7
8.	**Overall**		**INR 6400**	**8.5**

3.2 Battery

The configuration of the battery pack for the E-bike has been prepared on the basis of following factors [3]:

- Motor power consumption.
- Total weight of vehicle (including load).
- Range to be covered in a full charge mode (Fig. 4).

Voltage Calculation:

Required battery pack voltage = 48 V,
Nominal voltage of each cell = 3.7 V
Therefore,

$$\begin{aligned}\text{No. of cells in series} &= \text{required battery pack voltage/Nominal voltage of each cell}\\ &= 48/3.7\\ &= 12.97\ \text{cells(which is nearly 13)}\end{aligned} \tag{1}$$

Battery Current:

Required battery pack current = 25 Ah
Current from each cell = 2.5 Ah
Therefore,

$$\begin{aligned}\text{No. of cells in parallel} &= \text{required battery pack current/Current of each cell}\\ &= 25\ \text{Ah}/2.5\ \text{Ah}\\ &= 10\ \text{cells}\end{aligned} \tag{2}$$

From Eqs. (1) and (2);
Total No. of Cells Used = 13 * 10, i.e., **130 cells** used.

Weight estimation of battery pack

The weight of the total battery pack is calculated as total no. of cells in battery pack * Weight of one cell
= 130 * 48 = 6240 g, i.e., Approx. **7 kg** (Fig. 5).

3.3 Motor

A BLDC motor has been used as the main driving unit which is of 900 W heavy duty mainly used e-rickshaw. In compare to other motors, it has long life span and high efficiency. Being brushless, it hardly gets a chance to wear out [13] (Fig. 6).

Fig. 5 Battery pack used with the connections of motor

3.3.1 Mathematical modeling of BLDC Motor

Generally, the motors are driven by the balanced three-phase waveform. Equivalent circuit of each due to the rotation of the rotor. Figure 7 shows the DC phase consists of a winding inductance, induced emf voltage and a resistance.

Fig. 6 Cross-sectional and inside view of the BLDC motor used

Fig. 7 Phase equivalent circuit of brushless DC motor

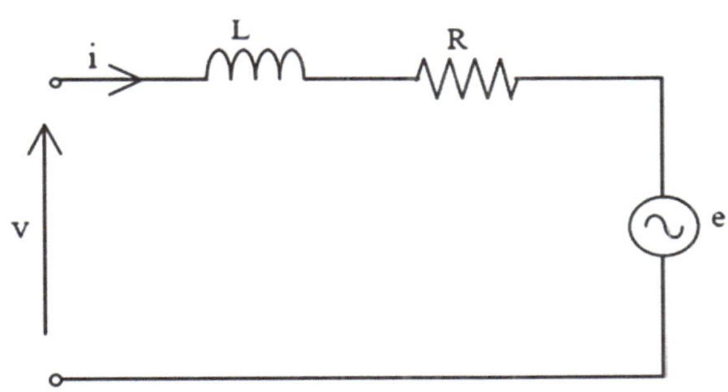

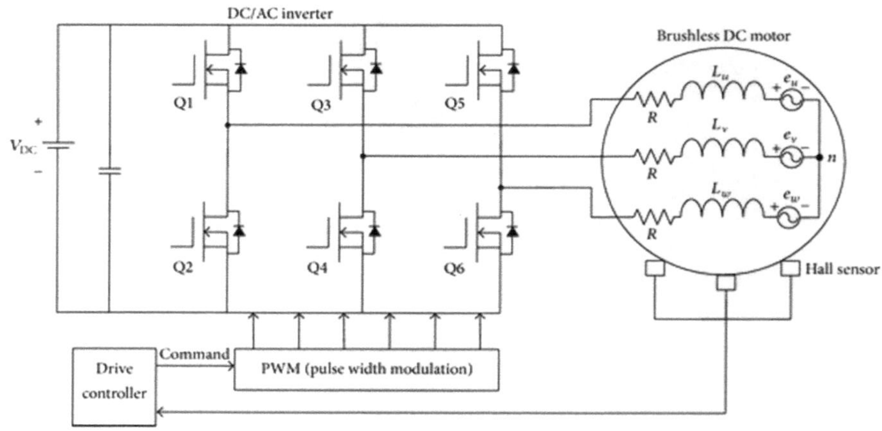

Fig. 8 Basic diagram of motor controller

Using the solar cell equivalent circuit, electrical equation can be obtained as in Eq. (3).

$$V = R_i + \left(\frac{Ld_i}{d_t}\right) + e \tag{3}$$

where

V is applied phase voltage,
i is phase current,
e is back emf voltage and
L is phase inductance.

By using the yield power of electrical motor, the electromagnetic torque that is produced can be gained. By the production of three-phase emf voltages and phase currents, output power of electrical motor is calculated. Power can be represented as output torque multiplied by the angular speed form mechanical point of view. Using these two definition, electromagnetic torque is defined in Eq. (4).

$$T = (e_a I_a + e_a I_b + e_a I_c)/w \tag{4}$$

where

w is mechanical speed of motor and
T is electromechanical torque.

Mechanical relation between the speed and the torque is represented in the Eq. (5).

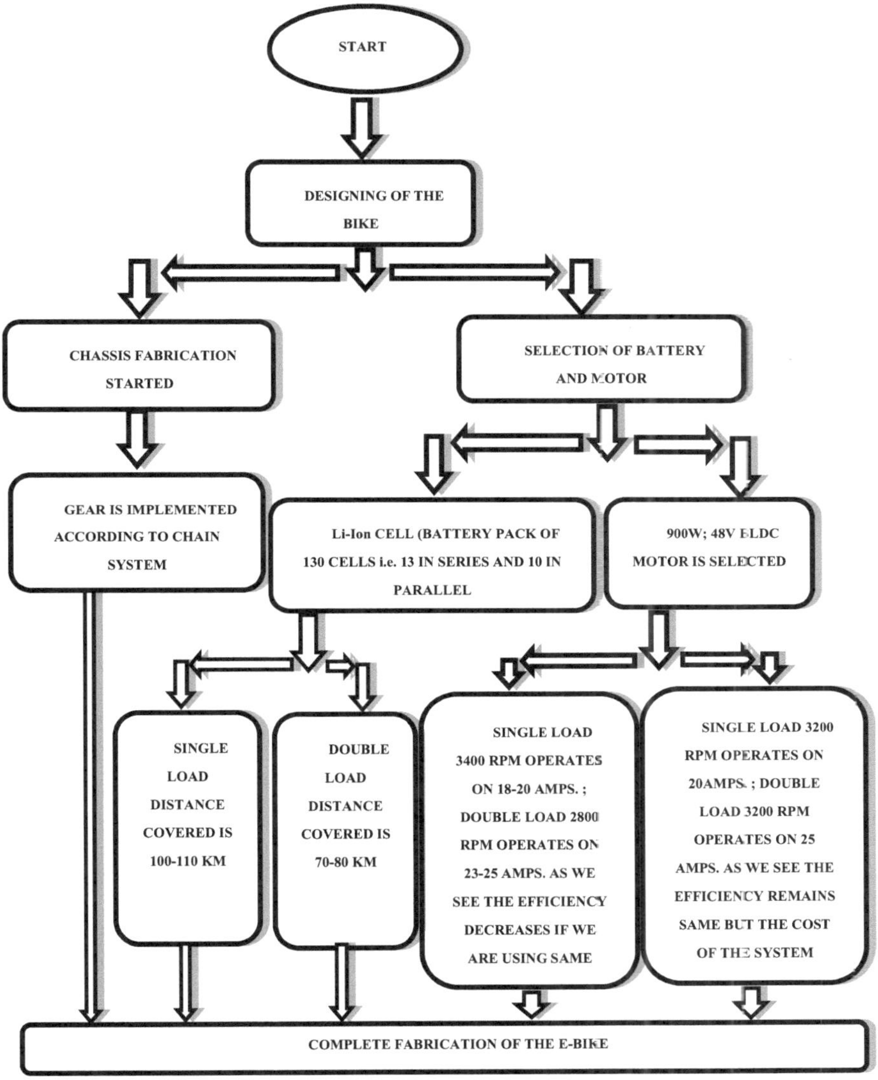

Fig. 9 Flow chart of system working through two-stage verifications

$$\frac{\mathrm{d}w}{\mathrm{d}t} = \left(\frac{P}{2}\right)(T_e - T_{\text{load}})/J \tag{5}$$

where $w = \mathrm{d}\alpha/\mathrm{d}t$

T Load is load torque,
J is rotating inertia and

α is rotor mechanical position and
P is number of pole.

3.4 Controller

To control a motor wrt the speed, start/ stop and rotation of the motor in a more proper way, controller has been used, instead of using simple mechanical switch due to limitation of a mechanical switch is the current limit [14]. In this E-bike, a compatible BLDC motor controller is used which has following rating:

- Supply voltage = 48 V DC
- Max current = 50 A
- Potentiometer voltage = 5 V (for throttle)
- No of MOSFETs = 24 (Fig. 8).

Inverter

A combination of MOSFET which create AC pulses in such a way that it makes three phases of AC power so conversion of DC power to AC power can take place. MOSFET is used for high frequency so better and smooth acceleration, multiple of 6 MOSFET has been used [15–17].

Drive Control Unit

This command has been proceeded manually by the throttle (potentiometer) and drive unit measures the required resistance and uses it as a signal for microcontroller.

Microcontroller

It has been used to change digital signal to analog signal when hall sensor detects the position of magnet, it gives digital signal in the form of 001 101 011 so micro-controller can decide PWM signal for the inverter so inverter can do its job [18–20]. The electrical parameters ratings, cost estimation and weight ratio have been listed in Table 3.

Table 3 Cost estimation of major component details

S. No.	Component	Ratings	Costing (approx.)	Weight (approx.) (kg)
1.	Motor	900 W	INR 4500	1.5
2.	Battery	48 V, 25 Ah	INR 10,000	7
3.	Controller	45 V, 25–46 Ah	INR 5500	0.5

4 Methodology

See Fig. 9.

5 Results and Discussion

Drawing from the above discussions, the E-bike performance calculation is potted in terms of different key factors. These include market tendencies and protocols, chances for improvement by special-purpose design to charm customers, identification of perhaps oversized components and drop of oversizing and identification of extents where advance research is required. Battery pack strength comparison for two different load conditions has been listed in Table 4. It is assumed that two loads as half load: 80 kg, single person and Full load: 150 kg, two persons. When speed is maximum (38–40 km/h) for both loads, battery is same, charging time of battery is same but for full load, bike running speed is reduced at 30%, whereas running time is also reduced to 40 %. The cost per kilometer has also been observed as INR 0.42 and INR 0.50/km at half load and full load, respectively. The detailed comparison has been shown in Tables 1 and 5. The comparison between the motor strength on half and full load.

The selection of motor according to uses or loading phenomena, a detailed listing has been shown in Table 6. Here, only rpm is changing with load variation as well as costing comparison, if the speed of bike has been kept constant or maximum. In motor choice 1, the cost is INR 4500 with RPM: 3400 and 2800 with half and full load, respectively.

Table 4 Battery strength comparison

S. No.	Half load (80 kg)	Full load (150 kg)
1.	Rating: 48 V, 25 Ah	Rating: 48 V, 25 Ah
2.	Discharging time/distance covered: Approx. 6 h/ 100–110 km	Discharging Time/Distance Covered: Approx. 3.5–4 h/70–80 km
3.	Charging Time: Approx. 8 h	Charging Time: Approx. 8 h
4.	Temperature almost remains constant throughout the riding	Temperature increases as the load increases because of increase in battery drain

Table 5 Motor strength comparison

S. No.	Half load (80 kg)	Full load (150 kg)
1.	Rating: 900 W, 48 V	Rating: 900 W, 48 V
2.	RPM: 3400	RPM: Approx. 2800
3.	Drawn current: 18–20 A	Drawn current: 23–25 A

Table 6 Motor strength comparison

S. No.	Motor choice 1		Motor choice 2	
	Half load	Full load	Half load	Full load
1.	Rating: 900 W, 48 V	Rating: 900 W, 48 V	Rating: 900 W, 48 V	Rating: 900 W, 48V
2.	RPM: 3400	RPM: Approx. 2800	RPM: 3200	RPM: 3200
			RPM remains same as load increases because of its high quality and its operating ratings since it needs heavy rating battery pack	
3.	Cost remains same, i.e., INR 4500 (Table 3)		Cost increases due to its different specifications, i.e., INR 10,000	

It is noted that for full load, the speed of bike is reduced due to addition load from half load, which is natural. Here, user has to compromise for speed. If any user is interested to make same speed with half load in full loading time, it is mandatory to upgrade the motor specification to RPM 3200 which cost is just double to the choice 1, which is not economical. But it gives fair speed as choice 1.

In present bike, smart phone has been used as speedometer, where the running speed and travelled distance have been shown and recorded. In Fig. 10, the distance covered as 5.6 km with the average speed of 9 km/h has been shown with the maximum speed of 38 km/h, i.e., it also shows the efficiency of the designed model of present proposed E-bike.

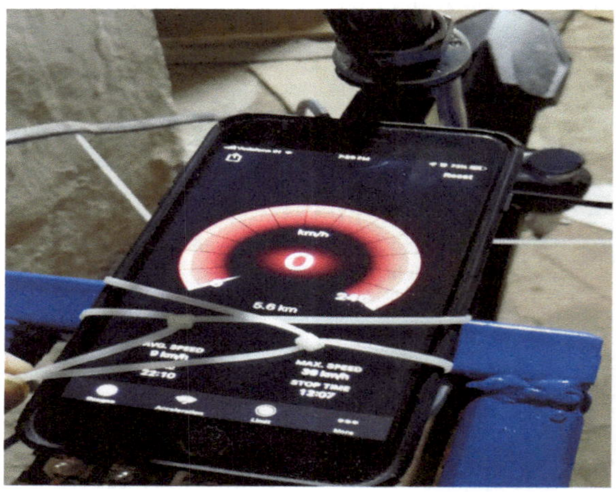

Fig. 10 Speedometer: Smart phone showing the different speed and distance measures

6 Conclusions

Some specific conclusions have been pointed out as following:

- The maximum speed of E-bike is observed as 40 km/h on natural Indian road condition.
- The efficiency is obtained as 88 and 67 % for half and full load, respectively.
- The cost of running is INR 0.42 and INR 0.50/km for half and full load, respectively.
- The cost of complete bike is INR 26,500 including all major and minor components, except speedometer.
- Such type of bikes can be used for delivery or courier boys, school or coaching going students and old age people for local uses.
- The biggest benefit of such bikes is as these are very economical as well as environment-friendly too.

References

1. J. Lee, J. Kim, B. Woo, Optimal design of in-wheel motor for an E-bike, in *Asia-Pacific (ITEC)*, June 1–4, 2016, pp. 441-443
2. R. Meireles, J. Silva, A. Teixeira, An E-bike design for the fourth generation bike-sharing services, in *EVS27 Barcelona*, Spain, November 2013, pp. 17–20
3. F. Dumitrache, M.C. Carp, G. Pana, E-bike electronic control unit, in *2016 IEEE 22nd International Symposium for Design and Technology in Electronic Packaging (SIITME)*, pp. 248–251
4. S.J. Lopes, A. Gattelu, A. Ghosalkar, S. Gonsalves, Environment friendly booster E-bike, in *2018 International Conference On Smart City and Emerging Technology (ICSCET)*
5. F. Pellitteri, V. Boscaino, A. O. Di Tommaso, F. Genduso, R. Miceli, E-bike battery charging: methods and circuits, in *2013 International Conference on Clean Electrical Power (ICCEP)*, pp. 107–114
6. E. Crisostomi, M. Faizrahnemoon, A. Schlote, R. Shorten, A Markov-chain based model for a bike-sharing system, in *2015 International Conference on Connected Vehicles and Expo (ICCVE)*, pp. 367–372
7. R. Tripathi, T.S. Bhatti, G.N. Tiwari, Effect of packing factor on electrical and overall energy generation through low concentrated photovoltaic thermal collector in composite climate condition. Mater. Today Proc. **31**(2), 449–453 (2020). https://doi.org/10.1016/j.matpr.2020.05.005
8. N. Pavan Kumar Reddy, K.V.S.S. Vishnu Prasanth, Next generation electric bike E-bike, in *IEEE International Conference on Power, Control, Signals and Instrumentation Engineering (ICPCSI-2017)*, pp. 2280–2285
9. R. Tripathi, G.N. Tiwari, Energy matrices, life cycle cost, carbon mitigation and credits of open-loop N concentrated photovoltaic thermal (CPVT) collector at cold climate in India: A comparative study. Sol. Energy **186**, 347–359 (2019)
10. T. Bhavani, K. Santhosh Kumar, K. Dhiraj Kumar, S. Udai, Novel design of solar electric bicycle with pedal assistance. Int. J. Magaz. Eng. Technol. Manag. Res. 104–108 (2015)
11. M.M. Trivedi, M.K. Budhvani, K.M. Sapovadiya, D.H. Pansuriya, Design & development of E-bike. Depart. Electr. Eng. 590–596 (2019)

12. R. Tripathi, G.N. Tiwari, V.K. Dwivedi, Overall energy and exergy performance of partially covered N-photovoltaic thermal (PVT)-compound parabolic concentrator (CPC) collectors connected in series, in *IEEE 1st International Conference on Power Electronics, Intelligent Control and Energy Systems*, pp. 1–6 (2016)
13. S. Nagwe, T. Shaikh, V. Sangale, J. Sapkale, A. Singh, Solar E-bike. IJARIIE 590-596 (2019). ISSN(O) 2395-4396
14. K.S. Mishra, S.V. Gadhawe, D.C. Chaudhari, B. Varma, S.B. Barve, Design and development of solar hybrid bicycle. Int. J. Current Eng. Technol. 377–380 (2016)
15. M. Reddi Sankar, T. Pushpave, V. Bhanu Prakash Reddy, Design and development of solar assisted bicycle. Int. J. Sci. Res. Pub. **3**(3), 1–6 (2013)
16. S. Pearlson, S. Durai Muragan, K. Vijay, Solar power bicycle. Int. J. Eng. Technol. Comput. Sci. Electron. **25**(5), 1-6 (2018). ISSN 0976-1353
17. R. Gupta, P. Patel, F. Dadhi, A. Pathan, V. Shastri, Design and development of electric bike. Int. J. Sci. Res. Develop. 889–893 (2017)
18. R. Tripathi, G.N. Tiwari, Annual energy, exergy, and environmental benefits of N half covered concentrated photovoltaic thermal (CPVT) air collectors, in *Advances in Smart Grid and Renewable Energy. Lecture Notes in Electrical Engineering*, vol. 435, eds. by S. SenGupta, A. Zobaa, K. Sherpa, A. Bhoi (Springer, Singapore). https://doi.org/10.1007/978-981-10-4286-7_12
19. S. Adhisuwignjo, S.M. Rifa'I, Development of a solar-powered electric bicycle in bike sharing transportation system. Earth Environ. Sci. **70**, 1–10 (2017)
20. R. Tripathi, G.N. Tiwari, Annual performance evaluation (energy and exergy) of fully covered concentrated photovoltaic thermal (PVT) water collector: An experimental validation. Sol. Energy **146**, 180–190 (2017)

A Novel Energy and Exergy Assessments of Solar Operated Combined Power and Absorption Refrigeration Cogeneration Cycle

Mohd Parvez, Mohammad Emran Khan, Faizan Khalid, Osama Khan, and Wasim Akram

Nomenclature

DNI	Direct normal irradiance [W/m^2]
$LiNO_3$-H_2O	Lithium nitrate-water
T_0	Ambient temperature [°C]
T	Working temperature [°C]
A_{feild}	Area of heliostat field [m^2]
$\dot{E}_e$	Exergy of the evaporator [kW]
$\dot{E}_P$	Exergy of heat [kW]
$\dot{E}_{in}$	Energy input [kW]
$\dot{E}_{x,out}$	Exergy output [kW]
$\dot{E}_{x,in}$	Exergy input [kW]
$\dot{Q}_A$	Energy of the absorber [kW]
$\dot{Q}_C$	The heat energy of condenser [kW]
$\dot{Q}_{CR}$	The heat energy of the central receiver [kW]

M. Parvez (✉)
Department of Mechanical Engineering, Al-Falah University, Faridabad, Haryana, India
e-mail: mparvezalig@rediffmail.com

M. E. Khan · F. Khalid · O. Khan
Department of Mechanical Engineering, Jamia Millia Islamia, New Delhi, India
e-mail: emran_jmi@yahoo.com

F. Khalid
e-mail: kuwarfaizan@gmail.com

O. Khan
e-mail: osamakhan6165@gmail.com

W. Akram
Department of Mechanical Engineering, Mewat Engineering College Palla, District Nuh Mewat, Haryana, India
e-mail: wasimakramkhan18@gmail.com

N. Patel et al. (eds.), *Electric Vehicles*, Green Energy and Technology,
https://doi.org/10.1007/978-981-15-9251-5_13

$\dot{Q}_g$	The heat energy of the generator [kW]
$\dot{Q}_{heat}$	Heat energy [kW]
$\dot{Q}_{lost,CR}$	Loss of heat energy from the central receiver [kW]
$\dot{Q}_{lost,heliostat}$	Loss of heat energy from heliostat [kW]
$\dot{Q}_P$	Process heat [kW]
$\dot{Q}_{salt}$	The heat energy of molten salt [kW]
$\dot{Q}_{solar}$	The heat energy of solar [kW]
$\dot{m}_{fuel}$	A mass flow rate of fuel [kg/s]
$\dot{m}_{cw}$	A mass flow rate of cold water [kg/s]
$\dot{m}_{water}$	A mass flow rate of water [kg/s]
$\dot{m}_{we}$	A mass flow rate of weak solution [kg/s]
$\dot{m}_{sto}$	Mass flow rate of strong solution [kg/s]
$\dot{m}_r$	Mass flow rate of refrigerant [kg/s]
$\dot{m}_s$	Mass flow rate of solution [kg/s]
$\dot{m}_{steam}$	Mass flow rate of steam [kg/s]
$\dot{m}_{salt}$	A mass flow rate of molten salt [kg/s]
$\dot{W}_P$	Pump work [kW]
$\dot{W}_T$	Turbine work [kW]
$\dot{W}_{net}$	Net power [kW]
$\dot{W}_{el}$	Electrical power [kW]
$\Delta_{energy,heliostat}$	The energy efficiency of heliostat
Δ_{gen}	Generator efficiency [%]
$\Delta_{ex,cogen}$	Exergy efficiency cogeneration cycle [%]
Δh	Change of enthalpy
Δs	Change of entropy
α	Constants
$\eta_{energy.cogen}$	The energy efficiency of cogeneration cycle [%]
a–h	State points of cogeneration cycle
1–17	State points of cogeneration cycle

1 Introduction

For a country to prosper, it is essential to groom sustainable energies which are not the only renewable in nature but ample in the atmosphere. The positive side of these sources can be interpreted by understanding the inherent nature of being eco-friendly to the environment. In the long run, conventional sources of energy lag in comparison with natural energies primarily due to excessive cost and availability issues which vary from country to country [1]. To fulfill various energy challenges, there is an immediate demand to develop and enhance a viable cost-effective energy alternative. In this context, concentrating solar thermal systems appears to progress as one of the most notable, promising and viable alternatives for achieving cleaner and cheaper power generation with the further benefit of reducing the level of pollution [2].

Solar thermal power plants eventually gained popularity and a certain boost when concentrated solar power (CSP) plant was successfully developed in California, U.S (1990). These plants utilize trough type solar concentrators in combination with a heat transfer device to generate steam, by applying a Rankine cycle at a low temperature greater than 150 °C [3]. An evolved version of this system is a combined cycle system that comprises a solar coupled steam-based generator, solar collector field in combined cycle operation. The general report after analyzing the various details associated with this cycle, that can be interpreted by the application of solar energy to produce high-quality steam at a particular temperature of 600 °C, which previously ensured a limit of 550 °C, is predicted to almost double up the cycle's efficiency in contrast to the conventional organic Rankine cycle which operates at somewhat alike temperatures; approximately from 9% to an exceptional 18% [4, 5].

To efficiently drive a solar-based combined power plant, a solar thermal collector is often considered an integral part of a generation of effective energy. The primary function associated with this device is to capture all the incoming packets of solar energy and eventually convert them into useful heat which is transferred into the working substance employed within the cycle which usually is fluids such as water/steam, pressurized air, atmospheric air, molten salt, and different types of oil [6, 7].

In past, solar heating was often employed in combination with various devices to derive energy. But recently, researchers are actively looking out for advanced techniques that may be applied to upgrade the solar absorptivity and further modify the output temperature range. The heat obtained by the application of solar thermal energy is obtained through special types of solar collectors that are available at the market in varying ranges such as flat plate, evacuated tube, parabolic trough, and solar tower technology. Among them, a flat plate is the most common and well-established collector for low-temperature applications in the range 50–70 °C [8–10].

In recent years, the commercialization of solar energy has compelled researchers to analyze and explore various concentrated solar collectors that provide high power at relatively low cost. Among them, solar tower technology has gained wide appreciation instinctively since it is seen to provide electric power at minimum cost as compared with the parabolic dish system. Furthermore, these solar collectors prove to furnish superior performance in comparison with other collector systems [11]. In a recent study [12], a mathematical code was developed to generate differential evolution optimization technique for a heliostat field that would be applied to calculate the optimum space requirement between heliostats. In another solar-based analysis [13], a proposal was designed which reduced the number of optimization steps involved for calculation of the accuracy of the solar tower system and further increased the accuracy of a cogeneration system that employed molten salt as working fluid. The project involved a solar power tower plant that consisted of 2650 heliostat aligned together inclined toward the sun and the data furnished from the analysis was established as a reference case. During the analysis, components such as the heliostat field, receiver, thermal storage system, and power block were developed by varying certain parameters to obtain a relationship between them. In one study [14], keeping the commercialization factor in mind, solar collector plants were designed which primarily included receiver dimensions, tower height as well as the layout of numbers heliostats used for power generation. Further in their study, a substantial amount of land was obtained to establish the plant named Noor III of capacity approximately around 150 MWe coupled with 7400 heliostats. The whole setup was designed on Campo code, which simplifies the optimization without sacrificing the accuracy. One of the researches [15] implemented an optimization technique that involved solar towers employing molten salt as working fluid coupled with cavity receivers. The whole setup was designed to integrate optical, thermal, hydraulic, and operational analysis of the system. The coupled system was predicted to furnish an additional 4% higher annual yield, in comparison with other conventional models based on static design considerations.

The growing potential of solar equipment has compelled the current generation to somehow exploit any solar thermal energy comprehensively and effectively present into the environment according to its potential. The global acceptance of this energy has substantially reduced the grid load besides safeguarding the atmosphere from harmful gases. Henceforth, considering the current zeal of the society toward solar energy, a hybrid combination of power and cooling cycles has been evolved to improve the overall energy conversion efficiency of the radiations on the PV modules. In this analysis, the cycles taken into consideration are a coalition of organic Rankine cycle and vapor absorption refrigeration cycle which retrieve waste heat from another resource and are further implemented to produce combined electric power and refrigeration simultaneously [16, 17].

The general idea behind the employment of waste heat recovery is to recover any type of waste heat issuing from various components that would have otherwise been discarded. This research focuses all its attention on providing a standard analysis on the recovery of this waste heat which is, later on, applied to drive an absorption cooling system placed in the cycle of an organic Rankine cycle [17, 18].

In another study, [19] analyzed the performance of first and second laws of a single effect, water lithium bromide absorption refrigeration system. The analysis predicted to have superior system performance for higher temperature ranges for both generator and evaporator. Simultaneously, the system performance was found to deteriorate when condenser and absorber temperatures are set for higher values. Among all components of the system zenith, exergy destruction occurred in the generator. Recently, [20] reviewed a solar operated refrigeration absorption system and in their study employed natural refrigerants as working substances for the production of cooling effect in the range of refrigeration and air conditioning. The system was found to have the ability to make effective utilization of waste heat lower temperature resources by deploying various configurations of the absorption refrigeration cycle on which past experimentations have been done. The uncomplicated design of this cycle has enabled it to be widely accepted among power plant users which besides are quite economic in operation too. In another study [21], modeling of solar operated absorption chiller system was analyzed to evaluate the effects of generator inlet temperature on the effectiveness of the solution heat exchanger, mass flow rate of the pump, COP, exergetic efficiency, exergy destruction, fuel depletion ratio, and improvement potential of the system. The results evaluated predicted solar collectors to be the origin of maximum exergy destruction since 71.9% of the supplied exergy was depleted. Further, this solar collector accounts for approximately 84% of the total exergy loss, and the overall exergetic enhancement capability of the system was estimated to be 84.7%. In another fruitful study [22], a solar-based combined power and cooling system were proposed to accomplish two functions simultaneously which are power production and cooling effect at two different temperature levels for cold. The proposed cycle was integrated with an internal whose heat source was obtained from a heat storage tank. The whole setup was further connected to evacuate which employed a combination of ammonia as a working fluid. The results obtained from the above research predict the thermal efficiency of the SCPC system to be close to 70.35%. Further, the exergy efficiency of all components combined is estimated to be around 45.52% with a renewable energy ratio of 18.1%. Another experiment on a solar-driven cogeneration cycle was performed in order to find out the effects of some influence parameters such as DNI, turbine inlet pressure, the mass flow rate of molten salt, and turbine entrance temperature was studied on power generation of the cycle. Each component's energy and exergy efficiency was analyzed to compute the final efficiency of the cycle [23].

The primary analysis presented in the current research is inclined toward examining energy, and exergy, of a combined power (Rankine cycle) and cooling (absorption refrigeration) cogeneration cycle which is driven by solar power tower. The solar power tower consists of the heliostat field and central receiver which further employ molten salt as a heat transfer fluid within the system. Therefore, the theoretical cycle is an attempt for the successful utilization of concentrated solar power (CSP) to satisfy the power and cooling to various energy requirement industries simultaneously like electricity, cold storages, space cooling, food processing, etc. Parallel to the advancement in refrigeration systems for the effective use of waste heat, exergy analysis along with the energy approach is highly desirable to make the theoretical analysis to identify and quantify the sources of losses for evaluating its true efficiency in the proposed cycle. Subsequently, the fluctuations in energy and exergy efficiency were studied when the characteristics of the associated system were varied. Finally, the exergy destruction of each components was evaluated, to find out the performance of the system. Numerical results are graphed and explained in the result and discussion section. The properties of the solar operated cogeneration cycle represented in Table 1.

2 System Description of Solar Operated Cogeneration Cycle and Assumptions

The theoretical cogeneration cycle comprises a Rankine cycle (RC) coupled with the absorption refrigeration cycle (ARC) by utilizing solar radiant energy for side

Table 1 Properties of solar operated cogeneration cycle [23, 25]

System properties values		
Heliostat field	Beam radiation (DNI)	800 W/m^2
	Overall field efficiency	75%
	Total heliostat aperture area	10000 m^2
Center receiver	Aperture area	12.5 m^2
	The internal temperature of molten salt	290 °C
	The outlet temperature of molten salt	565 °C
	View factor	0.8
	Tube diameter	0.019 m
	Tube thickness	0.00165 m
	Emissivity	0.8
	Reflectivity	0.04
	Wind velocity	5.0 ms^{-1}
	Passes	20
Heat recovery steam generator	The internal temperature of the water	239 °C
	The outlet temperature of the steam	552 °C
	Ambient temperature	20 °C
Absorption refrigeration cycle	Generator inlet temperature	90 °C
	Condenser outlet temperature	35 °C
	Evaporator outlet temperature	5 °C
	Absorption outlet temperature	35 °C
	The mixture used in ARS	$LiNO_3$-H_2O
Thermophysical properties of molten salt	Temperature range	220–600 °C
	Viscosity	25 (MPa-s)
	Density	863 (kg/m^3)
	Thermal conductivity	0.134(W. K^{-1} m^{-1})
	Specific heat	10 (kJ K^{-1} kg^{-1})
	Freezing temperature	238 °C
	Melting point	221 °C

by side production of power and cooling effect as depicted in Fig. 1. The molten salt used as working fluid enters the HRSG and generator, where it subsequently transmits all the thermal energy absorbed into the heat exchanger carrying water simultaneously on the other hand to produce steam. The steam is further transferred into the steam turbine where it is likely to be expanded to produce useful work and produce power by rotation of the shaft. The rejected steam after being utilized to produce work is condensed within the condenser chamber of the cycle, and saturated liquid is directed back to the HRSG of the RC cycle bypassing the pump in the

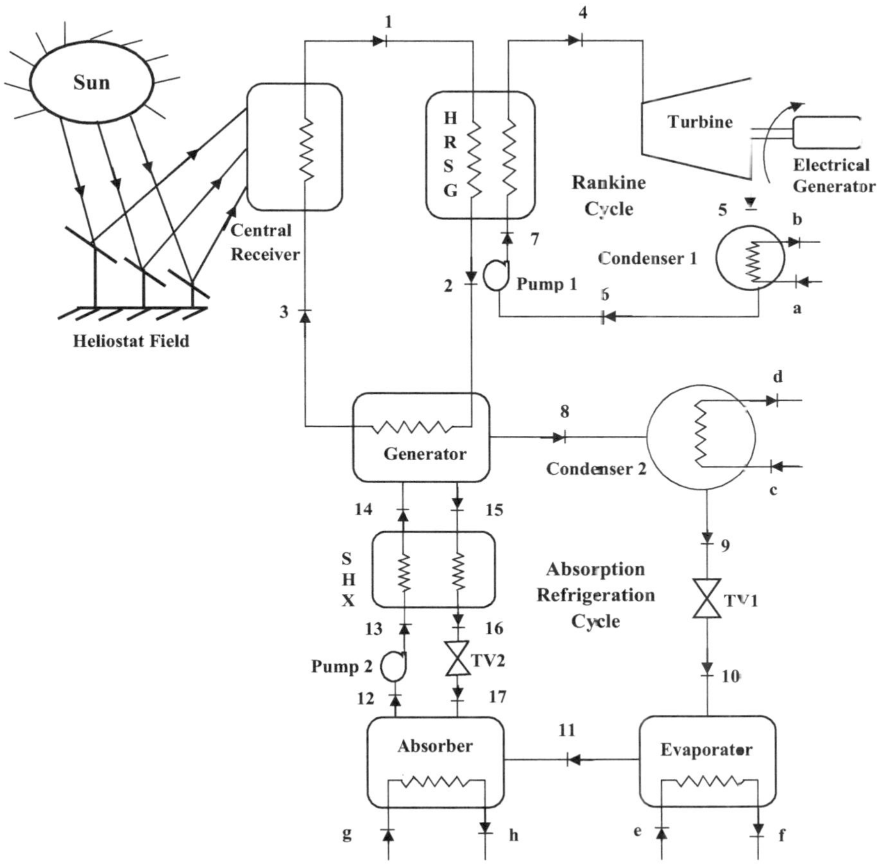

Fig. 1 Schematic diagram of a solar integrated cogeneration cycle

process. The stream being issued out of the HRSG proceeds into the generator and ultimately into the central receiver of the system. Also, a condenser is employed to receive all the superheated pure water vapor issued from the generator to condense the outgoing steam.

The system further consists of a throttle valve which yields saturated liquid at the same pressure found in the evaporator. Initially, the liquid enters the throttle valve in the saturated form at condenser pressure. The heat ejected from the evaporator is absorbed within the working substance in the absorber and hence transforms into saturated vapor. Eventually, the working fluid which is predominantly a blend of $LiNO_3$-H_2O enters the heat exchanger with the sole intention of rejecting its heat in the regenerator. Further, the substance flows into the throttle valve, and during its operation, its pressure value gets reduced to absorber pressure

Two different streams of fluid intermix with each other within the absorber and transform into a different mixture, which eventually proceeds into the pump and solution heat exchanger (SHX). The working substance is ultimately diverted into the generator of a cycle.

To ascertain the above hypothetical cycle and simplify the analysis, some assumptions have been prescribed below:

- All the units of the cycles have made a common assumption that all components are in a steady-state condition. Also, the receiver is perceived to have a constant supply of solar radiation irrespective of the time in the day.
- Any component (predominantly HRSG, steam turbine, condenser coils, and evaporators) which may have a sudden drop of pressure and ambient heat loss is to be neglected.
- As minimal changes may occur between incoming and outgoing enthalpy, hence the isenthalpic flow is to be considered neglecting the minute changes in enthalpy during its course of action in the throttle valve.
- The general state for refrigerant is assumed to be at a saturated liquid state exiting the condenser.
- At the end of the operation in the evaporator, the working substance is assumed to be saturated vapor.
- The analysis only includes physical exergies for the solar heat source and vapor flows
- Exergies such as kinetic, potential, and chemical of the given substance are not taken into consideration.
- The components such as generator and absorber enable lithium nitrate mixture to be in an equilibrium state concerning its temperature and pressure.
- The other refrigerant (water) is assumed to be at saturation condition while exiting the condenser and evaporator.
- The respective solutions (strong solution and weak solution) for generator and absorber while exiting the unit are presumed to be in a saturated state.
- A certain temperature of approximately 70 °C and above needs to be obtained in the throttle valve to evade crystallization within the mixture in ARC.
- Power consumption for both types of a pump (refrigerant pump and solution pump) needs to be assumed negligible to simplify the analysis.

3 Thermodynamic Analysis of Cogeneration Cycle

In this paper, a practical approach is adopted to approximate the exergy and energy efficiencies of all the components which eventually provide the overall performance of the cycle. The present prototype formulated for the thermodynamic analysis can speculate the overall performance and cycle efficiency. In this analysis, a method is provided which enables the researchers to minimize the optimization method by reducing the number of experimental tests conducted. These experiments are often considered to be quite expensive and further time-consuming too. Conversely, this cycle in the paper presents a new thermodynamic cycle technologically which appears to be extremely advanced and appraised. The evaluation involves exergetic analysis which involves evaluation of exergy loss for each component. From this data, the overall cycle performance is evaluated which is defined as "the highest feasible reversible work acquired in initiating the state of the system to equilibrium in accordance with that of atmosphere." Through the reversible transformation of an out flux of matter from the specified thermodynamic state to the hindered dead state where the specified out flux of matter will be in thermal and mechanical equilibrium with atmospheric conditions, the work obtained in the entire cycle can be interpreted to be as exergy, which can be explained from the equation given below as

$$E = \dot{m}[(h - h_0) - T_0(s - s_0)] \tag{1}$$

In a recent research article [24], the entropy generation for a space of control volume is expressed as

$$\dot{S}_{\text{gen}} = \frac{\mathrm{d}s}{\mathrm{d}t} - \sum_{i=0}^{n} \frac{Q_i}{T_i} - \sum_{\text{in}} \dot{m} + \sum_{\text{out}} \dot{m} \geq 0 \tag{2}$$

Often the Gouy–Stodola theorem is applied to evaluate the exergy destruction and entropy generation whose relationship is further expressed as

$$\dot{E}_D = T_o \dot{S}_{\text{gen}} \tag{3}$$

The amount of solar radiation input received from the heliostat field on the central receiver is expressed as

$$\dot{Q}_{\text{solar}} = \dot{A}_{\text{field}} q \tag{4}$$

The surface area of all the heliostat fields is linked with the aperture area and is often expressed as concentration ratio which is provided below:

$$C = \frac{A_{\text{field}}}{A_{\text{app}}} \tag{5}$$

A percentage of solar radiation energy is accepted by the heliostat and is delivered to the central receiver. The further remaining energy is lost within the environment is expressed as

$$\dot{Q}_{\text{solar}} = \dot{Q}_{\text{CR}} + \dot{Q}_{\text{lost,heliostat}} \tag{6}$$

The energy efficiency of the heliostat field is formulated as

$$\Delta_{\text{energy,heliostat}} = \frac{\dot{Q}_{\text{CR}}}{\dot{Q}_{\text{solar}}} \tag{7}$$

The total amount of radiation energy is accepted by a central receiver within the circulating molten salt, and the remaining part of the energy is perceived to be lost into the ambient environment. This statement in mathematical form can be expressed in the given equation:

$$\dot{Q}_{\text{CR}} = \dot{Q}_{\text{molten salt}} + \dot{Q}_{\text{lost,CR}} = \dot{m}_{\text{molten salt}}(h_1 - h_3) + \dot{Q}_{\text{lost,CR}} \tag{8}$$

$$\Delta_{\text{energy,CR}t} = \frac{\dot{Q}_{\text{salt}}}{\dot{Q}_{\text{CR}}} \tag{9}$$

The net amount of power produced by a proposed cogeneration cycle is given by

$$\dot{W}_{\text{net cogen}} = \dot{W}_T + \dot{Q}_E - \dot{W}_{P_1} - \dot{W}_{P_2} \tag{10}$$

The electrical power production from the complete process is evaluated by

$$\dot{W}_{\text{el}} = \Delta_{\text{gen}} \dot{W}_{\text{net}} \tag{11}$$

where Δ_{gen} is the electrical generation efficiency.

Often a cogeneration cycle whose prime function is to produce the electricity, therefore, the energy efficiency is explained as the ratio of useful energy output to the energy input:

$$\Delta_{\text{energy,cogen}} = \frac{\dot{W}_{\text{net}} + \dot{Q}_E}{\dot{Q}_{\text{in}}} \tag{12}$$

where $\dot{W}_{\text{net}}$ is the net power output, $\dot{Q}_E$ is the rate of heat supplied from the plant for space heating and $\dot{Q}_{\text{in}}$ is the rate of energy input to the cogeneration plant

The exergy efficiency of the cycle is presented as

$$\Delta_{\text{ex,cogen}} = \frac{\dot{E}_{x,\text{out}}}{\dot{E}_{x,\text{in}}} = \frac{\dot{W}_{\text{net}} + \dot{E}_{x,\text{heat}}}{\dot{E}_{x,\text{in}}} = 1 - \frac{\dot{E}_{x,\text{dest}}}{\dot{E}_{x,\text{in}}} \tag{13}$$

where $\dot{E}_{x,\text{dest}}$ is the rate of exergy destruction and $\dot{E}_{x,\text{heat}}$ is the exergy transfer associated with the transfer of heat, expressed as

$$\dot{E}_{x,\text{heat}} = \int \delta \dot{Q}_{\text{heat}} \left(1 - \frac{T_0}{T}\right) \tag{14}$$

Table 2 Energy balance equations of each component for the solar operated cogeneration cycle

Description	Energy balance equations
Central receiver	$\dot{Q}_{CR} = \dot{Q}_{salt} + \dot{Q}_{lost,CR} = \dot{m}_{salt}(h_1 - h_3) + \dot{Q}_{lost,CR}$
HRSG	$m_{ms}(h_1 - h_2) = m_{st}(h_4 - h_7)$
Turbine	$W_T = m_{st}(h_4 - h_5)$
Condenser 1	$m_c(h_b - h_a) = m_{st}(h_5 - h_6)$
Pump	$W_P = m_w(h_7 - h_6)$
Generator	$m_{ms}(h_2 - h_3) = m_s h_{14} - m_r h_{15} - m_s h_8$
Condenser 2	$m_{c2}(h_c - h_d) = m_r(h_8 - h_9)$
Throttling valve	$h_9 = h_8$
Evaporator	$m_e(h_f - h_e) = m_r(h_{11} - h_{10})$
Absorber	$m_{ws}h_{17} + m_r h_{11} - m_{ss}h_{12}$
Pump	$m_{ss}(h_{13} - h_{12})$
Solution heat exchanger	$m_{ss}h_{13} + m_{ws}h_{15} = m_{ss}h_{14} + m_{ws}h_{16}$
Expansion valve	$h_{16} = h_{17}$

The exergy rate of heating as the exergy may subsequently increase and is given by

$$\dot{E}_{x,heat} = \Delta\dot{E}_{x,water} = \dot{m}_{water}(\Delta h - T_0\Delta s)_{water} \tag{15}$$

where Δh and Δs are the enthalpy and entropy changes of the water, respectively. Then, the cogeneration exergy efficiency becomes

$$\Delta_{ex,cogen} = \frac{\dot{W}_{net} + \dot{m}_{water}(\Delta h - T_0\Delta s)_{water}}{\dot{E}_{x,in}} \tag{16}$$

The energy and exergy balance equations of each component for the solar operated cogeneration cycle are given in Tables 2 and 3.

4 Results and Discussion

The prime objective of this research is to investigate the effects of essential operating parameters on the performance of the hypothetical cogeneration cycle. Parameters such as DNI, variation in a mass flow rate of steam and molten salt on change DNI, energy-based efficiency, exergy-based efficiency, and exergy destruction of each component are evaluated for the solar-based cogeneration cycle [23, 25].

To understand the relationship between the change in DNI on the energy efficiencies of a Rankine cycle and cogeneration cycle coupled together is depicted in Fig. 2.

Table 3 Exergy balance equations of each component for the solar operated cogeneration cycle

Description	Exergy balance equations
HRSG	$\dot{E}_{D,\text{HRSG}} = T_0[\dot{m}_{\text{ms}}(s_1 - s_2) + m_{\text{st}}(s_4 - s_7)]$
Turbine	$\dot{E}_{D,\text{ST}} = \dot{m}_{\text{ST}}[(h_4 - h_5) - T_0(s_4 - s_5)] - W_T$
Condenser 1	$\dot{E}_{D,C1} = T_0\dot{m}_{\text{ST}}(s_5 - s_6)$
Pump 1	$\dot{E}_{D,P1} = T_0[\dot{m}_w(s_6 - s_7)]$
Generator	$\dot{E}_{D,g} = T_0[\dot{m}_r s_8 + (\dot{m}_s - \dot{m}_r)s_{15} - \dot{m}_s s_{14} - \dot{m}_{\text{ms}}(s_2 - s_3)]$
Condenser	$\dot{E}_{D,c} = \dot{m}_r[(h_8 - h_9) - T_0(s_8 - s_9)] + \dot{m}_{\text{cw}}$
Throttling valve	$\dot{E}_{D,\text{TV}_1} = \dot{m}_r T_0(s_{10} - s_9)$
Evaporator	$\dot{E}_{D,E} = \dot{m}_r[(h_{10} - h_{11}) - T_0(s_{10} - s_{11})]$
Absorber	$\dot{E}_{D,A} =$ $\dot{m}_r[(h_{11} - T_0 s_{11}) + (f - 1)(h_{17} - T_0 s_{17}) - f(h_{12} - T_0 s_{12})]$
Pump 2	$\dot{E}_{D,P_2} = T_0\dot{m}_s(s_{13} - s_{12})$
Solution heat exchanger	$\dot{E}_{D,\text{SHX}} = T_0[(m_s - m_r)(s_{15} - s_{16}) + m_s(s_{13} - s_{14})]$
Expansion valve	$\dot{E}_{D,\text{EV}} = (m_s - m_r)T_0(s_{17} - s_{16})$

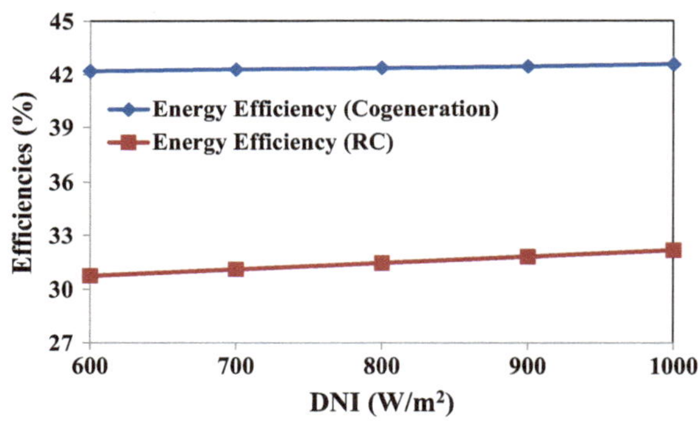

Fig. 2 Variation of energy efficiencies of the Rankine cycle and cogeneration cycle versus DNI

The graph predicts a slight gain in energy efficiencies for a substantial increase in the DNI ranged between 600 (W/m^2) and 1000 (W/m^2) for both the Rankine cycle and cogeneration cycle. The principal cause behind this considerable increase can be explained by the direct relationship between the DNI and the surface temperature of the receiver. Thereby, a gradual increase in the receiver temperature is often observed for a simultaneous increase in the rate of DNI.

Figure 3 shows the variation in DNI on the exergy efficiencies for the Rankine cycle and cogeneration cycle, and it is observed that both the efficiencies were increased as the DNI was increasing. The reason for such kind of trends is that, both the systems mainly depend on a central receiver. Furthermore, the potential exergy-based efficiency for the complete cogeneration cycle is more than a Rankine cycle [25].

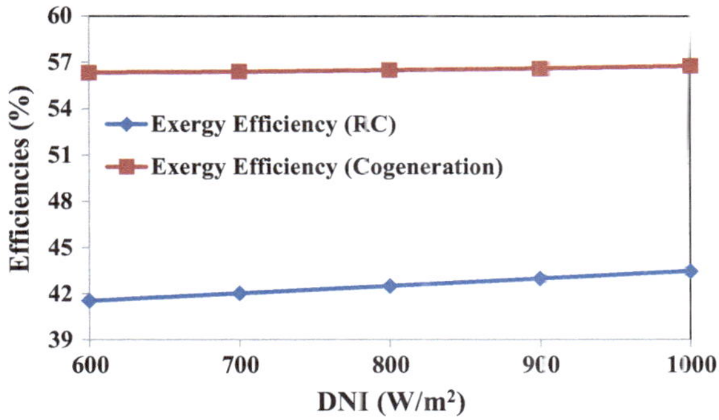

Fig. 3 Variation of exergy efficiencies of the Rankine cycle and cogeneration cycle versus DNI

The following Fig. 4 explains the interrelationship between energy and exergy efficiencies for subsequent variations occurring in DNI for a cogeneration cycle. The graph predicts a slight gain in the exergy efficiency for any substantial rise in the total DNI. The small gain in respective efficiencies can be explained by understanding the trend followed between DNI and surface temperature of the receiver. The trend shows that for a substantial rise in DNI value, the temperature of the receiver increases minorly. Furthermore, the portion of energy incident on the receiver often related as thermal energy is found to be far less than its exergy content, which eventually leads to a minute rise in the overall energy efficiency of the cogeneration cycle.

Figure 5 displays the variation of energy and exergy-based efficiencies for an eventual rise of pressure recorded at the entry of the turbine-based cogeneration cycle. The graphical representation predicts that both energy and exergy efficiencies were seen to increase significantly with a subsequent rise in pressure at the entry of the turbine. This rise is a direct outcome of the increase in refrigeration output of ARC and power output from the turbine of the RC cycle which particularly depends on pressure at the inlet. Furthermore, the rate of increase in the refrigeration output of ARC is substantially higher which eventually increases the power output from the turbine of the RC cycle. Henceforth, the above analysis proves that an eventual rise in pressure available at the turbine entrance increases the overall energy efficiency

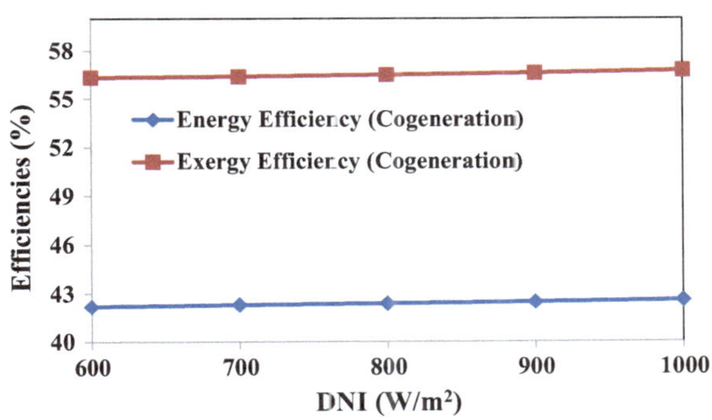

Fig. 4 Variation of energy and exergy efficiencies versus DNI

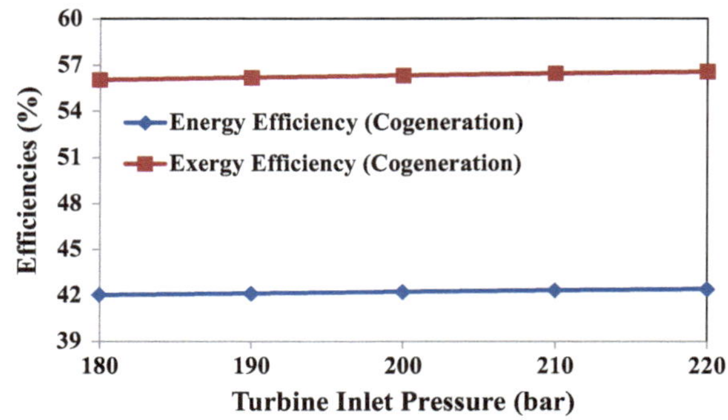

Fig. 5 Variation of energy and exergy efficiencies versus turbine inlet pressure

of the hypothetical cogeneration cycle. The above outcomes substantiate the benefits of integrating the ARC and RC cycle which can be exhibited by appreciating the significant increase in energy and exergy efficiency of the cogeneration cycle which would rather have decreased with variation in pressure at the entry of turbine.

Figure 6 predicts the variation of energy and exergy efficiencies with an increase in temperature obtained at the entrance of the turbine for the cycle. From the computed results, it can be comprehended that the energy and exergy efficiencies inflate significantly with a subsequent rise in the pressure at the entrance of the turbine. The probable reason behind this rise can be explained by the simultaneous increase in refrigeration output of ARC and power output for the turbine of the RC cycle. Also, the above statement provided for Fig. 5 holds good in this analysis also which states that the rate of increase in the refrigeration output of ARC. This finally leads to the conclusion that again the overall energy and exergy efficiencies of the theoretical cogeneration cycle may increase for a subsequent increase in temperature available at the entrance of the turbine.

Figure 7 explains the relationship between the total DNI and mass flow rate of steam or molten salt. The graph predicted an increase in the mass flow rate for both the fluids with a simultaneous raise in DNI value. Perhaps the main reason behind such kind of increasing trends is, due to the increase in DNI, the mass flow rate

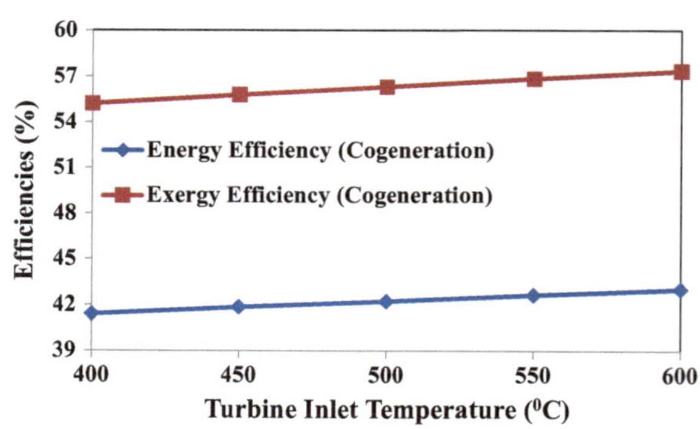

Fig. 6 Variation of energy and exergy efficiencies of cogeneration cycle versus turbine inlet temperature

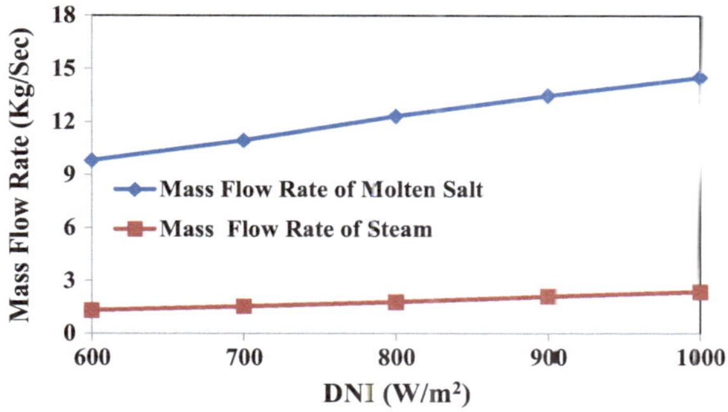

Fig. 7 Variation of a mass flow rate of cogeneration cycle versus DNI

furnishes a high rate of thermal energy to the HRSG which further reciprocates it in the form of higher steam flow rate in the Rankine cycle to increase the overall turbine output. Another valuable output derived from the analysis predicts the mass flow rate of steam to be more than that of the mass flow rate for molten salt. The obvious reason for such kind of trend is the increased flow rate of heat transfer fluid (molten salt) which in turn increases the mass flow rate of working fluid (steam) as it is seen after applying energy balance approach over the heat recovery steam generator (HRSG).

The thermodynamic irreversibility was presented in the form of exergy destruction in each component and exergy input to the cycle, and hence, it was expressed in kilowatt and is shown in Fig. 8. The result reveals that the central receiver and heliostat of the solar field compromise the higher thermodynamic irreversibility of 2495 kW and 1869 kW, respectively, in the whole cogeneration cycle. A considerable amount of exergy destruction occurs in the HRSG of approximately 586.61 kW. The exergy destruction in the rest of the cycle components was found to be 178 kW and 556 kW is exergy of heat. This analysis helps the designer to obtain the ranking of components among the proposed cogeneration cycle. The components of larger exergy destruction of cogeneration have the major scope of improvement for enhancing the thermodynamic performance of the overall cycle.

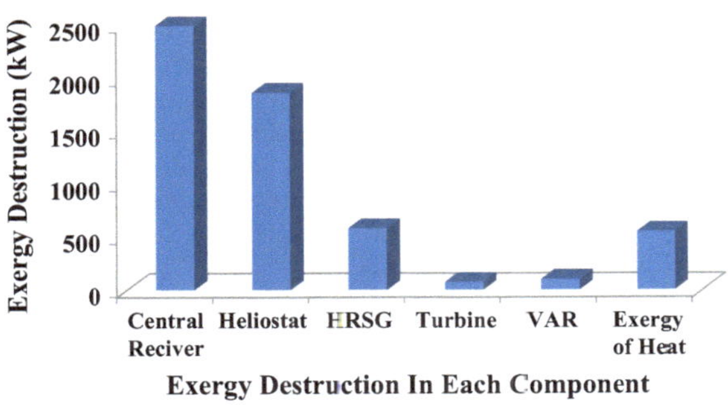

Fig. 8 Exergy destruction in each component in the cogeneration cycle

5 Conclusion

The vapor absorption-based cogeneration cycle running on solar energy provides a proper way to furnish the energy transformation through the efficient and eco-friendly thermal conversion of available solar thermal energy into the useful output of energy like electricity and cooling. The enhancement in the performance of an energy conversion system has shown drastic improvements after changing its mode from power generation to cogeneration by successful employment of the VAR cycle. From the above analysis, the following conclusion can be and is summarized below:

- A slight gain in energy efficiencies was observed after a substantial increase in DNI value was performed in the RC and cogeneration cycle.
- The gain in the exergy efficiencies was also observed for the RC and cogeneration cycle which was subsequently due to the rise in the DNI.
- The energy efficiencies of the Rankine cycle and cogeneration cycle lie between 31.45 and 42.38% and exergy efficiencies 42.51–56.5%.
- A significant increase was noted in the mass flow rate of steam and molten salt as the values of DNI were increased.
- A considerable amount of exergy destruction occurs in the central receiver and HRSG of around 2495 kW and 586.61 kW, respectively.

The development of a vapor absorption refrigeration system (VAR) at the bottom of the steam turbine-based cogeneration system driven by solar thermal energy resulted in the considerable enhancement of both energy and exergy efficiencies of cogeneration. This configuration not only enhances cogeneration efficiency but also can be utilized to refrigerate a specific thermal load of buildings where electric power and process heat are also desired. Results obtained after the combined application of energy and exergy of thermodynamics for the steam turbine-based cogeneration cycle utilizing the solar thermal energy are quite beneficial.

References

1. I. Dincer, M.A. Roshen, Thermodynamic aspects of renewables and sustainable development. Renew. Sustain. Energ. Rev. **9**, 169–189 (2005)
2. N.L. Panwar, S.C. Kaushik, S. Kothari, Role of renewable energy sources in environmental protection: a review. Renew. Sustain. Energ. Rev. **15**, 1513–1524 (2011)
3. A.M.D. Torres, L.G. Rodriguez, Analysis and optimization of the low temperature solar organic Rankine cycle (ORC). Energ. Convers. Manag. **51**, 2846–2856 (2010)
4. T.A. Yassen, Experimental and theoretical study of a parabolic trough solar collector. Anbar J. Eng. Sci. **5**(1), 109–125 (2012)
5. Y.S. Sanusi, P. Gandhidasan, E.M.A. Mokheimer, Performance analysis of integrated solar combined cycle power plant for Dhahran, Saudi Arabia. Appl. Mech. Mater. **293**, 568–573 (2014)
6. M.S. Acar, O. Arslan, Energy and exergy analysis of solar energy-integrated, geothermal energy-powered organic Rankine cycle. J. Therm. Anal. Calorim. **137**(2), 659–666 (2019)

7. P.A. Gonzalez–Gomez, F. Petrakopoulou, J.V. Briongos, D., Santana Steam generator design for solar towers using solar salt as heat transfer fluid, in *Solar PACES 2016. AIP Conference Proceedings*, vol. 1850, pp. 030020-1–030020-8 (2016). https://doi.org/10.1063/1.4984363
8. M.S. Hossain, R. Saidur, H. Fayaz, N.A. Rahim, M.R. Islam, J.U. Ahamed, M.M. Rahman, Review on solar water heater collector and thermal energy performance of circulating pipe. Renew. Sustain. Energ. Rev. **15**, 3801–3812 (2011)
9. K.C. Manjunatha, A.K. Bhoi, K.S. Sherpa, Design and development of buck-boost regulator for DC motor used in electric vehicle for the application of renewable energy, in *Advances in Smart Grid and Renewable Energy* (Springer, Singapore, 2018), pp. 33–37
10. H. Yagli, C. Karakus, Y. Koc, M. Cevik, I. Ugurlu, A. Koc, Designing and exergetic analysis of a solar power tower system for Iskenderun region. Int. J. Exergy **28**(1), 96–112 (2019)
11. Y. Ma, T. Morosuk, J. Luo, M. Liu, J. Liu, Superstructure design and optimization on supercritical carbon dioxide cycle for application in concentrated solar power plant. Energ. Convers. Manag. **206**, 112290 (2020)
12. M. Atif, F.A. Al-Sulaiman, Development of a mathematical model for optimizing a heliostat field layout using differential evaluation algorithm. Int. J. Energ. Res. **39**(9), 1241–1255 (2015a)
13. Y. Leo, T. Lu, X. Du, Novel optimization design strategy for solar power tower plants. Energ. Convers. Manag. **177**, 682–692 (2018)
14. F.C. Collado, J. Guallar, Quick design of a regular heliostat field for commercial solar tower power plants. Energy **178**, 115–125 (2019)
15. P. Schottl, G. Bern, D.W.V. Rooyen, J.A.F. Pretel, T. Fluri, P. Nitz, Optimization of Solar Tower molten salt cavity receivers for maximum yield based on annual performance assessment. Sol. Energ. **199**, 278–294 (2020)
16. E.M.A. Mokheimer, Y.A. Dabwan, Performance analysis of integrated solar tower with a conventional heat and power co-generation plant. J. Energ. Res. Technol. **141**(2), 021201-1–021201-13
17. H.A. Dhahada, H.M. Hussena, P.T. Nguyenb, H. Ghaebic, M.A. Ashrafd, Thermodynamic and thermoeconomic analysis of innovative integration of Kalina and absorption refrigeration cycles for simultaneously cooling and power generation. Energ. Convers. Manag. **203**, 112241 (2020)
18. N. Priyadarshi, F. Azam, A.K. Bhoi, A.K. Sharma, A multilevel inverter-controlled photovoltaic generation, *Advances in Greener Energy Technologies* (Springer, Singapore, 2020), pp. 149–155
19. M. Kilic, Kaynakli, Second law-based thermodynamic analysis of water-lithium bromide absorption refrigeration system. Energy **32**(8), 1505–1512 (2007)
20. M.U. Siddiqui, S.A.M. Said, A review of solar powered absorption systems Renew. Sustain. Energ. Rev. **42**, 93–115 (2015)
21. E.D. Kerme, A. Chafidz, O.P. Agboola, J. Orfi, A.H. Fakeeha, A.S. Al-Fatesh, Energetic and exergetic analysis of solar-powered lithium bromide-water absorption cooling system. J. Clean. Prod. **151**, 60–73 (2017)
22. Y. Chen, D. Xu, Z. Chen, X. Gao, W. Han, Energetic and exergetic analysis of a solar-assisted combined power and cooling (SCPC) system with two different cooling temperature level. Energ. Convers. Manag. **182**, 497–507 (2019)
23. M. Parvez, F. Khalid, O. Khan, Thermodynamic performance assessment of solar-based combined power and absorption refrigeration cycle. Int. J. Exergy **31**(3), 232–248 (2020)
24. A. Bejan, Fundamentals of exergy analysis, entropy generation minimization and the generation of flow architecture. Int. J. Energ. Res. **26**, 545–565 (2002)
25. C. Xu, Z. Wang, X. Li, F. Sun, Energy and exergy analysis of solar power tower plants. Appl. Therm. Eng. **31**, 3904–3913 (2011)

The Effectiveness of Smart Grids V2G and Integration of Renewable Energy Sources

Partha Pratim Das, Awasthi Aditya Bachchan, and Vijay Chaudhary

1 Introduction

Natural resources are on the verge of their depletion. By each day passing the resources are getting lesser and lesser. We are cutting forests to meet our energy requirement, which is not only bad for the fauna by demolition of their habitat but also with the increase in the level of pollution which is eventually harming we humans. Decreasing natural resources is also affecting our economy. With Increase in the price of resource like natural gas coal crude oil etc. is affecting the lives of common people and making it hard for the people to bear this economic burden, hence resulting in a drastic effect in their lifestyle and livelihood. There are various methods people incorporate generally such as using solar panels, solar water heaters, windmills etc., but they have their own pros and cons.

Smart Grid (SG) are those networks for circulation of energy that provide assurance of increment in the performance of energy supply by using by directional flow of energy and distributed generation. "The first energy grid was established in Great Barington, Massachusetts USA" which was centralized unidirectional system and provided demand driven control [1]. It is the incorporation of communication the control technology, sensing and computers all working along with electric Power-Grid so that its efficiency could be enhanced and decreasing the economic burden over the customers [2].

SG is a smart system or network which is environmentally friendly in nature and supplies energy according to the nature of consumption of the consumer and their

P. P. Das (✉) · A. A. Bachchan · V. Chaudhary
Department of Mechanical Engineering, Amity University, Noida, Uttar Pradesh 201313, India
e-mail: das1.parthapratim@gmail.com

A. A. Bachchan
e-mail: awasthiadityabachchan@gmail.com

V. Chaudhary
e-mail: vijaychaudhary111@gmail.com

N. Patel et al. (eds.), *Electric Vehicles*, Green Energy and Technology,
https://doi.org/10.1007/978-981-15-9251-5_14

behavior [4–6]. "We know that how the conventional grid works" but we need to know how different they are from the SG. SG handle public safety better than the conventional one and there is relatively less energy loss they have different quality of power according to their consumption behavior. Smart grid can be connected to renewable energy readily, however connecting conventional grid with renewable energy result in problem. Last but not the least there is participation of customer in SG whereas in conventional grid there is nothing as such. Various studies have been made already over the smart grid on its effectiveness. The application and control of smart grid technology is emerging as an important aspect nowadays [3]. It is an essential tool for a flourishing society. The flow of energy has become a major issue in the world in the present days. Increasing business and day by day, the increase in population has increased the demand of establishments which has caused hike in the cost of utilities including energy. This initiative to make energy grid smart is what making a smart grid is. This transformation is very much profitable to companies. Distribution through smart grid can decrease the consumption of energy by 5–10%, 13–25% less carbon emission and 87% less cost of disturbance related to power. It can decrease the use of electricity by 10–15% and the critical peak load to 43% [7]. The smart grid technology is a solution of increase in demand of energy. It helps in delivering energy at low cost with the same exceptional quality [5]. It uses the information communication technology to collect data from its customers. Advanced metering of establishments include collecting using and storing of data which are main equipments smart grid uses in its operation. Many other tools like phasor measurement measure the condition of power grid and increases the capacity of transferring power [8]. The lacking and basics of SG's and presented the challenges and prospects of SG in 3 connected areas. Emphasis must be made on the fact that integration of the future development of these focal areas must be done [9]. It is represented by a loosely coupled series of micro grid clusters and each cluster incorporating one or greater number of machines such as a wind turbine, cogeneration systems etc. and numerous direct power injection system power electronic compensator etc. A design approach holistic in nature can be incorporated to distribute the global optimization tasks into subtasks for clusters established locally for the convenience of formation of global control strategy and designing of convertors could be done to answer to local subtracts and activate global control which is hierarchical and distributed [8, 9] (Fig. 1).

According to US SG report 2018 using SG is beneficial in following ways [10]:

- Advanced transmission system technologies for control and visibility to a wide area.
- Advanced distribution systems technologies or voltage optimization health monitoring and self-healing automation.
- Advanced metering infrastructure
- Smart customer devices and management systems that activate response of demand.

Smart grid is coming as an emerging technology for future electrical system. It is an advanced system of grid that sustainably economically and reliably manner,

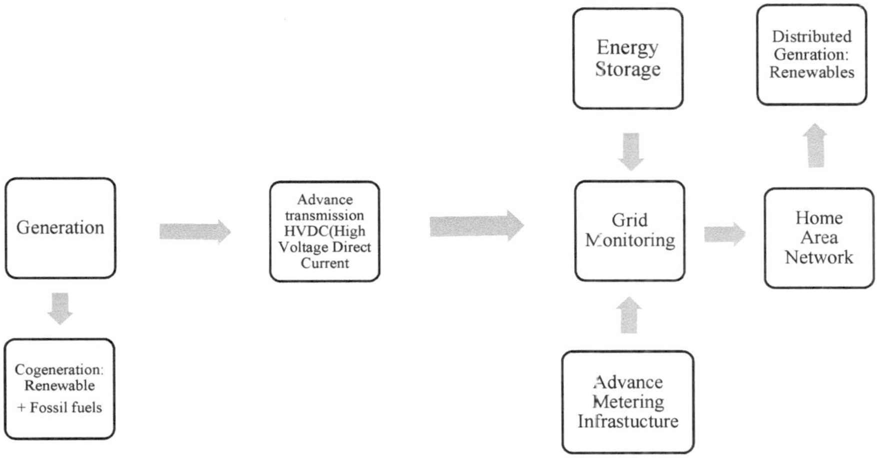

Fig. 1 Block diagram showing the smart grid concept [22]

established on infrastructure advanced in nature. They can fulfill the rising demand by supplying the adequate amount of energy [11]. The time of use program is one of the advantageous tools to enhance SG. The amount and time of usage of energy get monitored by smart meter and the user can program their usage accordingly. The demand of energy is variable throughout the day being lowest in the midnight and highest in noon and it varies according to the weather also. Power plant perform maximum to meet the need of its consumers. The time of use program enables us to use the energy at the time of lowest demand hence giving lowest price at that time. We can monitor that data from our personal computer and smart phones [11].

A smart grid incorporates bilateral electricity flow and information to make a widely distributed energy delivery network. Smart protection systems, smart management, and smart infrastructure systems are the major three subsystems. We evade the smart energy subsystems the smart communication subsystems and smart information subsystems for smart infrastructure systems [12]. Numerous management objectives like reducing cost maximizing utility profiling demand and improving energy efficiency are also studied for smart protection systems and smart grid management systems [12] (Table 1).

Table 1 Shows the overall benefits of using smart grid with respect to its utility and consumers benefits [42–44]

S. No.	Benefits based on utility	Benefits to the customers
1	The Reduction in losses due to Renewable energy sources integration	Prosumers (Producer and Consumer) enablement
2	Increased Grid stability	User friendly and transparent interface with utilities
3	Peak load management	Improves the quality of power supply
4	Self-healing grid	Reduction of electricity bills by load changes from peak hours to non-peak hours
5	Reduced Capital & operational cost	Opportunity to communicate with the electricity markets through the home area network and synchronization of smart meters
6	Increased employee safety	Opportunity to buy energy from renewable resources, further increasing pressure for transition from a carbon-based to a "green economy"

2 Challenges That Obstruct the Path of Smart Grid Development Are

2.1 Making Grid Stronger

In spite being a revolution, a smart grid has its weak points which are points to ponder upon. The unprecedented occurrence in the smart grid are:

1. Blackouts
2. Authenticating smart meter
3. Corrosion
4. Faulty control system
5. Weak base
6. Cyber theft attack.

2.2 Compatible Equipments

To cover the program controlling the whole system are sophisticated and are compatible.

2.3 Advance Technology Avoiding Capacity

Hi tech devices are used all over the network. Technology is never stagnant and is always dynamic. It must absorb the latest technology.

2.4 Cyber Security and Privacy

Granular and frequent information on consumption of and digital communication networks make us ponder upon misuse of data and cyber security.

2.5 Assessment of Cost

It may be higher than projected because there is a state of flux in the protocol and standards that operates and design smart systems.

2.6 Education of Consumers

To procure maximum benefit from the smart grid the consumer must be able to engage with the technology and for that the consumer must be educated.

2.7 Government Resource

Willingness of government plays a very important role as well as effective energy policy is needed for successfully implementing the smart grids.

2.8 The Financial Resource

Significant amount of investment is required to begin the journey of smart grid. We need an adequate financial resource to create important distributed networks.

The remaining challenges for the establishment of smart grids are as follows [13, 14]:

- The information technology industry has begun to improve the standard for communication control device interface, sensing and message interface. Sensing and message interface in spite much work is left to be done.

- Multiscale sensing telecommunication network and power grid are highly interdependent on each other which is a major source of complexity. This increases issue on each stage.
- The ability for a widely distributed system with great penetration and sources renewable in nature that show variable generation and dispatchable nature constitutes of opportunities and challenges for smart grid.

3 The Vehicle to Grid (V2G) System

V2G is a reciprocating flow of energy between a vehicle and the receiving end that could be a building, a low voltage micro grid or a grid. Along with shifting demand and reduction and reduction in electricity cost by making user aware of the peak costs during high demands, it also provides financial incentive to the customer by offering facilities of regulation of frequency and energy storage. V2G helps in the use of localized renewable resources. However, they raise concern about lithium ion battery degradation [15]. V2G involves the generation of electricity from vehicles that are parked. According to V2G every vehicle is an underutilized source of production of electricity. A car is on the road for 5% of the day and during peak tariff hours it is standing still at some parking lot. However, the power generated is limited to the amount of energy stored in the car [16, 43].

V2G demonstrates various strategies for connecting the power system and the transported system to provide reciprocal benefits for them. It consists of treat plug-in controllable lead electric vehicles with the unidirectional flow of electricity such as smart charging or pricing time of usage. VGI carries in V2G and highly advanced view with bidirectional flow electricity in between the power grid and the car, using PEV as storage. Such VGI systems will help minimize transportation carbon emissions, incorporate smart grid renewable resources, and raise capital for car owners. It was found that most of the VGI studies emphasize the technical aspects of VGI especially on the notable of V2G systems to enable balancing of load or to reduce the electricity cost and also achieving environmental goals and benefits. However, only a few studies investigate the role of acceptance by customer and behavior of the driver in these systems rarely any study the need for capacity of the institute [17].

The bidirectional flow of electricity between electricity grid and vehicle gives the possibility to join alternating electricity generation with uneven availability of electric vehicle at charging station V2G is an option for balancing of the grid in a specific regions with a high possibility of penetrating renewable energy. It can eradicate the need for storage station by investing more on the EV battery in parking. It may be astonishing that V2G has not been widely deployed yet. For the lacking, the status of development and deployment of V2G and found that the return benefits of V2G are not so great for deployment in the present scenario however it could be made possible in the after the launching of other EV [18]. A technology road map for distributing smart V2G technology was developed in the pacific northwest. The research emphasized on V2G application in the residential chargers. It makes us aware of the technical analysis, products and market drivers which will influence the market in the upcoming years [19]. The V2G development from a perspective co evolutionary in nature, consisting of interactions between the EV diffusion and grid system upgradation. The logic behind development of V2G is made in accordance with relation between development and upgrading of EV. Then the characteristic of every phase including the management hierarchy relation, physical trade structures as well as the measure for management were taken into consideration for making the proposal [20]. The V2G transition could facilitate vehicles to enhance the efficiency of grid decrease greenhouse effect from transport, establish low carbon source of energy and provide saving to the customers. To comprehend the latest state if this research field. The technical side of V2G including storage of renewable energy batteries and balancing of load to lower down the cost. A very small proportion articles emphasized on making use of the environmental features of transition to V2G [21].

The electric vehicle to grid technology is the noise-free, efficient, and clean means of transport can be achieved by electrification of vehicle. Numerous sections, for example, charging infrastructure, communication technology, the methodology adopted, signalling was studied. It was concluded that using high voltage battery results in less fuel consumption and greater range which was unlike a hybrid electric vehicle, offboard charging of the battery must be done [22]. Various aspects like battery degradation are primary topics to ponder upon. The investigation was also done upon factors like voltage regulation, control of reactive power, valley filling, load shifting etc was also done which are an ideal approach for making the active and reactive EV integrated power design, able. The present status and potential scenario of electric vehicles are achieving a great potential in the present scenario. It is possible to distinguish plug-in electric vehicles (PEVs) into battery electric vehicles and plug-in hybrid electric vehicles. The electric vehicle plug is expected to dominate the mobility market by 2030. Adapting to electric vehicle plug-in would offer cultural, social, and future benefits [23]. To manage the increase in the price of fossil fuels and meet the environmental goals it is necessary to pay more emphasis on the electric vehicle venture. Due to unavailability of charging infrastructure, operation cost and capital the future of EV cannot be predicted. Although there are lots of incentives provided by the government still there is no interest taken by the market. The market is not able to trust on the PEV venture despite the strong and attractive predictions. However, the major manufacture is already ready with their plan of action for the launching of the PEV venture. Using this technology for mobility has both market and technological impact. There is a lot of research-based work done to study the impact of this technology over factors like environment, economy, market etc. Adequate work has also been done to eradicate the challenges with the infrastructure fracture of charging stations and the cost of operation. In the coming future, it can be expected that the PEV venture can achieve market awareness and can meet the environment goals [21–23].

The PEV also has an impact on the power system. A probabilistic Plug-in electric vehicle charging model was studied over its impact on a distribution grid. Agent-based techniques such as Monte Carlo methodology, queuing theory and probabilistic variable were used to form this model. The PEV charging can be characterized into private charging and public charging. The private charging point consists of variables like arrival time, distance travelled and the number of trips. All these data when fed into a probabilistic distribution function gives a probabilistic model. The Public charging points are used by the public to increase the autonomy of the vehicle succeeding this is a case study to study the impact on the power network. A Probabilistic grid impact consists [44] of probability to increase maximum drop in the voltage with current and transformer saturation. Saturation of line suffer most of the impact, we can control the private charging points as per our usage but controlling the public charging points makes no sense. Some challenges are presented by fast chargers for grid integration since they are 50 KW power rate charging points. To increase the autonomy of the vehicles fast chargers are used and they arrive on the station in stochastic manner. Queuing theory is incorporated for the fast charger's probabilistic arrival and the evaluation of electricity demand is done [24].

4 Smart Grid: Integration with Renewable Energy Sources

Renewable energy sources are the sources which provide a never-ending supply of energy. Whether it is tidal energy hydel energy, wind vanes or the ultimate source of energy, the sun all come under renewable energy. Renewable source of energy are those which need not be conserved, and which can renew itself in a very short span of time. Renewable source of energy is not only used to produce electricity but is also used for processes like the boiling of water. Among all the renewable sources of energy, the wind and solar power plants are one of the mostly used in different application [25].

The production of energy from these resources is not constant and is variable. One cannot predict the amount of energy derived from these sources. In case of the wind power plant on can't predict the direction, intensity, duration, the velocity and frequency of the wind and this can be only installed in places with adequate wind flow and can't be installed in areas with the inadequate wind. In India states like Gujrat, Maharashtra etc have adequate and frequent wind supply. In the case of the solar power plant, the setup location should have adequate daylight. Fortunately, in India, most of the states have adequate daylight and therefore solar plant can be used in every state except for states like Meghalaya. There have been numerous works being performed for the integration of renewable energy to eradicate the backdrop of renewable energy such as the non-frequent generation of electricity [45]. The systems and control opportunities in integrating renewable energy with smart grid is also an important aspect. The biggest hurdle that come in way is the uncertain, dispatchable and variable characteristics of the renewable energy. The electricity grid is expected to suck up these energies via solution portfolio. This consists of distributed demand response, local generation, storage, operating reserve, curtailment, and aggregation of variable generation. Dynamic coordination of the elements in this portfolio must be done according to the data stored in the electricity grid framework, which will have need of essential technologies and techniques derived from control, modeling, and optimization [25]. The opportunities and hurdles that come in the way of integrating smart grid with renewable energy is also important to be considered. As we are experiencing rapid climate change and increase in cost of natural resources like petrol there is decrease in the price of renewable energy. There is a need of an efficient strategy for the management of the system for the widespread commercialization. The price allocation is also an important challenge in the promotion of electric energy. The unique characteristics associated with the renewable electricity production must be focused upon to get the insight [26].

On the other hand, the reliability on smart grid for the security of energy and renewable energy is also another factor that must be considered. Producing electricity from renewable energy can result in both direct and indirect benefits which includes achieving environmental goals. The implementation of renewable energy sources should be done for sustainable development and carbon-free energy. By the year 2050, about fifty seven percent (57%) of the total energy would be made from renewable energy resource. Integration of renewable energy with smart grid can be the best alternative for a future energy security [26, 27]. Smart grid system focuses on degradation of energy source and information technology for improving of the efficiency. Smart grid will invent new market for private organizations develop smart and energy efficient appliances. Smart grid can be used to replace the conventional methods of energy generation. Smart grid can benefit the environment as it makes use of renewable resources distribution.

In the year 2019, the solar grid integration technologies come into existence and highlighted the benefits of integrating solar grid with electric grid. Due to the depletion level of energy resources it is necessary to integrate the smart grid with renewable resources and integrating smart grid with solar energy might be the first step in the revolution. The importance of renewable energy resources is as such that it removes the dependence of us over the non-replenish able resources and over fossil fuels. The PV cell are utilizing the solar power in the form of ac or dc power, hence leading this revolution. Various researchers and scientists understand the prominence of renewable energy resources integration with smart grid due to its advantages over non-renewable resources [28].

Although integrating a renewable energy source is beneficial, but there are problems that could hamper the integration of smart grids with renewable energy sources. The increasing reliance on renewable energy resources would require concerted effort to integrate them into future grid technology. The key technology is the communication between the electronic devices although there are specific models for the energy systems according IEC61850, i.e., intelligent electronic devices. Today's fragmented devices could be integrated by the migration concept [29].

Further studies are being started to implement the smart grid and renewable energy sources in other countries [30]. The effect of solar thermal plant's integration with smart grid was studied by Camacho et al. [31]. There has been a rapid increase in the market of solar energy in recent years. Solar energy is irregular therefore its integration with smart grid is complicated. Modification of an electric bulb was done for the optimization problem encountered while integrating solar energy with smart grid. Mixed integer programming algorithm was use in a scattered way [31].

The market for grid-connected photovoltaic (PV) systems has recently grown rapidly and, therefore, grid-connected PV inverters are more in demand. Various researchers are now working on cost optimization and tracking of PV system efficiency, thereby increasing the power quality of the grid-connected PV system [46]. Significantly, renewable energy sources (RE) are also becoming more common in the traditional power system. A traditional vertical power grid is seen as needing to be changed to meet the strong demands of uninterruptible power, rising energy consumption in rural and urban areas, minimizing voltage fluctuation problems for far-reaching loads and the overall costs of building massive long-range resources transmission networks for remote feeding areas [47].

5 Smart Grids Technologies

Various smart grid technologies are being used in the present scenario and are applied in many systems and in many forms. However, some have not yet been deployed in the market. Some are used in specifically in the electrical system and some have their cross over in information technology also [32] (Table 2).

6 Conclusion

In Present scenario, smart grid V2G and its integration with renewable energy has gaining a lot of potential in electric vehicles (EVs). Several points regarding the Smart Grid, V2G and its integration with renewable energy for making EVs as more energy efficient mentioned below:

- The smart grid is penetrating in the market as we head toward future. Its integration with renewable energy might result in increase in the influence of renewable energy resource over the market unlike now which consists of shallow penetration in spite of promising performance.
- The generation of electricity from renewable energy sources also delivers more direct and indirect economic benefits than costs, as well as environmental benefits by reducing CO2 emissions.
- A smart grid will turn the power grid of the twentieth century into a smarter, more versatile, more efficient, more self-balanced and more responsive network that enables the economic growth, environmental regulation, operational capacity, energy protection and increased customer control.
- The environmentalists still need renewable energy in the hope of producing a cleaner and more efficient electricity generation.

Table 2 Shows some smart grid technologies [33–41]

S. No.	Technologies	Description	References
1	SCADA (Supervisory control and data acquisition system)	The use of the program logic controller is done for control management and can manage a few to a more than a million devices. The software consists of two layers namely client layer and data server layer. Client layer is for personal human use whereas data server layer is for controlling the process of the system. SCADA is based on the real-time database system and can perform multitasking. It requires constant commanding and monitoring due to its large-scale operation. Various open source SCADA is employed by operators however it consists of many loopholes	[33]
2	Smart meters	Smart meters are multi-dimensional technical devices which can be upgraded with the help of integration of new information. The bidirectional and automatic communication between the operator and the utility can be done with the help of smart meters. Conventional meters can only show the amount of energy spent whereas the smart meter can also pass data back and forth to and from the operator. A Smart meter enables the user for accessing the data remotely, offers connection between the device and the premise-based network and use real-time data	[34]
3	IED—(Intelligent electronic devices)	IED are the devices that can process data and can interact with other external devices. They are the component of automation which are automatic in nature. They are used to perform jobs that were done by humans previously. They are more likely to commit error in large systems complex in nature	[35]
4	Wide area protection systems	WAPS works on two motives namely relay protection and security with stability control. In security and stability WAPS can stop long term voltage collapse. It does not need real-time data exchange. It requires more complex calculation and has a wider geographical scope as compared to conventional strategies	[36]

(continued)

Table 2 (continued)

S. No.	Technologies	Description	References
5	PMU (Phasor Measurement unit)	It serves the purpose of providing frequent positive sequence of voltage and current in the matter of microseconds. It makes use of GPS and uses the data processing technique used for computer relaying purposes. These systems are also used to measure the rate of change of frequency along with local frequency. It can also be modified for measuring harmonics, zero and negative sequencing and phase current as well as individual voltages and currents also	[37]
6	FACTS (Flexible AC transmission system)	It is developed based on power electronics and offers a chance to develop the ability to control, stabilize and transfer power of AC transmission system. It was pioneered by Hingoraniin 1987 and has been a subject to numerous studies ever since. It was initially used for controlling transmission network in remote places. Distribution grids consisting of lower voltage and power rating and are near to the customers and have fast power switches can be incorporated into FACTS devices	[38]
7	DMS (Distribution management system)	DMS is a method of monitoring the flow of energy in a smart grid. As there is also a development in the smart grid the DMS should also evolve. The adaptations that a DMS should make are as follows bidirectional power flow consideration, load estimation and management, forecasting customer behavior and tools for incorporating field device data	[39]
8	EMS (Energy management system)	It performs the aggregation of energy resource distribution which helps the smart grid participation in the open market	[40]
9	WASA (Wide area situational awareness)	It is one of the areas of priority for standardization of communication and is an important aspect of wide area monitoring and control system. It is one of the methods to counter instability in power systems and eradication of phenomenon like blackouts	[41]

- EVs emerges as one of the most significant emerging technologies in terms of green and sustainable transport network.
- The implementation of EVs would certainly have a positive impact and far more benefits in terms of lower fuel dependency, major GHG (Greenhouse Gasses) reduction and various emissions of toxic contaminants that are very harmful to the environment.

References

1. H. Gabbar, Design and planning support tool for interconnected micro energy grids. Br. J. Appl. Sci. Technol. **12**(6), 1–15 (2016). https://doi.org/10.9734/bjast/2016/21968
2. M. Amin, North America's electricity infrastructure: are we ready for more perfect storms? IEEE Secur. Priv. **1**(5), 19–25 (2003). https://doi.org/10.1109/msecp.2003.1236231
3. J. Choi, G. Jang, Guest editorial special section on smart grid technology in South Korea. IEEE Trans. Smart Grid **4**(1), 351–352 (2013). https://doi.org/10.1109/tsg.2013.2245531
4. R. Siqueira de Carvalho, P. Kumar Sen, Y. Nag Velaga, L. Feksa Ramos, L. Neves Canha, Communication system design for an advanced metering infrastructure. Sensors **18**(11), 3734 (2018). https://doi.org/10.3390/s18113734
5. R. Mudumbai, S. Dasgupta, B. Cho, Distributed control for optimal economic dispatch of a network of heterogeneous power generators. IEEE Trans. Power Syst. **27**(4), 1750–1760 (2012). https://doi.org/10.1109/tpwrs.2012.2188048
6. H. Gharavi, R. Ghafurian, Smart grid: the electric energy system of the future [scanning the issue]. Proc. IEEE **99**(6), 917–921 (2011). https://doi.org/10.1109/jproc.2011.2124210
7. E. Bijami, M. Farsangi, K. Lee, Distributed control of networked wide-area systems: a power system application. IEEE Trans. Smart Grid, 1–1 (2019). 10.1109/tsg.2019.2961111
8. A. Nigam, I. Kaur, K.K. Sharma, Smart grid technology: a review. Int. J. Recent Technol. Eng. (IJRTE) **7**(6S4) (2019). ISSN: 2277-3878
9. X. Yu, C. Cecati, T. Dillon, M. Simões, The new frontier of smart grids. IEEE Ind. Electron. Mag. **5**(3), 49–63 (2011). https://doi.org/10.1109/mie.2011.942176
10. P. Das, R. Choudhary, A. Sanyal, Review report on multi-agent system control analysis for smart grid system. SSRN Electron. J. (2020). https://doi.org/10.2139/ssrn.3517356
11. K.M. Ravi Eswar, Smart grid-future for electrical systems. Int. J. Electr. Electron. Res. **3**(2), 603–612 (2015)
12. D.M. Souran, H.H. Safe, B.G. Moghadam, M. Ghasempour, P.T. Heravi, Smart grid technology in power systems, in *Advances in Intelligent Systems and Computing*, vol. 357 (2016). 10.1007/978-3-319-18416-6_109
13. M.G. Rahman, M.F. Chowdhury, M.A. Al Mamun, M.R. Hasan, S. Mahfuz, Summary of smart grid: benefits and issues. Int. J. Sci. Eng. Res. **4**(3), 1–5 (2013). ISSN 2229-5518
14. S. Massoud Amin, Smart grid: overview, issues and opportunities. Advances and challenges in sensing, modeling, simulation, optimization and control. Eur. J. Control **2011**(5–6), 547–567 (2011). https://doi.org/10.3166/EJC.17.547-567
15. Kotub Uddin, Matthieu Dubarry, Mark B. Glick, The viability of vehicle-to-grid operations from a battery technology and policy perspective. Energy Policy **113**(2018), 342–347 (2018). https://doi.org/10.1016/j.enpol.2017.11.015
16. R. Deivanayagam, Vehicle-to-grid technology: concept, status, and challenges (2017). https://doi.org/10.5210/jur.v10i1.8013
17. Benjamin K. Sovacool, Jonn Axsen, Willett Kempton, The future promise of vehicle-to-grid (v2g) integration: a sociotechnical review and research agenda. Annu. Rev. Environ. Resour. **2017**(42), 377–406 (2017). https://doi.org/10.1146/annurev-environ-030117-020220

18. D. Lauinger, F. Vuille, D. Kuhn, A review of the state of research on vehicle-to-grid (V2G): progress and barriers to deployment, in *Conference: European Battery, Hybrid and Fuel Cell Electric Vehicle Congress* (2017)
19. Tugrul U. Daim, Xiaowen Wang, Kelly Cowan, Tom Shott, Technology roadmap for smart electric vehicle-to-grid (V2G) of residential chargers. J. Innov. Entrepreneurship **5**, 15 (2016). https://doi.org/10.1186/s13731-016-0043-y
20. L. Shi, T. Lv, Y. Wang, Vehicle-to-grid service development logic and management formulation. J. Mod. Power Syst. Clean Energy **7**(4), 935–947 (2019). https://doi.org/10.1007/s40565-018-0464-7
21. B.K. Sovacool, L. Noel, J. Axsen, W. Kempton, The neglected social dimensions to a vehicle-to-grid (V2G) transition: a critical and systematic review. Environ. Res. Lett. **13**(1), 013001 (2018). https://doi.org/10.1088/1748-9326/aa9c6d
22. S. Shariff, D. Iqbal, M. Saad Alam, F. Ahmad, A state-of-the-art review of electric vehicle to grid (V2G) technology. IOP Conf. Ser. Mater. Sci. Eng. **561**, 012103 (2019). https://doi.org/10.1088/1757-899x/561/1/012103
23. K. Ramalingam, C.S. Indulkar, *Overview of Plug-in Electric Vehicle Technologies* (2015). 10.1007/978-981-287-299-9_1
24. P. Olivella-Rosell, R. Villafafila-Robles, A. Sumper, *Impact Evaluation of Plug-in Electric Vehicles on Power System* (2015). https://doi.org/10.1007/978-981-287-299-9_6
25. E. Bitar, P. Khargonekar, K. Poolla, Systems and control opportunities in the integration of renewable energy into the smart grid. IFAC Proc. Volumes **44**(1), 4927–4932 (2011). https://doi.org/10.3182/20110828-6-it-1002.01244
26. N. Phuangpornpitaka, S. Tia, Opportunities and challenges of integrating renewable energy in smart grid system. Energy Procedia **34**(2013), 282–290 (2013). https://doi.org/10.1016/j.egypro.2013.06.756
27. M.A. Islam, M. Hasanuzzaman, N.A. Rahim, A. Nahar, M. Hosenuzzaman, Global renewable energy-based electricity generation and smart grid system for energy security. e Sci. World J. (2014). http://dx.doi.org/10.1155/2014/197136
28. K.N. Nwaigwe, P. Mutabilwa, E. Dintwa, An overview of solar power (PV systems) integration into electricity grids. Mater. Sci. Energy Technol. **2**(2019), 629–633 (2019). https://doi.org/10.1016/j.mset.2019.07.002
29. Antonello Gaviano, Karl Weber, Christian Dirmeier, Challenges and integration of PV and wind energy facilities from a smart grid point of view. Energy Procedia **25**(2012), 118–125 (2012). https://doi.org/10.1016/j.egypro.2012.07.016
30. P. Barua, T. Chowdhury, H. Chowdhury, S. Hossain, M. Dey, A study on renewable energy in smart grid, in *National Conference on Energy Technology, and Industrial Automation 2018 (NCETIA2018)* 13th December 2018
31. E.F. Camacho, A.J. del Real, C. Bordons, A. Arce, Solar thermal plants integration in smart grids, in *Proceedings of the 18th World Congress the International Federation of Automatic Control Milano* (Italy), 28 August 28–2 September 2011
32. A.H. Bagdadee, L. Zhang, Smart grid: a brief assessment of the smart grid technologies for modern power system. J. Eng. Technol. **8**(1), 122–142 (2019)
33. M. Eissa, Editorial (Thematic issue: "renewable resources and challenges for smart grid"). Recent Adv. Commun. Netw. Technol. **4**(2), 68–68 (2016). https://doi.org/10.2174/221508110402160406105328
34. G. Shamim, M. Rihan, Multi-domain feature extraction for improved clustering of smart meter data. Technol. Econ. Smart Grids Sustain. Energy **5**, 1–8 (2020). https://doi.org/10.1007/s40866-020-00080-w
35. A. Friedanian, L. Cameron, A framework for implementation of adaptive autonomy for intelligent electronic devices. J. Appl. Sci. (2008). https://doi.org/10.3923/jas.2008.3721.3726
36. L. Luo, N. Tai, G. Yang, Wide-area protection research in the smart grid. Energy Procedia **16**(2012), 1601–1606 (2012). https://doi.org/10.1016/j.egypro.2012.01.249
37. K.S. Chandarani Sutar, Verma: calibration of dynamic line rating model for phasor measurement unit based wide area measurement system in smart grid application. Int. J. Recent Technol. Eng. **8**(4), 1879–1883 (2019). https://doi.org/10.35940/ijrte.c4621.118419

38. C. Bhatia, M. Rao, P. Kumar, A. Khare, Intelligent computer control of flexible AC transmission system (FACTS). IETE J. Res. **41**(2), 135–141 (1995). https://doi.org/10.1080/03772063.1995.11437238
39. J. Leite, J. Mantovani, Development of a smart grid simulation environment, part II: implementation of the advanced distribution management system. J. Control Autom. Electr. Syst. **26**(1), 96–104 (2014). https://doi.org/10.1007/s40313-014-0158-y
40. C. Cecati, C. Citro, P. Siano, Combined operations of renewable energy systems and responsive demand in a smart grid. IEEE Trans. Sustain. Energy **2**(4), 468–476 (2011). https://doi.org/10.1109/tste.2011.2161624
41. C. Alcaraz, J. Lopez, Wide-area situational awareness for critical infrastructure protection. Computer **46**(4), 30–37 (2013). https://doi.org/10.1109/mc.2013.72
42. R. Abe, H. Taoka, D. McQuilkin, Digital grid: communicative electrical grids of the future. IEEE Trans. Smart Grid **2**(2), 399–410 (2011). https://doi.org/10.1109/tsg.2011.2132744
43. S. Vadi, R. Bayindir, A.M. Colak, E. Hossain, A review on communication standards and charging topologies of V2G and V2H operation strategies. Energies **12**(19), 3748 (2019). https://doi.org/10.3390/en12193748
44. H. Lund, Corrigendum to "Renewable heating strategies and their consequences for storage and grid infrastructures comparing a smart grid to a smart energy systems approach" [Energy 151 (2018) 94–102]. Energy **153**, 1087 (2018). https://doi.org/10.1016/j.energy.2018.04.159
45. A. Almehizia, H. Al-Masri, M. Ehsani, Integration of renewable energy sources by load shifting and utilizing value storage. IEEE Trans. Smart Grid **10**(5), 4974–4984 (2019). https://doi.org/10.1109/tsg.2018.2871806
46. N. Priyadarshi, F. Azam, A.K. Bhoi, A.K. Sharma, Dynamic operation of grid-connected photovoltaic power system, in *Advances in Greener Energy Technologies. Green Energy and Technology* (Springer, Singapore, 2020). https://doi.org/10.1007/978-981-15-4246-6_13
47. N. Patel, D. Porwal, A.K. Bhoi, D.P. Kothari, A. Kalam, An overview on structural advancements in conventional power system with renewable energy integration and role of smart grids in future power corridors, in *Advances in Greener Energy Technologies. Green Energy and Technology* (Springer, Singapore, 2020). https://doi.org/10.1007/978-981-15-4246-6_1

A New Series-Parallel Switched Capacitor Configuration of a DC–DC Converter for Variable Voltage Applications

B. Hemanth Kumar, A. Bhavani, C. V. Jeevithesh, Sanjeevikumar Padmanaban, and Vivekanandan Subburaj

1 Introduction

DC–DC converters are designed to supply a fixed output voltage from a fluctuating or a variable input DC voltage. DC–DC converters are classified into two types, linear regulators and switch converters. The output current in a linear regulator is fed directly from the power source, hence the performance is expressed as fraction of the output to the input voltages. It is clear that, lower efficiency is achieved, if the output voltage is much smaller than the supply voltage. Due to intercepted energy transmission, switching converters are more powerful than linear regulators. This is achieved by constantly swapping components from energy storage to transfer a portion of power from the input source to the load end. DC–DC converters may be classified into two broad classes (except for resonant converters): capacitive and inductive.

Owing to the broad range of potential in voltage and current conditions, inductive converters utilizing one or many inductors based on the load requirement for different applications. As inductors are usually large and not able to fabricate on-chip for IC applications. Two severe demerits in DC–DC converters based on inductors can be addressed as follows: one problem is due to high voltage spikes in these converters, which either need to be damped or repaired, but the power electronic switches that are not capable for these limitations will burst, while the rest of the circuit may be destroyed. The other inductive converter problem is the input current is pulsating in nature in these DC–DC converters, which may create EMI from other lines of equipment and conductors. This intrusion can penetrate sensitive devices and trigger

B. Hemanth Kumar · A. Bhavani · C. V. Jeevithesh · V. Subburaj (✉)
Department of Electrical Engineering, Sree Vidyanikethan Engineering College, Tirupathi, A.P, India
e-mail: vivekeee810@gmail.com

S. Padmanaban
Aalborg University, 6700 Aalborg, Denmark

N. Patel et al. (eds.), *Electric Vehicles*, Green Energy and Technology,
https://doi.org/10.1007/978-981-15-9251-5_15

irregular action. So, a special filter and often shielding is needed for the pulsating input present. Both these considerations increases the size of the PCB and the expense of inductive converters.

DC–DC converters based on capacitors is said to be capacitive converters, which are commonly used in applications needing low power and does not require any isolation. These circuits have fairly minimum noise, minimally radiated EMI, and are easily fabricated in ICs, which makes these type of DC–DC converters are highly used in mobile device power source management and Electric Vehicles (EVs). In EVs switched capacitor converter [1] plays a vital role for controlling the speed of the vehicle and supporting the electronics loads such as LEDs, head lamps and audio systems. An additional property of these converters is the no loaded operating choice with no need for dummy loads. On the other hand, these circuits experience from inbuilt power loss while discharging and charging of capacitors, which are in parallel to the other capacitor or source voltage. Theory assumes that the power loss is equal to the difference in square voltage that happens prior to the related circuit is configured. Therefore, these DC–DC converters show a very high output in compared to components that hold identical voltages are the capacitors pre-charged to those voltages.

The well-known DC–DC capacitive based converter is said to be as charge pump; it is also seen, for historical purposes, as a boost converter made from diodes and capacitors along with power electronic switches. Currently, by using transistor switches and capacitors, charge pump circuits are designed, it's design in principle does not vary from switching DC–DC capacitor converters. The foundation pillar in all circuits is the reconfigurable series of capacitors and switches commonly referred to as flying capacitors. All such capacitors are charged through input voltage, and it is an excellently-known fact that while a capacitive DC-DC converter works at the desired voltage ratios (output voltage/source voltage), the efficiency of such converters is good and is almost 90%. It is since the capacitors do not have appreciable voltage fluctuations at such voltage levels. The efficiency decreases significantly when the same capacitive converter works under the target voltage ratios or above them.

Inverting switched capacitive DC–DC converters (ISCC) were used to monitor white light emitting diodes (LEDs) in black light [2]. Capacitive DC–DC circuit is a low-power converter capable to produce a negative or positive boost or buck voltage. SCC functions as a Low Drop-out (LDO) or a voltage regulator in certain systems. White LEDs has a perfect show of black light for mobile phones and computer screens. These drivers may either be introduced using switched capacitor or inductor DC–DC converters. Certainly, the inductive converter offers higher efficiency. However, inductors are larger in size, allowing them undesirable for PCB's [3]. Conversely, for all low-power applications, SCC has smaller bulk, low cost and less battery consumption. Inverting switched capacitive circuits plays a key function in wireless micro-power applications. It is primarily used for black-lights operating the WLEDs under various voltage conditions.

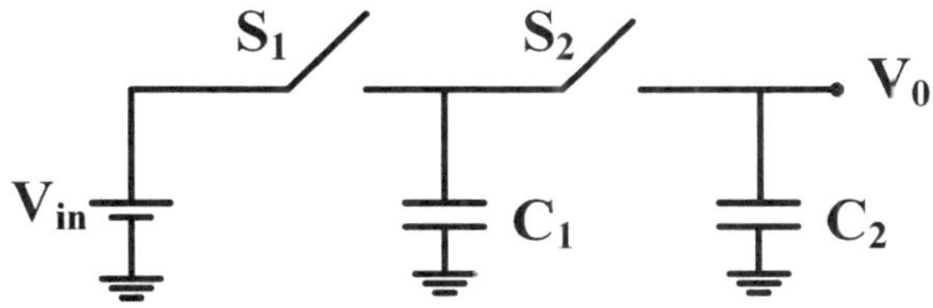

Fig. 1 Basic charge pump circuit

1.1 Charge Pump Simplified Analysis

Basically, Charge Pump (CP) is an electrical circuit, which transforms input DC source (V_{in}) into the output DC voltage, which is multiple times higher or lower than V_{in}. Unlike conventional DC–DC inductive converters, CPs are composed of switches (or diodes) and capacitors only, thus facilitating easily in PCB circuits [4]. Initially CP's, are used in power management IC's [5] and non-volatile memories [6]. CPs were often used in a broad variety of embedded devices, including switched capacitor circuits, voltage regulators, mechanical amplifiers, SRAMs, LCD drivers, RF antenna transfer controls, piezoelectric actuators, etc., due to continuous reduction of integrated circuit power supplies.

1.2 One-Stage Charge Pump

A simple load pump circuit illustrates a CP circuit, which has two switches that are switched OFF and ON successively is shown in Fig. 1. Once the switch S_1 is switched ON, the input voltage is appeared across the capacitor C_1 (Charging state) and when the switch S_2 is switched OFF, the energy of capacitor C_1 passes to capacitor C_2 and the switches are alternately controlled by using PWM signals. This process is termed transferring the energy among many capacitors and the output voltage approaches the input voltage point over many cycles [7]. For example, if 10 V is given as input to V_{in}, then the output is obtained as 10 V since the charge pump circuit charges and just transfers the input voltage to the output by charging and discharging. The pulses for the switches is given by PWM signal S_1 is in ON state from 0° to 180°. Whereas, S_2 conducts from 180° to 360°.

1.3 Doubler Circuit

The circuit for two-phase SC voltage doubler, simply called as a doubler circuit is shown in Fig. 2. The Capacitor voltage (V_{C_1}) is charged up to V_{in} in the first step, as shown in Fig. 2. In the second step, capacitor C_1 is positioned with the source input in cascade, typically doubling the level of output voltage [8]. Where 10 V is given

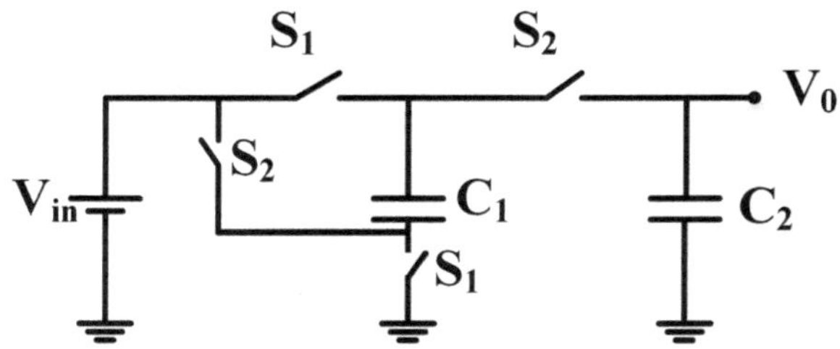

Fig. 2 Doubler circuit

as input to V_{in} and the output is obtained as 20 V since the doubler circuit doubles the input voltage applied to the circuit. The pulses for the switches is given by PWM signal S_1 is in ON state from 0° to 180°. Whereas, S_2 conducts from 180° to 360°.

The doubler circuit may be coupled in cascaded, in order to get higher output voltages, which is shown in the Fig. 3. Here, each stage's output voltage is twice the input voltage to the respective stage. So, the overall circuit V_o is four times the V_{in}. The output voltage for the doubler circuit connected in the series if the voltage applied to the input terminal is 10 V and therefore the output obtained is 40 V since the circuit boost the voltage four times the input value. The pulses for the switches is given by PWM signal S_1 is in ON state from 0° to 180° degree. Whereas, S_2 conducts from 180° to 360°.

Based on the knowledge of charge pump and voltage divider circuits, several topologies of SC based DC–DC converters are presented in the literature [9–13]. Detailed review of existing topologies of various SCC circuits are explained in [14]. These topologies are aimed for low-power applications. The performance limitations of such converters are presented in [15].

These SCC DC–DC topologies can also used for high and medium power applications. The output of such DC–DC converter is given to DC–AC inverters to obtain AC voltage by controlling the inverter with certain PWM techniques [16–18] for better conversion of DC to AC supply. High and medium power DC–DC converters are used for integration of renewable energy sources to the grid.

In this chapter, an SCC DC–DC converter is proposed which is capable of utilizing the input source to generate different output voltage ratios. The uses for the converter

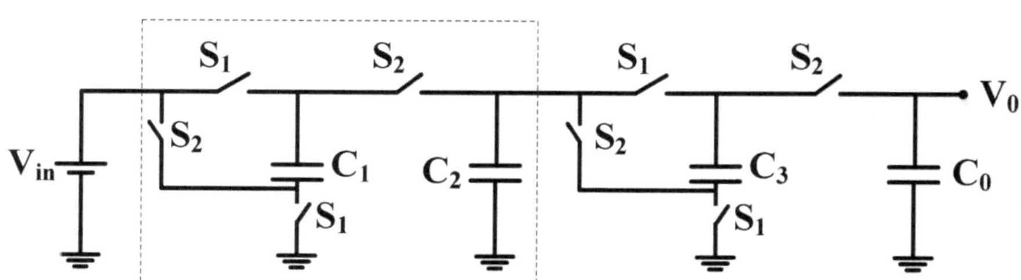

Fig. 3 TPVD circuit connected in series

proposed are alternative energy sources such as solar PV or a mix of battery and solar panels. The converter has the potential to reconfigure its benefit by selectively enabling converter switches, utilizing variable circuit structure. In a predetermined pattern, similar capacitors and switches are reused and accordingly joined in order to generate the appropriate output voltage with less number of the circuit elements. The proposed SCC converter utilizes four capacitors and 14 power electronic switches (MOSFET) to obtain the necessary output voltage, which has a capacity to produce 21 different output voltage ratios based on the proper switching sequence.

The Chapter is organized as the following sections. The proposed topology of the Switched Capacitor DC–DC circuit and its switching sequence of different modes of performance gain are covered in Sect. 2. The theoretical and conceptual basis for establishing the analogous resistance validation are discussed in Sect. 3 for the proposed SCC. Section 4 describes the simulated analysis of the proposed SCC by using PSIM software. Followed by the conclusion and references.

2 Proposed Converter

The SCC circuit shown in Fig. 4 is the presented capacitive DC–DC converter, which accepts the broad range of input voltages, which uses 14 bidirectional switches, four flying capacitors and one output capacitor. The load is connected at the output side. The proposed circuit provides 21 voltage conversion ratios such as 5/1, 4/1, 3/1, 2/1, 1/2, 2/3, 1/5, 1/11, 1/21, 1/31, 1/41, 1/16, 3/43, 2/7, 3/13, 1/6, −1/2, −1, −2/1, −3/1, and −4/1 which are as shown in Table 1. The 14 active switches of MOSFETs are used to achieve the required output with only one input voltage. Such switches are bidirectional, conducting current in all directions, the switch is OFF when the gate pulse is low and ON when the gate pulse is high, regardless of the switch's voltage bias conditions.

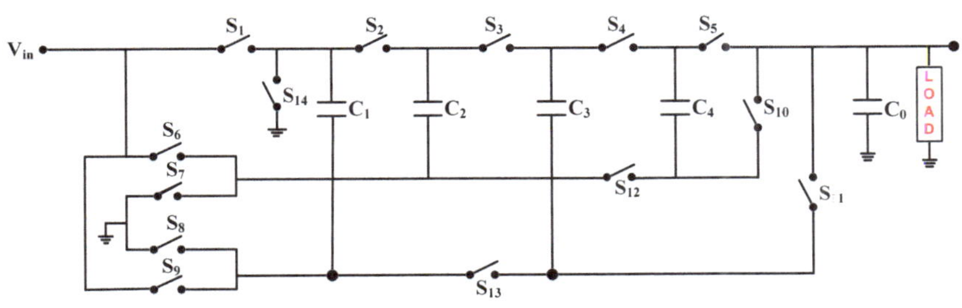

Fig. 4 Proposed SC based DC–DC converter

Table 1 Switching logic of the proposed circuit

Conversion mode	S_1	S_2	S_3	S_4	S_5	S_6	S_7	S_8	S_9	S_{10}	S_{11}	S_{12}	S_{13}	S_{14}
5 up	C_1	C_2	C_1	C_2	C_1	C_1	C_2	C_1	C_2	X	X	O	O	X
4 up	O	C_2	C_1	C_2	C_1	C_1	C_2	C_1	C_2	X	X	O	O	X
3 up	O	O	C_1	C_2	C_1	C_1	C_2	C_1	C_2	X	X	O	O	X
2 up	O	O	O	C_2	C_1	C_1	C_2	C_2	C_2	X	X	O	O	X
1	O	O	O	O	O	X	X	X	X	X	X	X	X	X
3/2 up	C_1	C_1	C_2	O	C_2	C_2	X	C_1	C_2	C_2	X	C_1	O	X
1/2 down	C_1	C_1	C_2	C_2	C_2	X	C_2	X	X	C_1	X	C_1	X	X
2/3 down	C_1	C_1	C_2	C_1	C_1	X	C_2	X	C_1	C_1	C_2	C_1	C_1	X
1/5 down	C_1	C_1	C_2	C_2	X	X	C_2	X	X	O	X	C_1	X	X
1/11 down	C_1	X	X	X	X	X	X	C_2	X	X	C_1	X	C_1	X
1/21 down	X	X	O	O	X	C_1	C_2	X	X	C_1	X	X	C_1	X
1/31 down	X	O	X	X	C_1	X	X	C_2	C_1	X	X	O	C_1	X
1/41 down	X	O	X	O	X	C_1	C_2	X	X	C_1	X	X	C_1	X
1/16 down	X	O	O	O	X	X	X	C_2	C_1	C_1	X	O	C_1	X
3/43 down	X	O	O	O	X	X	X	C_2	C_1	C_1	C_1	O	X	X
2/7 down	X	O	O	O	C_1	C_1	C_2	C_2	$C1$	X	X	O	O	X
3/13 down	X	X	O	O	C_1	C_1	C_2	C_2	$C1$	X	X	O	O	X
1/6 down	X	X	X	O	C_1	C_1	C_2	C_2	$C1$	X	X	O	O	X
−1/2 down	X	C_2	C_1	C_1	C_1	C_2	C_1	C_2	X	X	C_1	X	C_1	C_1
−1	X	C_1	C_1	C_1	C_1	X	X	C_1	C_2	X	X	X	X	C_2
−2 up	X	C_1	C_2	C_2	C_2	C_1	C_2	C_1	C_2	X	X	X	X	C_2
−3 up	X	C_1	C_2	C_1	C_1	C_1	C_2	C_1	C_2	X	X	X	O	C_2
−4 up	X	C_1	C_2	C_2	C_2	C_1	C_2	C_1	C_2	X	X	O	O	C_2

2.1 Circuit Description

The proposed circuit uses four flying capacitors of 4.7 μF each for charging and discharging the voltages and one output capacitor which is ten times of flying capacitors value is used for filtering any ripples in the output voltage. The 14 bidirectional semiconductor switches in the proposed SCC are operated individually by using PWM signals for generating different voltage ratios. While voltage conversion is performing by the four capacitors which charges and discharges respectively in the SC converter.

Table 1 shows the switching logic for the proposed circuit, which provides 21 voltage ratios. Out of 24, 4 are up modes i.e., which boosts the input voltage to a higher voltage based on the corresponding switching logic, 12 are down ratios which bucks the input voltage and the remaining 5 are of negative ratios.

Table 2 Clock signal states

States	Duty cycle	Phase delay
C_1	0.5	0
C_2	0.5	180
O	1	0
X	0	0

The proposed switched capacitor converter circuit has the potential to reconfigure circuit's structure such that the output voltage varies by selectively triggering switches through adjusting the pulses provided to the switches, which in effect generates both positive and negative voltage ratios. In a fixed pattern, similar capacitors and switches are repeated and are joined to obtain the necessary output voltage ratio with optimizing the usage of elements in the SCC.

The switching mode is chosen to limit the voltage at the output received from the output terminal (V_o) to a possible value according to the input voltage point. As seen in Table 2, the clock signal notes that state C_1 means switching PWM signals, and the C_2 is the inversion of the C_1 PWM signal. The notation X indicates that particular switch is OFF completely and notation O indicates that the particular switch is ON.

The PWM signal can be either from the IGBT, MOSFET, Gate driver circuit. In these we have taken from the MOSFET switch. Two phases 1 and 2 are used to control bidirectional MOSFET switches. Phase 1 and phase 2 are 50% non-overlapping duty cycle clock pulses, whereas phase 2 is discharging phase and phase 1 is considered to be the charging phase.

2.2 *Modes of Operation*

2.2.1 Working Operation for Positive Voltage Ratios

The Working operation for the 5 V up mode voltage ratio is discussed in detail as follows, where phase 2 is the discharging state and the phase 1 is considered to be the charging phase. During the phase 1 charging phase as shown in the Fig. 5. In SCC Circuit diagram during phase 1 for 5 V up mode the switches S_1, S_3, S_5, S_6, S_8, S_{12} and S_{13} are in ON state and the switches S_2, S_4, S_7, S_9, S_{10}, S_{11} and S_{14} are in OFF state, i.e., it doesn't conduct during the phase 1. During the phase 2 discharging phase as shown in the Fig. 6. SCC circuit diagram during the phase 2 for 5 V up mode the switches S_1, S_3, S_5, S_6, S_8, S_{12} and S_{13} are in non-conducting state (OFF) and the switches S_2, S_4, S_7, S_9, S_{10}, S_{11} and S_{14} are in conducting state (ON).

When the switches of phase 1 are ON, the capacitor C_1 appeared parallel to input voltage (V_{in}). Therefore, the capacitor voltage (V_{C_1}) is same as the V_{in}. When the switches of phase 2 are on, the capacitor C_2 is placed in the series with C_1 and V_{in}. Then the current direction is reversed and capacitor C_1 will deliver the charge to the capacitor C_2. This cycle repeats itself by discharge of capacitor C_2 to capacitor C_3

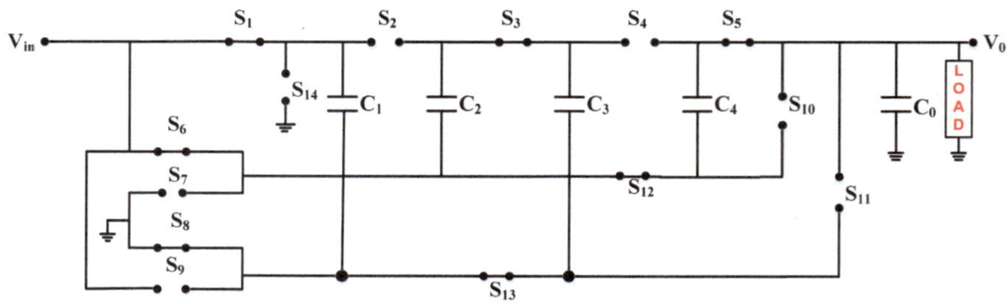

Fig. 5 SCC Circuit diagram during phase1 for 5 V up mode

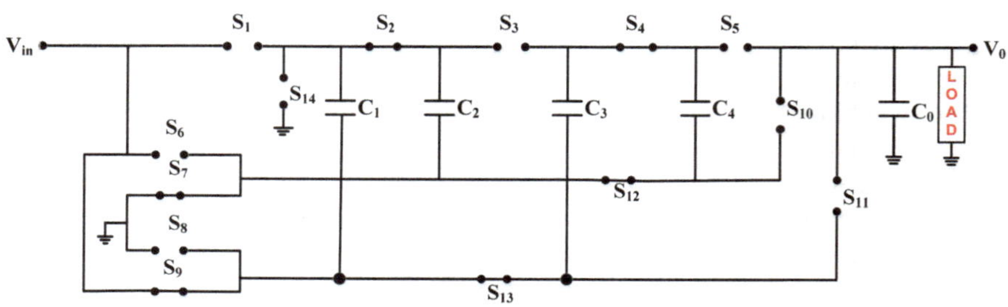

Fig. 6 SCC circuit diagram during phase 2 for 5 V up mode

in phase 1 and the discharge of C_3 to C_4 in phase 2. As a result, the voltage gets multiplied and appears across the output capacitor.

The increase in voltage is explained by the following terms. During phase 1, V_{in} is appearing across C_1 is shown in Fig. 7.

$$\Rightarrow q = C \times V$$

$$\Rightarrow q_1 = C_1 \times V_{\text{in}}$$

Let $V_{\text{in}} = 1$ V and $C_1 = 1\ \mu\text{F}$

$$\Rightarrow q_1 = 1\ \mu \times 1 = 1\ \mu\text{C}$$

During phase 2 (Fig. 8).

As, $q_1 = 1\ \mu\text{C}$, $C_1 = 1\ \mu\text{F}$

Fig. 7 Capacitor connected across the voltage source

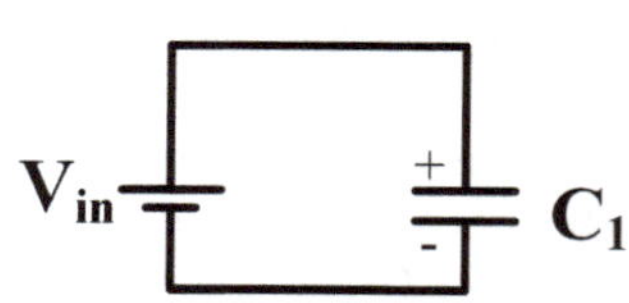

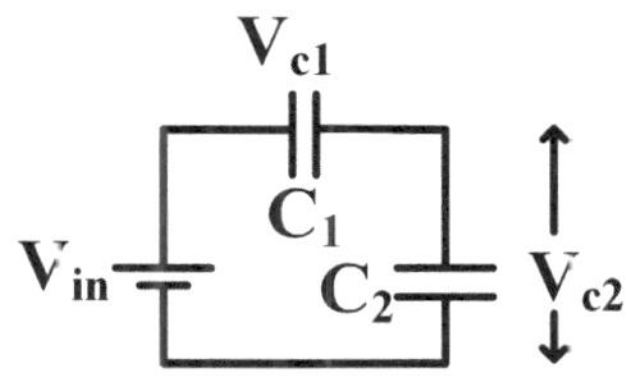

Fig. 8 Circuit during phase 2

$$\Rightarrow V_{c_1} = \frac{q_1}{C_1} = 1\text{ V}$$

By applying KVL we get,

$$V_{c_2} = V_{\text{in}} + V_{c1}$$

$$= 1 + 1$$

$$= 2V$$

As V_{C_2} is obtained 2 V, therefore the voltage across the C_2 is appeared as 2 V, then the capacitor C_2 will be charged to 2 V. Thus, the stored energy in the capacitor C_1 is discharged to C_2. The cycle repeats accordingly, and the capacitor C_4 will be charged to 4 V. As a result $V_{\text{in}} + V_{C4}$ will be appeared across the output capacitor. Which will be 5 V at the output terminal. Therefore the output obtained is five times the input voltage value.

2.2.2 Working Operation for Negative Voltage Ratios

The capacitors are charged via negative terminals for the negative voltage ratios and discharged through positive terminals to produce negative voltage conversion ratios. For each VCR, an individual gate is selected and used to generate signals to control bidirectional switches. For ease, $-2 * V_{\text{in}}$ Voltage conversion ratio is considered for designing the converter.

The switches are connected to charge (S_3, S_4, S_5, S_7, S_9 and S_{14}) and discharge (S_2, S_6 and S_8) during both Phases respectively. The Equivalent circuit of SCC for both Phases is depicted in the Figs. 9 and 10. Figure 9 shows the SCC circuit diagram during phase 1 for −2 V and the Fig. 10 shows the SCC circuit diagram during phase 2 for −2 V. As explained in the above calculations for 5 V up mode how the input voltage occurs at the output terminal similarly, it occurs for the negative voltage ratios also.

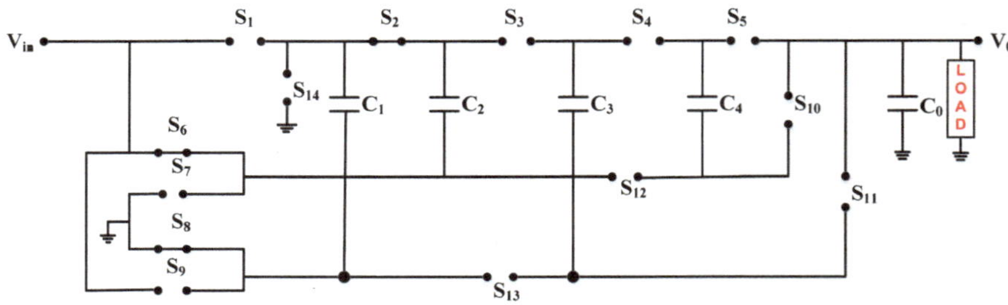

Fig. 9 SCC circuit diagram during phase 1 for −2 V

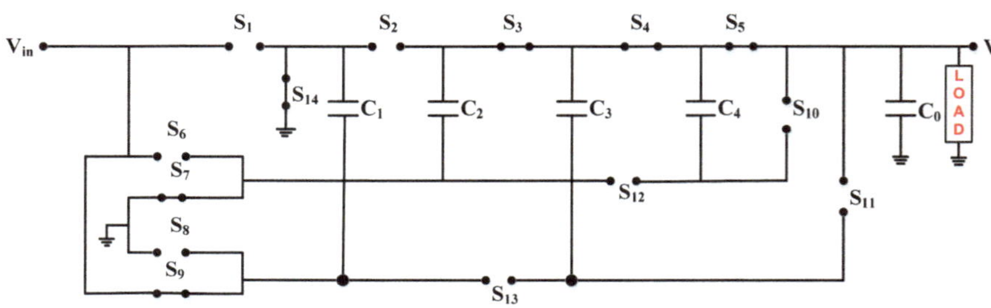

Fig. 10 SCC circuit diagram during phase 2 for −2 V

3 Analysis of the Proposed SCC

R_{eq} is helpful in measuring ohmic and conduction losses in power electronic switches used in the proposed SCC. In [19, 20], presented the analogous ISCC circuit model for an average voltage output ($V_{out} = -n * V_{in}$). Here n is desired voltage ratio and V_{in} is the supply voltage.

Measures to be taken to establish a linear equation to describe the function of the charge flow on SCC at each time of the switches are

1. Apply Kirchhoff current law (KCL).
2. Consider the steady state error value as zero.
3. The cumulative charge is applied to numerical calculations and is equated.

In every SCC operating phase I, which corresponds to a T_i time period., If the frequency of switching pulse is low, the corresponding time duration is high, otherwise the capacitors may be completely discharged and charged, this condition is considered as CC (Complete Charge) or SSL (slow switching limit) mode. Alternatively, as the frequency of switching pulse is high, which means time interval is low and thus the capacitor can never be charged fully, this condition is considered as NC (No charging) or FSL (fast switching limit mode. Likewise, if the charging and discharged is partly done, it must work in the condition of the PC (Partial Charging) mode. In any mode, i.e., PC, NC or CC, operating sub circuit equivalent resistance is dependent on R_iC_i and T_i, where C_i is the equivalent capacitance value in farads and R_i is the equivalent resistance value of capacitance ESR and all the active switches with respect to the time period T_i during the respective ith operating mode.

The analysis of R_{eq} is examined in the literature [20], but they produced estimated results using infinite filter capacitances.

The most popular way to obtain the R_{eq} mathematically as given below,

$$R_{\text{eq}} = \left[(\text{RSSL})^2 + (\text{RFSL})^2\right]^{1/2}$$

where RFSL and RSSL are fast switching and slow switching R_{eq}. On the contrary, another way to obtain R_{eq} mathematically is given as,

$$R_{\text{eq}} = \left[(\text{RSSL})^{2.545} + (\text{RFSL})^{2.545}\right]^{1/2.545}$$

According to above equations, R_{eq} calculations are fairly accurate. Accurate equal resistance is measured and verified using SCC developed for smartphone devices to address these problems.

Many researchers in the literature have presented the work on integrated SCC. In [20], they converted the higher order RC circuits to the first order and obtained R_{eq} for the following VCR.

The mathematical approach is illustrated for 5 V up the VCR for the calculation of R_{eq} is same to calculate R_{eq} for the other VCR's. From the Table 1, Switching logic of the proposed work, in state 1 for a 5 V voltage conversion ratio, switches S_1, S_3, S_5, S_6, S_8, S_{12} and S_{13} are in the charging state (CS) and the switches S_2, S_4, S_7, S_9, S_{10}, S_{11} and S_{14} conduct in the transfer state (TS). In the following the SCC circuits during charging, transfer states and the respective switches and flying capacitors in the charging state (CS) and the transfer state (TS) are shown in Fig. 11.

As the two systems have separate circuit configurations, KCL has to consider a proportionality factor φ_i responsible for the load balance in capacitors.

The major assumptions in [20] are be considered while analyzing the equivalent circuits. Consider R_1, R_2, R_3, R_4 are capacitor equivalent series resistance. Since all the flying capacitor values are same i.e., $C_1 = C_2 = C_3 = C_4$

$$\Rightarrow R_1 = R_2 = R_3 = R_4$$

Let

$$C_1 = C_2 = C_3 = C_4 = C$$

$$R_1 = R_2 = R_3 = R_4 = R$$

From CS,

By applying charge balance and KCL for the flying capacitors C_1, C_2, C_3 and C_4 (1), (2) and (3) can be formulated as,

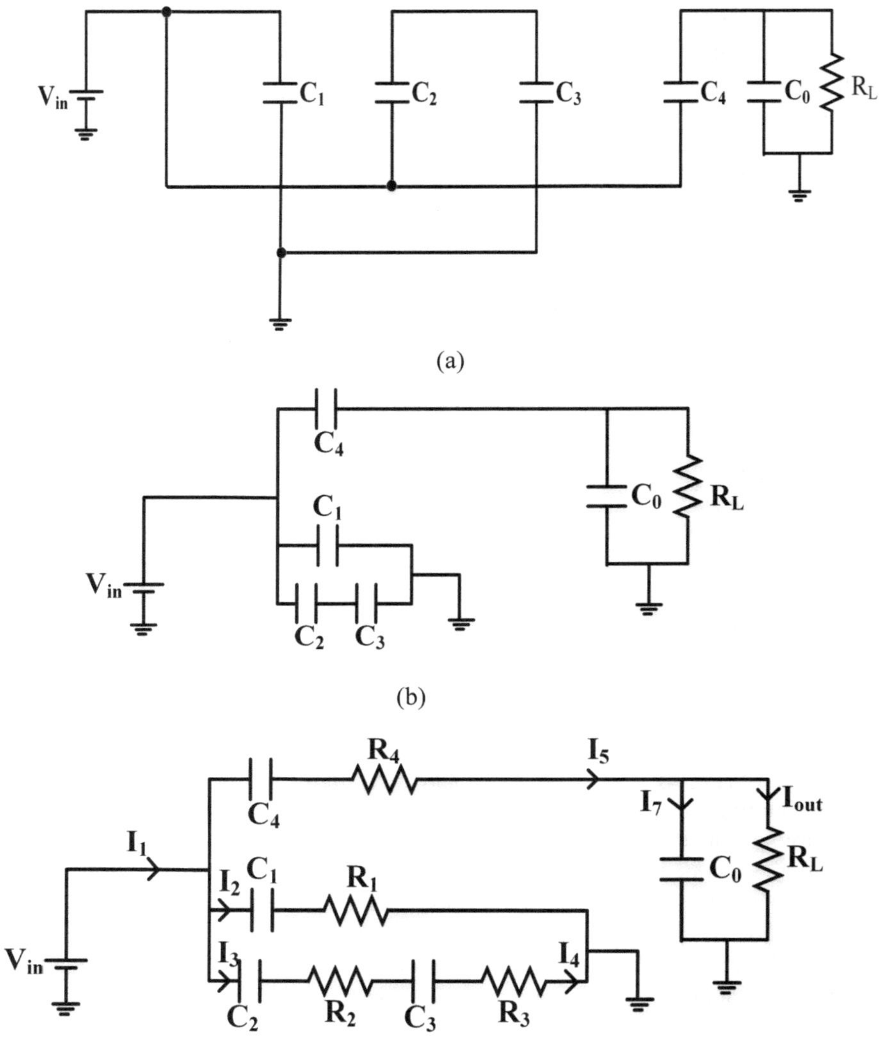

(c)

Fig. 11 Corresponding circuits in state 1 for a VCR of 5 V. **a** Actual circuit in CS. **b** Redrawing the actual circuit in CS. **c** Corresponding circuit in CS. **d** Actual circuit in TS. **e** Corresponding circuit in TS. **f** Modified corresponding circuit in CS. **g** Modified corresponding circuit in TS

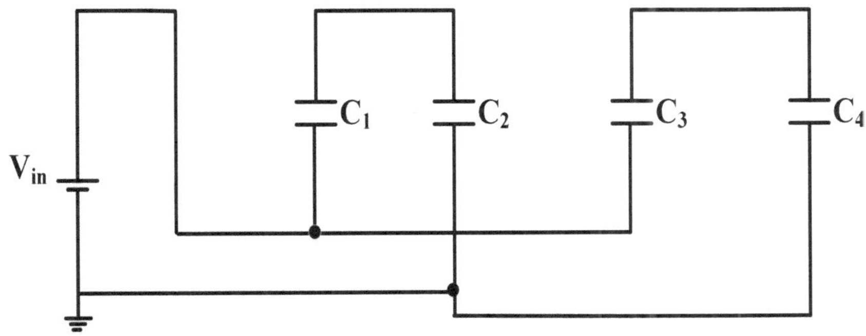

(d)

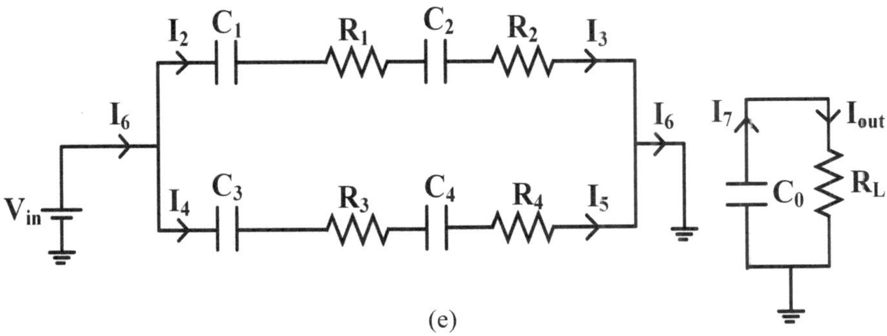

(e)

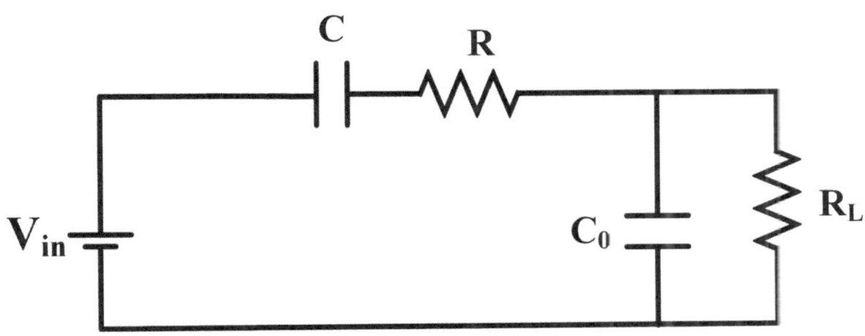

(f)

Fig. 11 (continued)

Fig. 11 (continued)

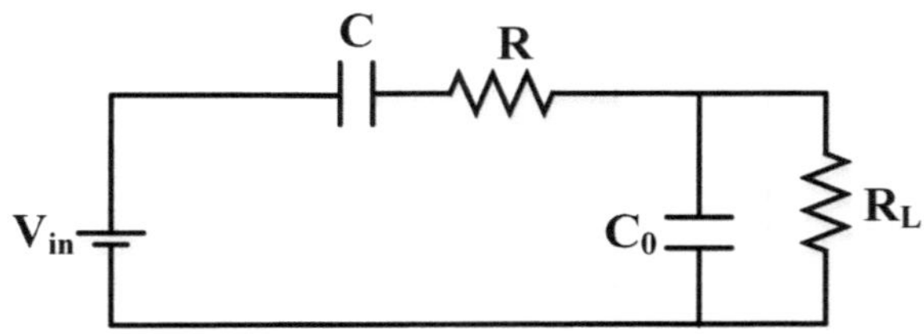

(g)

$$I_1 - I_2 - I_3 - I_5 = 0 \tag{1}$$

$$I_3 - I_4 = 0 \tag{2}$$

$$I_5 - I_7 = I_{out} \tag{3}$$

From TS,
By using charge balance and KCL (4), (5), (6) and (7) can be written as,

$$I_6 - I_2 - I_4 = 0 \tag{4}$$

$$I_2 - I_3 = 0 \tag{5}$$

$$I_4 - I_5 = 0 \tag{6}$$

$$I_7 = I_{out} \tag{7}$$

By solving all the above (1)–(7) equations,

$$I_1 = 6I_{out}$$

$$I_6 = 4I_{out}$$

Arranging the Eqs. (1)–(7) equations in matrix form as,

$$\begin{bmatrix} 1 & -1 & -1 & 0 & -1 & 0 & 0 \\ 0 & 0 & -1 & -1 & 0 & 0 & 0 \\ 0 & -1 & 0 & -1 & 0 & 1 & 0 \\ 0 & 1 & -1 & 0 & 0 & 0 & 0 \\ 0 & 0 & 0 & 1 & -1 & 0 & 0 \\ 0 & 0 & 0 & 0 & 1 & 0 & -1 \\ 0 & 0 & 0 & 0 & 0 & 0 & -1 \end{bmatrix} \begin{bmatrix} I_1 \\ I_2 \\ I_3 \\ I_4 \\ I_5 \\ I_6 \\ I_7 \end{bmatrix} = \begin{bmatrix} 0 \\ 0 \\ 0 \\ 0 \\ 0 \\ I_{\text{out}} \\ I_{\text{out}} \end{bmatrix} \tag{8}$$

$\phi_{\text{CS}}, \phi_{\text{TS}}$ corresponds to both CS and TS respectively.

$$\phi_{\text{CS}} = \frac{I_1}{I_{\text{out}}} = \frac{6I_{\text{out}}}{I_{\text{out}}} = 6 \tag{9}$$

$$\phi_{\text{TS}} = \frac{I_6}{I_{\text{out}}} = \frac{4I_{\text{out}}}{I_{\text{out}}} = 4 \tag{10}$$

The proportionality factor $\phi_{\text{CS}} = 6$ and $\phi_{\text{TS}} = 4$ are determined. Since, $T_i = T_{\text{CS}} = T_{\text{TS}} = 0.5$, then

$$\phi_{\text{CS}} = 3 \tag{11}$$

$$\phi_{\text{TS}} = 2 \tag{12}$$

From Fig. 11f, the equivalent capacitance C_{CS} and equivalent resistance R_{CS} during Charging state (CS) are given by (13) and (14),

$$R_{\text{CS}} = R \tag{13}$$

$$C_{\text{CS}} = \frac{53C}{22} \tag{14}$$

From Fig. 11g, the equivalent capacitance C_{TS} and equivalent resistance R_{TS} during Transfer state (TS) are given by (15) and (16),

$$R_{\text{TS}} = R \tag{15}$$

$$C_{\text{TS}} = C \tag{16}$$

If T_{CS} and T_{TS} are the time period during the charging state and transfer state, then,

$$\lambda_{\text{CS}} = \frac{T_{\text{CS}}}{R_{\text{CS}} \cdot C_{\text{CS}}} = \frac{T_{\text{CS}}}{R \cdot \frac{53C}{22}} \tag{17}$$

$$\lambda_{\text{TS}} = \frac{T_{\text{TS}}}{R_{\text{TS}} \cdot C_{\text{TS}}} = \frac{T_{\text{TS}}}{R \cdot C} \tag{18}$$

Then the output resistance becomes,

$$\begin{aligned} R_{\text{outCS}} &= (\phi_{CS})^2 \cdot \frac{1}{2f.C_{CS}} \cdot \coth\left(\frac{\lambda_{\text{CS}}}{2}\right) \\ &= 9.\frac{1}{2f.\frac{53C}{22}}.\coth(\frac{\lambda_{CS}}{2}) \end{aligned} \tag{19}$$

$$\begin{aligned} R_{\text{outTS}} &= (\phi_{\text{TS}})^2 \cdot \frac{1}{2f.C_{\text{TS}}} \cdot \coth\left(\frac{\lambda_{\text{TS}}}{2}\right) \\ &= 2 \cdot \frac{1}{f \cdot C} \cdot \coth\left(\frac{\lambda_{\text{TS}}}{2}\right) \end{aligned} \tag{20}$$

Hence the equivalent R_{out} of SCC as in Fig. 4 Proposed switched capacitor converter circuit is obtained by combining the Eqs. (19) and (20),

$$R_{\text{eq}} = R_{\text{outCS}} + R_{\text{outTS}}$$

$$R_{\text{eq}} = \frac{9}{2 \cdot f \cdot \frac{53C}{22}} \coth\left(\frac{\lambda_{CS}}{2}\right) + \frac{2}{f \cdot C} \coth\left(\frac{\lambda_{TS}}{2}\right) \tag{21}$$

By substituting all the values in the Eq. (21),

$$\begin{aligned} f &= 100 \times 10^3, \\ R &= 100 \times 10^{-3}, \\ C &= 4.7 \times 10^{-6} \end{aligned}$$

Then, $\lambda_{\text{CS}} = \frac{T_{\text{CS}}}{R_{\text{CS}} \cdot C_{\text{CS}}} = 441589.723$

$$\coth\left(\frac{\lambda_{CS}}{2}\right) = \coth(220794.862) = 1$$

$$\coth\left(\frac{\lambda_{\text{TS}}}{2}\right) = 1$$

Finally,

$$\begin{aligned} R_{\text{eq}} &= \frac{9}{2 \cdot f \cdot \frac{53C}{22}} \coth\left(\frac{\lambda_{\text{CS}}}{2}\right) + \frac{2}{f \cdot C} \coth\left(\frac{\lambda_{\text{TS}}}{2}\right) \\ &= 3.9743 + 4.2553 \end{aligned}$$

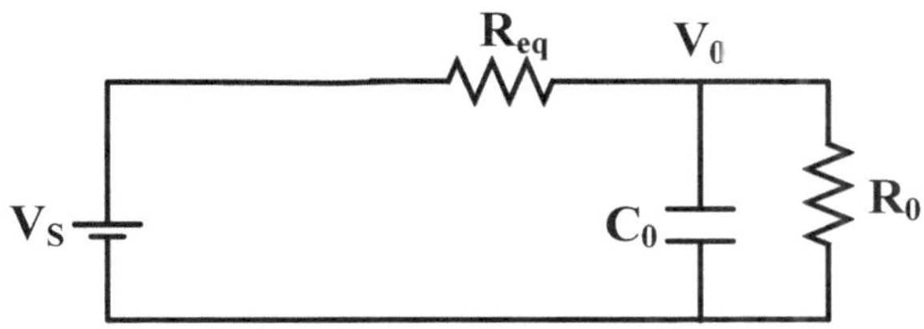

Fig. 12 Equivalent circuit for an average output voltage

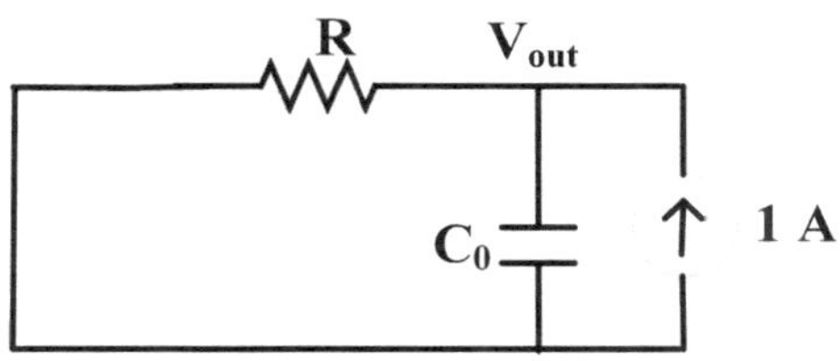

Fig. 13 Circuit with a zero DC source and 1 A, load as current source

$$= 8.2296\ \Omega$$

3.1 R$_{eq}$ *Validation*

From [19], the mean output voltage corresponding circuit as shown in Fig. 12. By neglecting the capacitor losses, the equivalent resistance provides an idea of an average loss of the converter which is an ohmic loss due to power electronic switches resistance and capacitance ESR. By shortening the DC voltage source and connecting 1 A current source in place of the load, which is shown in Fig. 13 is an appropriate way to validate the calculated equivalent resistance for the particular VCR. Then the magnitude of measured output voltage is same as the value of R_{eq} in ohms, which is shown in Fig. 14.

4 Simulation Results and Discussions

Simulations results of the proposed switched capacitor converter circuit for different voltage ratios are presented in this Section. The parameters of the proposed circuit and the simulation results are as follows. For simulating the proposed SCC, PSIM software has been used. The switches S_1–S_{13} used in the circuit for theoretical modeling

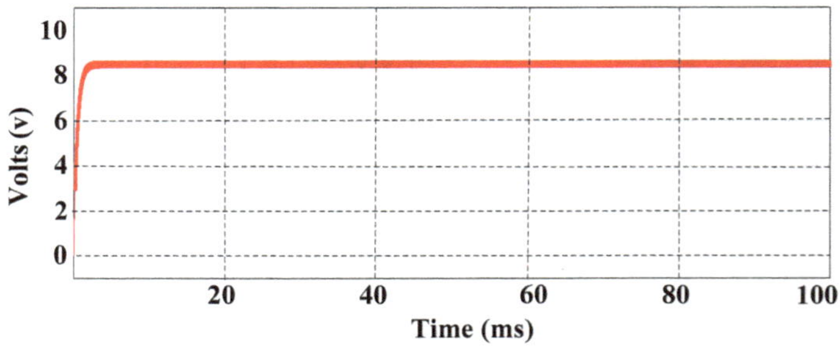

Fig. 14 Output voltage waveform for the zero dc source and 1 A current source as load

and simulation are of bidirectional switches. Ideally, the switches has zero internal resistance, whereas in the mathematical calculations considering the internal resistance of each switch has 1.2 Ω. These 14 active switches of MOSFETs are used to achieve the required output with only one input voltage. In the switch, a quick transition pushes the current in both directions. It is ON when the gate pulse to the switch is high, and OFF when the gate pulse to the switch is low, regardless of the switch's voltage bias conditions.

The proposed circuit uses four flying capacitors or charge pump capacitors of 4.7 μF each for charging and discharging the voltages and one output capacitor which is known as filter capacitor C_o is ten times of the flying capacitors value, i.e., 47 μF is used for filtering any ripples in the output voltage. The load resistance R_L is taken as 12 Ω.

Initially, the conversion ratio is 5 V up mode is considered. The simulation condition is taken as $V_{in} = 10$ V (V_{in} in Fig. 4). The flying capacitor values $C_1 = C_2 = C_3 = C_4 = 4.7$ μF and output capacitor value $C_o = 47$ μF. As shown in Fig. 15, the output voltage waveform for 5 V up mode, the input is given as 10 V, so, the output voltage V_{out} is obtained as 50 V based on the switching logic for 5 V up mode voltage ratio. It shown in Fig. 16, the voltage waveform for 4 V up mode, the input voltage V_{in} is taken as 10 V and the corresponding output obtained is 40 V which

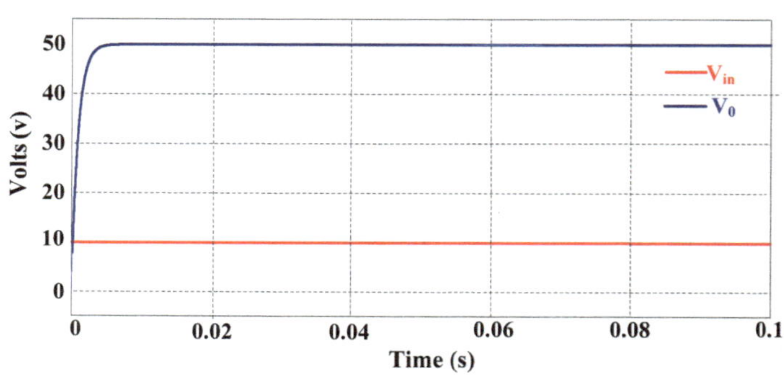

Fig. 15 Voltage waveforms for 5 V up mode

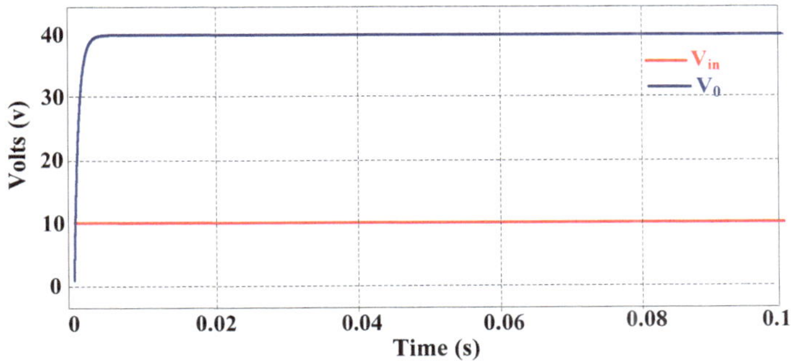

Fig. 16 Voltage waveforms for 4 V up mode

boosts the output voltage by 4 times the V_{in}. The output obtained in this particular ratio is based on the switching logic for 4 V up the mode ratio given in the Table 1 Switching logic for the proposed circuit. The voltage across each capacitor for 4 V up mode is as shown in the following Fig. 17. Voltage across capacitors for 4 V up mode. The capacitor voltage V_{C_1} is 20 V as the explained in the above section and the capacitor voltage V_{C_2} is 30 V. The voltage across capacitor C_3 and C_4 is 40 V but the ripple in the voltage across Capacitor C_4 is reduced as compared to the ripple in the voltage across the capacitor C_3. Figure 18 shows the output voltage for 2 V up mode, the output obtained is 20 V, which is twice the input voltage value i.e., 10 V.

The output voltage waveforms for the down mode voltage ratios are as follows. It is shown in Fig. 19, the output voltage waveform for 1/2 V down mode, the input voltage V_{in} is taken as 10 V and the corresponding output obtained is 5 V which is half of the input voltage. The output obtained in this particular ratio is based on the switching logic for 1/2 V down mode ratio given in the Table 1. The voltage across each capacitor for 1/2 V down the mode ratio is as shown in Fig. 20.

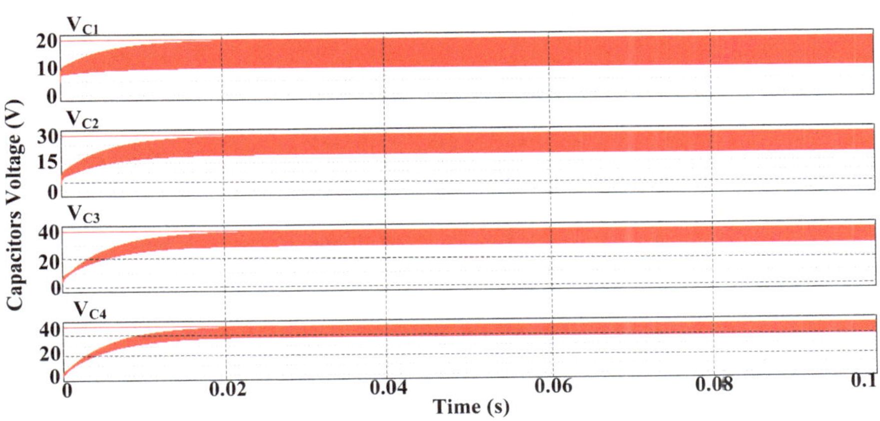

Fig. 17 Voltage across capacitors for 4 V up mode

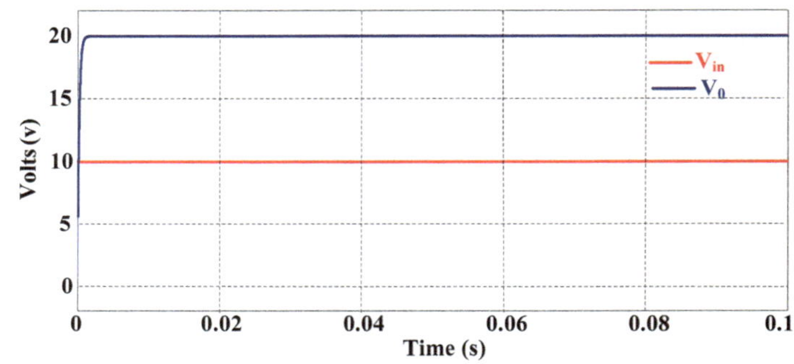

Fig. 18 Voltage waveforms for 2 V up mode

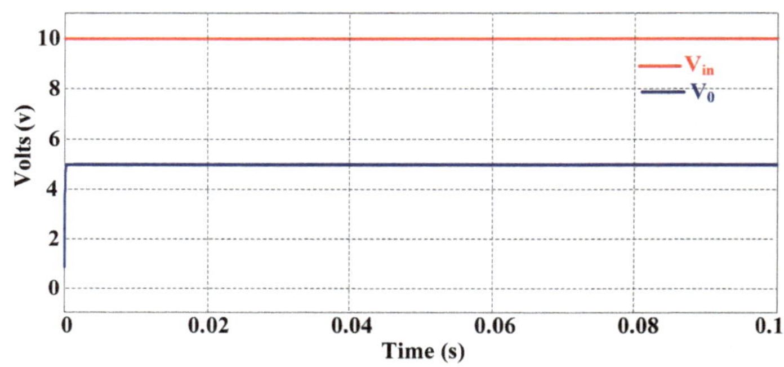

Fig. 19 Voltage waveforms for 1/2 V down mode

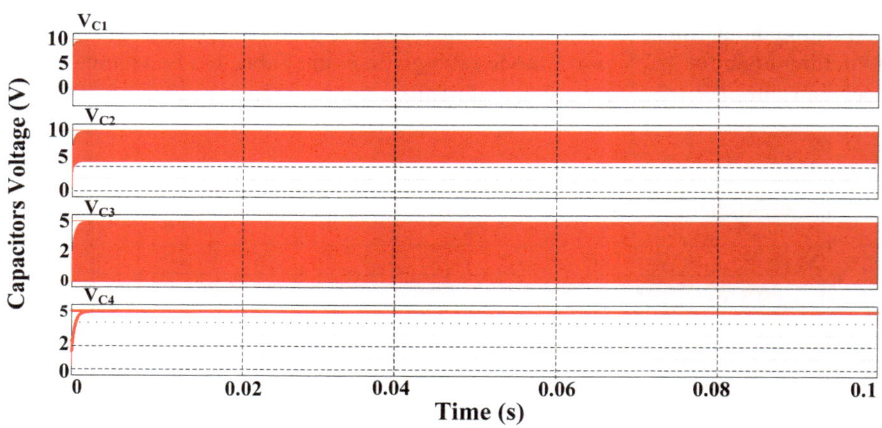

Fig. 20 Voltage across capacitors for 1/2 V down mode

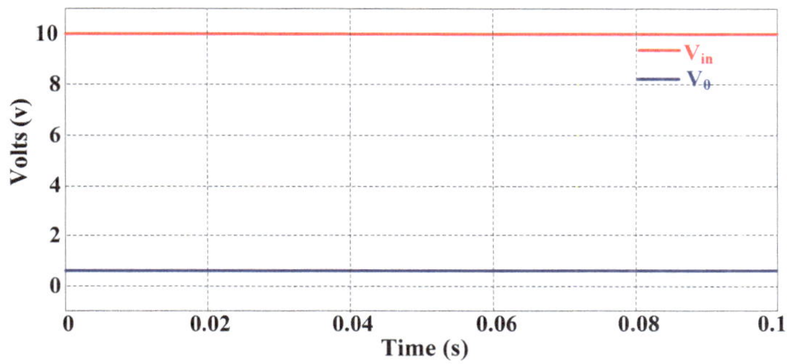

Fig. 21 Voltage waveforms for 2/3 V down mode

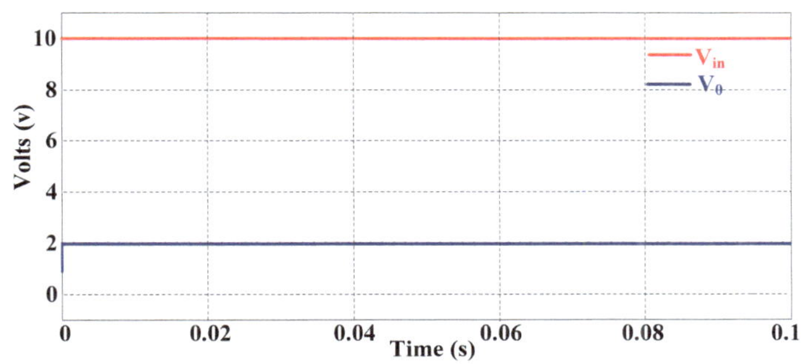

Fig. 22 Voltage waveforms for 1/5 V down mode

The simulation result of voltage ratio 2/3 V down mode is shown in Fig. 21, which bucks the input voltage and the output obtained is according to the switching logic 2/3 V down mode is shown in the Table 1. Similarly, as shown in the Fig. 22, the output voltage waveform for 1/5 V down mode, the input voltage is taken as 10 V and the output voltage obtained is 2 V.

As the proposed circuit is able to produce the negative voltage ratios. The output voltage waveforms for negative ratios are as shown below. Figure 23 shows the output voltage waveform for −2 V input is taken as 10 V and the output voltage obtained is −20 V which is according to the switching logic for −2 V. Similarly, Fig. 24 shows the output voltage waveform for −3 V, the output voltage is obtained as −30 V where input is taken as 10 V.

Similarly the output voltage waveform for voltage ratio of −1/2 V is as shown in Fig. 25. Output voltage waveform for −1/2 V where the input is taken as 10 V then output voltage is obtained as −5 V based on the switching logic for −1/2 V down mode in the Table 1.

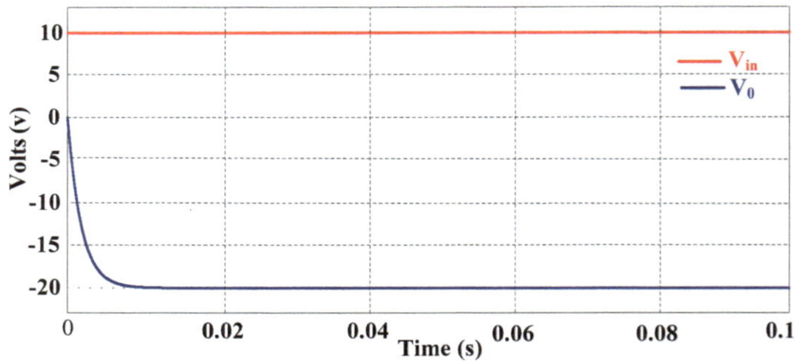

Fig. 23 Voltage waveforms for −2 V ratio

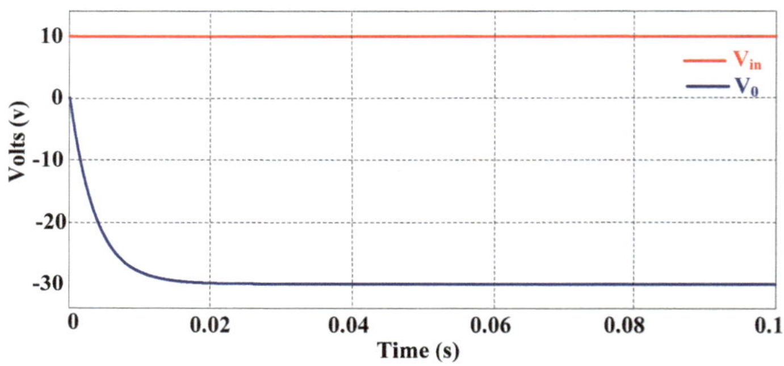

Fig. 24 Voltage waveforms for −3 V ratio

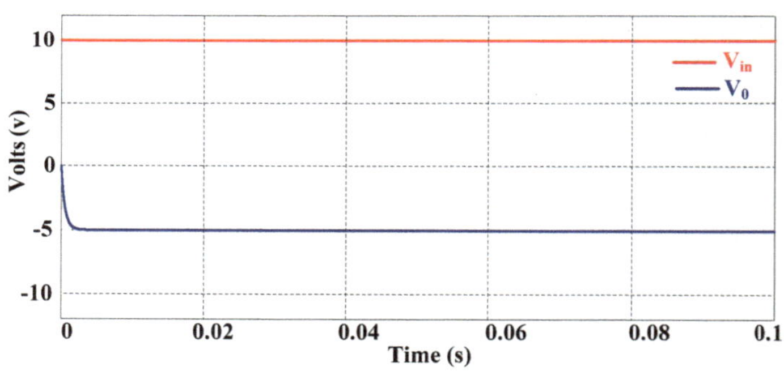

Fig. 25 Voltage waveforms −1/2 V down mode

5 Conclusion

This chapter proposed a new topology of the switched capacitor (SCC) circuit. The converter has a basic construction that does not require any magnetic elements which is powerful enough to produce 21 specific voltage conversion ratios. Out of 21 voltage ratios, 4 ratios are of up ratios (which lift the voltage), 12 ratios are of down ratios up ratios (which down the voltage) and the remaining 5 ratios are negative ratios. One possible application for the proposed SCC is the use of a combined configuration of energy storage devices such as batteries and alternative energy sources such as solar etc. The proposed SCC is evaluated and analyzed to measure and simulate the exact output voltage depending on the equivalent resistance method. The simulation results for the proposed SCC is verified using the PSIM program. The simulation results are confirmed well with theoretical conditions. In future work, the efficiency of the proposed work can be addressed in both theoretical and mathematical analysis. And also, there is a possibility to develop the new SCC's with the lower switch count without decreasing the number voltage ratios.

References

1. H.V. Gandhodi, K.T.R. Konkala, R.K. Bandameeda, J. Chaviti, P.K. Yenna, V. Subburaj, A new single input-multiple output switching converter using circular capacitors for e-vehicle applications, in TENSYMP 2020-2020 IEEE Region 10 Conference (IEEE, 2020), pp. 1696–1699
2. C. Abraham et al., Reconfigurable highly efficient CMOS-based dual input variable output switched capacitor converter for low power applications. Electron. Lett. **54**(2), 89–91 (2018)
3. J. Dickson, On-chip high-voltage generation MNOS integrated circuits using an improved voltage multiplier technique. IEEE J. Solid-State Circ. **SC-11**(3), 374–378 (1976, June)
4. C. Richardson, *LED Applications and Driving Techniques* (National Semiconductor Corporation, CA, USA, 2007). [Online]. Available www.national.com/onlineseminar/2007/led/national_LEDseminar.pdf
5. R. Gariboldi, F. Pulvirenti, A monolithic quad line driver for industrial application. IEEE J. Solid State Circuits **29**(9), 957–962 (1994)
6. G. Palumbo, D. Pappalardo, M. Gaibotti, Charge pump with adaptive stages for non-volatile memories. IEE ProcCirc. Devices Syst. **153**(2), 136–142 (2006)
7. G. Palumb, D. Pappalardo, Charge pump circuits: An overview on design strategies and topologies. IEEE Circuits Syst. Mag. **10**(1), 31–45 (2010) (First Quarter)
8. J.A. Starzyk, J. Ying-Wei, F. Qiu, A DC-DC charge pump design based on voltage doublers. IEEE Trans. Circuits Syst. I, Fundam. Theory Appl. **48**(3), 350–359 (2001, March)
9. F.L. Luo, Switched-capacitorized DC/DC converters, in *Proceedings of IEEE 6th International Power Electronics and Motion Control Conference*, 2009, pp 1074–1079
10. D. Cao, W. Qian, F.Z. Peng, A high voltage gain multilevel modular switched-capacitor DC-DC converter, in *Proceedings of. IEEE Energy Conversion Congress and Exposition (ECCE)*, Pittsburgh, PA, USA, 2014 September, pp. 5749–5756
11. S. Li, K. Xiangli, Y. Zheng, K.M. Smedley, Analysis and design of the ladder resonant switched-capacitor converters for regulated output voltage applications. IEEE Trans. Ind. Electron. **64**(10), 7769–7779 (2017, October)
12. M.D. Seeman, S.R. Sanders, Analysis and optimization of switched capacitor DC-DC converters. IEEE Trans. Power Electron. **23**(2), 841–851 (2008, March)
13. Ioinovici, Switched-capacitor power electronics circuits. IEEE Circuits Syst. Mag. **1**(3), 37–42 (2001)
14. W. Li, X. He, Review of nonisolated high-step-up DC/DC converters in photovoltaic grid-connected applications. IEEE Trans. Ind. Electron. **58**(4), 1239–1250 (2011)

15. M. Makowski, D. Maksimovic, Performance limits of switched-capacitor DC-DC converters. Proc. IEEE Power Electron. Specialists Conf. (PESC) **2**, 1215–1221 (1995)
16. B. Hemanth Kumar, M.M. Lokhande, An enhanced space vector PWM for nine-level inverter employing single voltage source, in Transportation Electrification Conference (IEEE ITEC-India) (2017, December), pp. 1–6
17. B. Hemanth Kumar, M.M. Lokhande, Raghavendra Reddy Karasani, V.B. Borghate, An improved space vector pulse width modulation for nine-level asymmetric cascaded H-bridge three-phase inverter. Arab. J. Sci. Eng. **44**(3), 2453–2465 (2019, March)
18. B. Hemanth Kumar, M.M. Lokhande, R.R. Karasani, V.B. Borghate, A modified space vector PWM approach for nine-level cascaded H-bridge inverter. Arab. J. Sci. Eng. **44**(3), 2131–2149 (2019, March)
19. Y. Mustafa, A. Zhaikhan, A. Ruderman, SCC Equivalent Resistance: The relationship for complementary buck and boost and accurate calculation for 2-phase converters, in Proceedings of IEEE 18th International Power Electronics and Motion Control conference (PEMC) (Budapest, Hungary, 2018), pp. 188–193
20. C. Abraham, B.R. Jose, J. Mathew, M. Evzelman, Modelling, simulation and experimental investigation of a new two input, series-parallel switched capacitor converter. IET Power Electron. **10**(3), 368–376 (2017)

A Comprehensive Study on Electrical Vehicle in Charging Infrastructure, Challenges and Future Scope

Piyush Sharma, Ashok Kumar Sharma, Neeraj Priyadarshi, Farooque Azam, Sajeevikumar Padmanaban, and Akash Kumar Bhoi

1 Introduction

Today world facing mostly rising of atmospheric pollutions because of the exponential development of number, on-road vehicles which emit carbon emission and pollute our atmosphere also affect our greenhouse. To tone down this growing issue there is direct need to execute of some changes in transportation system. Electric vehicles (EVs), Plug-in Electric Vehicles (PEVs), and Plug-in Hybrid-Electric Vehicles (PHEVs) [1–5] are fetching admired unconventional to gasoline powered or internal combustion engine (ICE) owned vehicle to minimize greenhouse gases which reduce take of the edge of atmospheric pollutions [6–10].

P. Sharma · A. K. Sharma
Department of Electrical Engineering, Rajasthan Technological University, Kota, Rajasthan 324010, India
e-mail: piyushsharma034@gmail.com

A. K. Sharma
e-mail: aksharma@rtu.ac.in

N. Priyadarshi (✉) · S. Padmanaban
Department of EnergyTechnology, Aalborg University, Esbjerg 6700, Denmark
e-mail: neerajrjd@gmail.com

S. Padmanaban
e-mail: san@et.aau.dk

F. Azam
School of Computing and IT, REVAUniversity, Bangalore, Karnataka 560064, India
e-mail: farooque53786@gmail.com

A. K. Bhoi
Department of Electrical and Electronics Engineering, Sikkim Manipal Institute of Technology, Sikkim Manipal University, Sikkim 737136, India
e-mail: akashkrbhoi@gmail.com

N. Patel et al. (eds.), *Electric Vehicles*, Green Energy and Technology,
https://doi.org/10.1007/978-981-15-9251-5_16

PHEVs have healthier performance and high efficiency than ICE vehicles. These EVs are fully equipped by electric motors, higher battery bank voltage storing equipment and transmission of energy from one side to another side. In worldwide most of countries use these types of EVs, PEVs, and PHEVs but major concern in these vehicles is charging of battery. Acceptance is very low of these EVs because due to less EV Charging station as compare to refueling station of internal combustion engine vehicles so charging infrastructure is major concern of this era. There are different methods to charge EVs are known as conductive charging or inductive charging. Conductive charging scheme shows physical connection of power supply and EVs and an inductive charging scheme shows there is no physical connection in between power supply and EVs. Last few years we use conductive charging scheme but we facing some major problems in it like overcrowding at charging station, queuing of vehicle at charging station, charging time of battery and also a major concern about is location of charging station within a range of EVs [11–15].

Inductive charging scheme has a potential to limit above these major problems. This type of charging infrastructure gives world to exponentially growth in field of EVs. That scheme has a several advantages w.r.t. above one conductive charging scheme. While EVs is at rest or in motion in both cases it has several advantages and it includes also simplicity, reliability, user friendliness, security and convenience of charging.

Non conductive charging sanctions the robotized charging of EVs which can be accomplished by 3 distinct modes,

(1) Static Wireless Charging (SWC)
(2) Quasi-Dynamic Wireless Charging (QDWC), and
(3) Dynamic Wireless Charging (DWC).

In Static Wireless Charging scheme have a major benefits with old one wired technology charging scheme in terms of eliminate electrical Distress hazard and capability to implement under suitable areas such as quarters or parking more.

We know some EVs are not working for lesser period as toll, traffic lights in this manner we need more battery energy storage requirement because these circumstances makes decreases energy storage of EVs. So the Quasi- Dynamic Charging (QDC) system provides charging to EVs in these circumstances.

The Dynamic Wireless Charging (DWC) system continuously charges the EVs energy storage battery system while en-route on the road through specified charging lane. This technology reduces size of energy storage battery and also increases driving range in all categories of EVs.

For charging of battery for all EVs there are various categories of charging station are made through AC EVSE Level I, AC EVSE Level II, AC EVSE Level III, DC Rapid Charging, Superchargers.

Nowadays researchers are move to wireless charging scheme because of its advantages whish was disused above earlier. Capacitive contact less power transfer (CPT), Magnetic gear contact less power transfer (MGWPT),Inductive contact less power transfer (IPT), Resonant contact less power transfer (RPT) are techniques to transfer of power from one end side to another end of side without circuit connection and researches are also growing to increase the air gap for WPT.

In this paper we are wrapping an assortment of EVs without wire charging of, as well as charging stages, charging effect of EVs on Economic growth of country, sustainable and environmental impact of charging without wire, infrastructure, standards, challenges, economic analysis and future aspects. Finally, an upcoming promising concept of Vehicle to Grid (V2G) flow with wireless power transfer has been discussed.

Conductive analysis and inductive battery charging scheme for all types of EVs, PEVs, and PHEVs are outline in segment 2. In Sect. 3, several stages of EVs, PEVs, and PHEVs with their battery charging level, charging capacity of battery, size of battery and range and efficiency (EPA) are compared and reviewed. In segment 4, factor affecting in Wireless Charging in EVs is discussed. With Wireless charging technology, battery charging circuit discussed in segment 5. In segment 6 Standards of contactless charging systems are reviewed. Segment 7 reviewed the financial examination of remote charging and social ramifications, maintainability and wellbeing issues. Future of V2G applications discussed in segment 8. Finishing up comments are introduced in segment 9.

2 Classification of Conductive and Inductive Charging Systems of EVs

For enlargement of EVs, the charging scheme of all variety of EVs playing a vital role. Charging of any kind of EVs require more time as compare to ICE type vehicle refueling and another drawback is low no of dedicated charging stations and less no of charging station. So, improving for charging capacity of all EVs researchers designed a new type of charging station which was make charge vehicle 50% in 3 min or 80% in 15 min. Battery charging is not a very simple task. In battery charging scheme, circuit have charger converters and its controlling algorithms. Through this we can control current and voltage and that was deliver to battery for fast charging adoption. For fast charging adoption two methods are discussed here.

2.1 Conductive Charging Scheme (CCS)

As we discussed earlier Conductive Charging involves physical connection of electrical power supply and EVs. It consists of AC to DC converter or connected to low frequency AC to high-frequency AC converter with power factor correction (PFC). Conductive charging scheme is also further classified into two parts viz ON board (ON-BC) or OFF board chargers (OFF-BC). In ON-BC scheme battery current flows inside a vehicle and OFF-BC scheme battery current flows outside a vehicle. Because

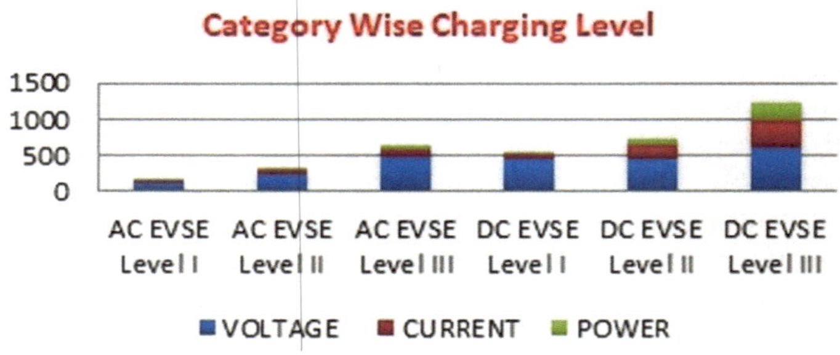

Fig. 1 Charging level infrastructure on EVs

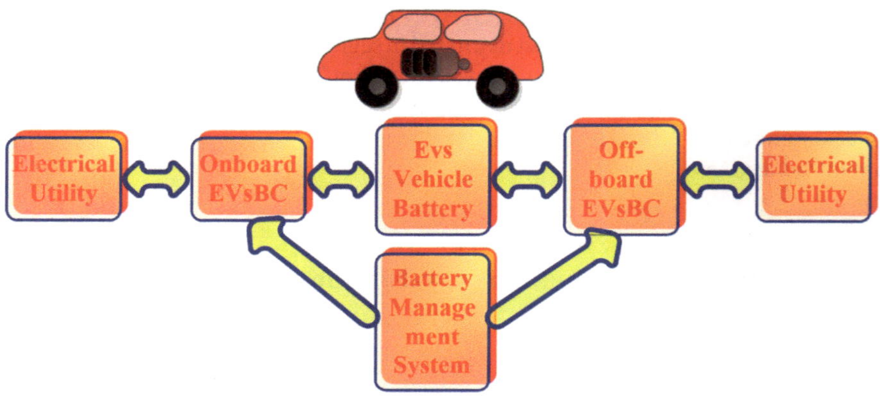

Fig. 2 ON board and OFF board charger scheme

of charging level, conductive charging also classified into different level of AC and DC charging level. AC EVSE Level I chargers, charges battery less than 3 kW, AC EVSE Level II chargers, charges battery 4–20 kW, AC EVSE Level I chargers, charges battery up to 50 kW, Fast DC Power level 1, 2 and 3 chargers, charges battery more than 30–240 kW (Figs. 1 and 2).

2.2 *Inductive Charging Scheme (ICS) or Wireless Charging Scheme (WCS)*

Nikola Tesla was invented power transfer by wireless mode. It has no physical electrical connection in between power supply and EVs. Power was transfer from primary to secondary side is very tedious task and that will be resolved by researchers. In year of 1999 distance between primary and secondary winding was only 3 mm and power was transferred up to 8.3 kW with an efficiency of 97%. So air gap through

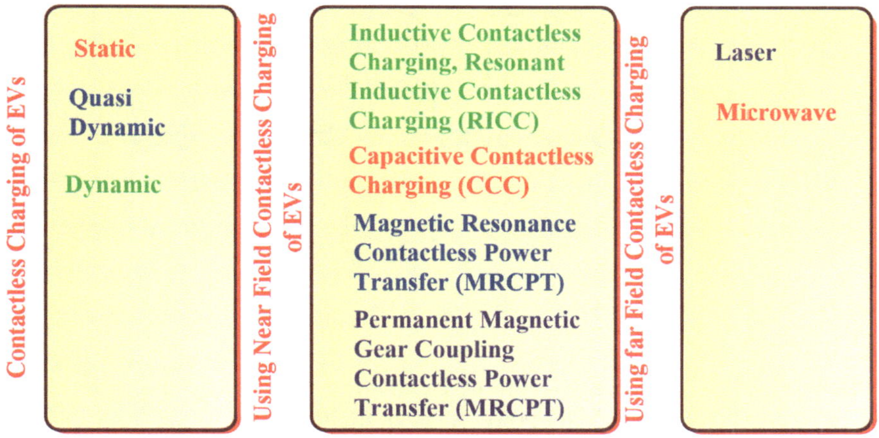

Fig. 3 Classification of EVs Battery Charging System

this researchers conclude that we can increases air gap between of two windings. Nowadays researchers are working to enhance a gap between primary and secondary charging structure side into meters.

Classification of WCS is discussed earlier in different ways such as Static, Dynamic, and Quasi-dynamic Wireless Charging Scheme, Near and Far Field charging infrastructure. Near field classify such as Inductive Contactless Charging, Resonant Inductive contactless charging (RICC), capacitive contactless charging (CCC), Magnetic Resonance Contactless Power Transfer (MRCPT), Permanent magnetic gear coupling contactless power transfer, and for Far filed Laser and Microwave wireless charging (Fig. 3).

2.2.1 Inductive Contactless Charging (ICC)

Inductive contactless charging scheme is one of the oldest scheme of battery charging. In year of 1914 Nikola Tesla invented how can power transfer from one side to another side without connection. Basic Fig. 4 is shown below It is a basic block diagram of inductive contactless charging which have transmitter side and another is receiver without physical connection.

Based on survey Inductive contactless charging scheme it has been tested wide range of power transfer in milliwatts to kilowatts in EVs application. In 1996, General Motors introduced Chevrolet S10 eV which was charge with AC EVSE Level II and AC EVSE Level III fast charging. Paddle was inserted to EVs that was primary coil and secondary coil received power to charge EVs battery. University of Georgia developed new ICC scheme for charging of EVs with a range of 200–400 V batteries at operating frequency 77 kHz for 6.6 kW AC EVSE Level II. Main advantage of ICC is no metal connection between primary and secondary side so due to this sparking

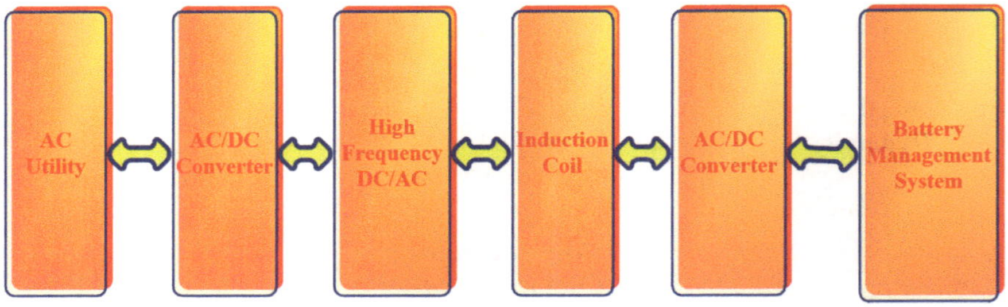

Fig. 4 Diagram of Inductive contactless charging (ICC)

issue is resolved and efficiency will increased up to 96%. With increase air gap in between primary and secondary power transfer efficiency is low.

2.2.2 Resonant Inductive Contactless Charging (RICC)

It was advanced version of previous ICC. In which researchers include some solid state devices and contactless transformer coils. Basic Fig. 5 is shown below for EVs. RIPT scheme have only few changes, it have some series/parallel compensation network add both side of RPIT structure.

So power flows source to battery in this way like, firstly AC mains voltage to converted to High-frequency AC voltage, then HF AC voltage send to primary side which was link with compensation network and then power transfer to secondary side through air gap which was equipped by additional compensation network. That received ac power at secondary side is converted to DC and store at battery management system (BMS). For maximum power transfer resonant frequency (f_{res}) should be same at primary and secondary side. That operating frequency has ranges from 10 to 100 kHz. Coupling coefficient is low due to absence of magnetic core. Normally EVs have low height clearance which was 150–300 mm so for that height coupling coefficient (k_{cc}) range is 0.2–0.3. So for improvement of coupling coefficient (k_{cc}), structure of magnetic ferrite improved for contactless transformer design. Due to

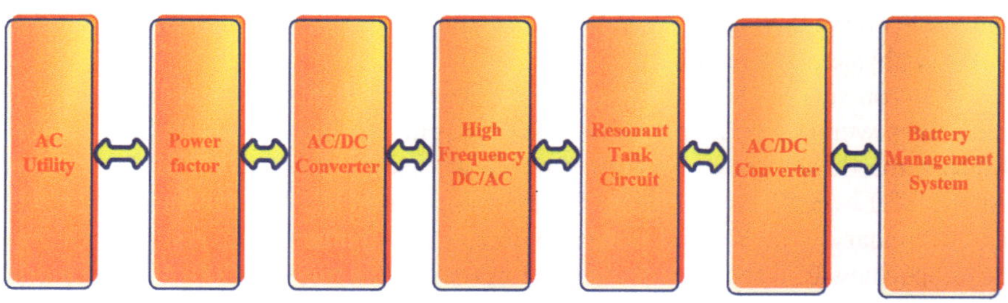

Fig. 5 Diagram of Resonant Inductive contactless charging (RICC)

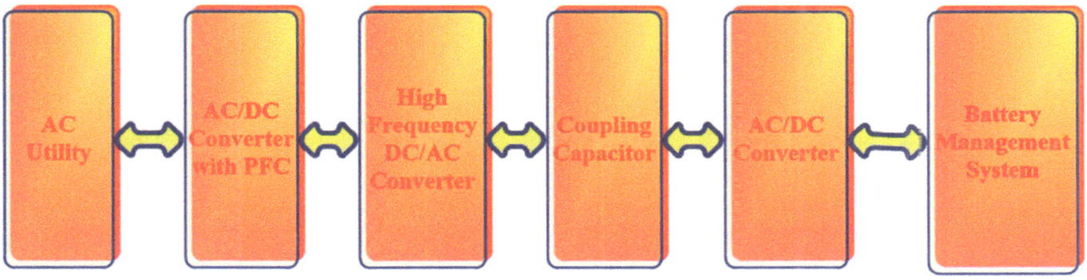

Fig. 6 Diagram of Capacitive contactless charging (CCC)

high frequency some losses appears that reduced transfer power efficiency so that will be improved by parasitic component then Quality factor (Q_{f}) will be improved.

$$f_{\mathrm{res(pri,sec)}} = \frac{0.159}{\sqrt{L_{\mathrm{pri,sec}} * C_{\mathrm{pri,sec}}}} \tag{1}$$

$$\mathrm{k_{cc}} = \frac{L_m}{\sqrt{L_{\mathrm{pri}} L_{\mathrm{sec}}}} \tag{2}$$

$$Q_{\mathrm{f}} = \frac{2\pi f L_{\mathrm{pri,sec}}}{R_{\mathrm{pri,sec}}} \tag{3}$$

2.2.3 Capacitive Contactless Charging (CCC)

Capacitive contactless charging are mostly use in low power application like cellular phone charger but nowadays researchers are focused on this method in EVs for power transfer from primary to secondary. CCC has two metallic sheets which was insulated, one plate work as primary and second as secondary. Mains connected to converter through PFC circuit and it generates HF AC and the power is transferred to secondary side through capacitor plates shows in Fig. 6. To reduce its impedance a series inductor is connected in coupling capacitor primary and secondary side of plate. At BMS side a filter circuit is added for remove AC contents.

2.2.4 Magnetic Resonance Contactless Power Transfer (MRCPT)

It is work on basis of electromagnetic induction principle for to charge batter of EVs through magnetic phenomenon. In this circuit it has a two coil for power transfer. Mains connected to converter and current flows to primary coil then it induce a magnetic field, that induced magnetic field transmit to secondary side for to charge battery of EVs. By its magnetic induction method power transfer from primary side fixed and secondary side moving. Due to magnetic coupling effect air gap is high. So efficiency is low.

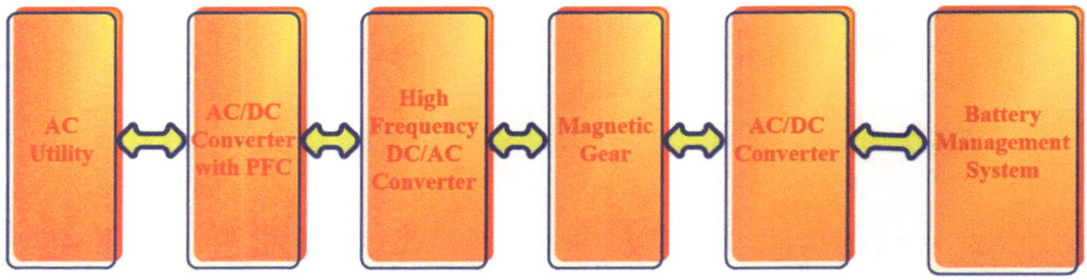

Fig. 7 Permanent magnetic gear coupling contactless power transfer

2.2.5 Permanent Magnetic Gear Coupling Contactless Power Transfer

It is relatively very different from above methods. As like their name it has two permanent magnet at both side primary as well as secondary and both are synchronized. When circuit get a supply from mains then permanent magnet gets energize and produce mechanical torque at transmitter side winding of EVs battery charging system. Through mechanical torque permanent magnet at primary side rotates and due to this secondary side placed permanent magnet also rotates. By this method power will transfer from primary to secondary side without circuitry connection then power transfer to battery is an arrangement of power electronics circuits (Fig. 7).

2.2.6 Laser and Microwave or Radio Wave Contactless Power Transfer

In this type of energy power transfers from one side to another side without circuit connection have some important points consider. First power is converted on radio frequency; microwave or laser then transmits that power from primary to secondary side without circuit interconnection. There are mainly two type of this type of energy or power transfer technique Microwave contactless power transfer and Laser contactless power transfer.

Microwave Contactless Power Transfer

Firstly Nikola Tesla invented in 1904, and developed a model at level of 150 kHz using radio wave for power transfer with wireless concept. After that many researcher work in this field. In 1964, researcher invented and glided a contactless helicopter without a battery, with 2.45 GHz. Now increasing distance is very tedious task for transfer power with high efficiency so researchers make a system with Antenna and Klystron with a distance of 1.6 km at frequency 2.45 GHz.

Laser Contactless Power Transfer

Power transfer from laser technology was vey less entertain in EVs battery charging system because it has very less efficiency. It has very complex technique to transfer power from primary to secondary. Firstly power convert into laser beam then that is focus on photovoltaic cell. In this we have to check also radiation spectrum because if radiation is high then that is become a dangerous for human life also.

3 Various Electrified EVs, PEVs, and PHEVs Specification

In this section we discuss about all type of EVs, PEVs, and PHEVs specifications in terms of EPA range, type of battery size, full time charge of battery and charging rate. Normally electric cars have such verity sedan, SUV, Wagon/Van and Coupe type structure. Electric car are very efficient instead of a gasoline tank and a electric motor instead of a internal combustion engine (ICE) cars. PHEVs are a blend of gas and electric vehicle so that have battery, electric motor, gas tank and ICE. The quantity of miles an EV will go before the battery should be energized is regularly not exactly the separation your gas vehicle can go before being refueled, however normally is sufficiently still to achieve the normal individual's day by day driving needs. An electric vehicle's mileage is accounted for as far as miles per gallon of gas identical (MPGe). Think about this as being like MPG, however as opposed to introducing miles per gallon of the vehicle's fuel type, it speaks to the quantity of miles the vehicle can go utilizing an amount of power with a similar vitality content as a gallon of gas. This permits you to contrast an EV and a gas vehicle despite the fact that power isn't administered or consumed as far as gallons. PHEVs typically have driving ranges that are comparable to gasoline vehicles. PHEVs have two fuel economy values: one for when the vehicle operates primarily on electricity and one for when the vehicle operates only on gasoline (listed as MPG). EPA range combined estimates (55% of city + 45% on highway). Charging of battery are discussed above like various categories AC EVSE Level I, AC EVSE Level II, AC EVSE Level III, DC Rapid Charging, Superchargers. Range of battery size for all categories of EVs extended to 9–95 kWh and there charging time may be varies by its internal features. Charging rate also major parameter for all EVs that have a range of 3–17.2 kW (Fig. 8).

4 Factor Affecting in Wireless Charging in EVs

Electrical vehicle technology is a growing industry in all over world it have several advantages over a IC engine. Innovation factors are firmly identified with attributes of the EV, for example, driving separation, charging time, EV buy outlays. Constrained driving reach, long charging time and EV's high buy outlays are the impediments to

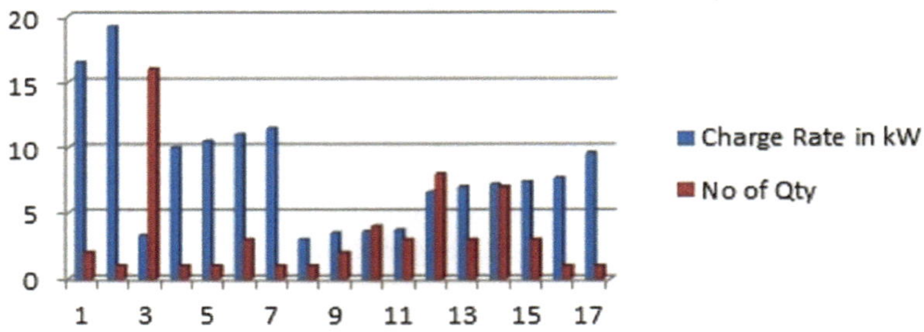

Fig. 8 Tabular chart of charge rate vs no of quantity of vehicle

EV appropriation and dispersion. Outlay, battery charging, replacement of battery, charging station of EV is major issue in current era. It has take 15 min to charge fast dc charging scheme but for filling full tank fuel it takes only 4 min. Installation for batter charging SWC is much outlay then DWC.

5 Charging Circuitry Design for EVs

In the previous decade, the difficulties of decreasing ozone depleting substance librations and appropriate intends to give cleaner vitality are pushing the car business toward presenting another age of EVs. With headways in power gadgets, machine plan, and battery advancements a noteworthy open door is accessible to accomplish the progress toward cleaner wellsprings of vitality inside the car business. Worldwide worries of vitality effectiveness and eco-friendliness further catalyzed the car business in developing PEVs (Fig. 9).

A front-end AC/DC converter is utilized to change over the gracefully AC voltage to a middle of the road DC-connect voltage and to shape the info current for both PFC

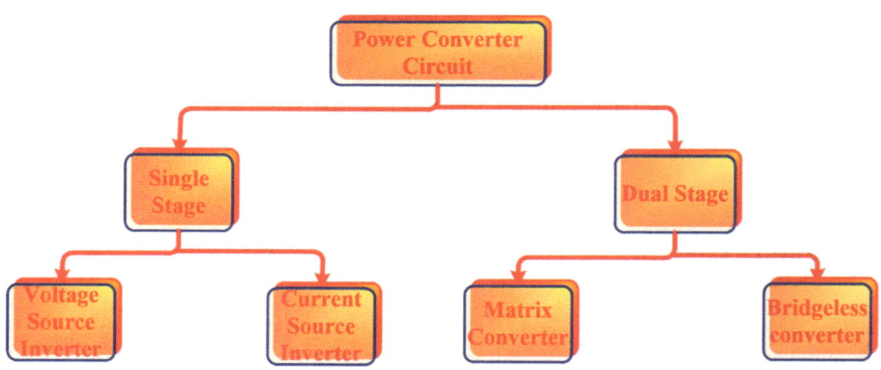

Fig. 9 Circuit topology for EVs

and symphonious decrease. Inductive contactless charging scheme two CSI topology half bridge, full bridge and push-pull. Single stage AC/AC converter widely used in ICC technology, for this matrix converter is used in it. Several type matrix converter including buck, half bridge and full bridge power electronics converter are used in ICC technology. Major difference in between those technology are single stage conversion means from utility supply AC supply is converted in AC with matrix converter and dual stage AC utility converts AC/DC/AC conversion.

6 Standards of Contactless Charging Systems

It is essential to explain the contrasts among Standards and Regulations. Measures are deliberate reports delivered by specialized panels and endorsed by the delegates of the concerned and intrigued substances with regards to the countries (in particular the IEC National Committees or ISO national norms organizations). Then again, guidelines are legitimately restricting records, which are given by government. In any case, our cutting edge lifestyle would not be conceivable without guidelines (and numerous gauges at long last are received has guideline by the nations). From an electrical perspective, it does the trick to figure what might occur in the event that we had a few unique plugs in every nation, with various voltage levels and frequencies. The equivalent occurs with the EVs, and at this phase of the principal wave of present day EVs arriving at the market, there are numerous parts of EVs that despite everything can and should be normalized as, the fittings for charging, the chargers voltages, the correspondence between the vehicle and the chargers, quick charging frameworks and particular charging, safety efforts for vehicle wellbeing and security of people against electric stun, electromagnetic similarity and wellbeing impacts issues, on-board electric vitality stockpiling for drive and vehicle vitality execution and vitality estimations.

EVs, are relied upon to contribute authoritatively to the feasible portability. In any case, so as to permit a quicker market infiltration, normalization is critical. Without a doubt, for agreeable and secure vehicle use, there are numerous parts of EVs that need to be normalized as, the fittings for charging, the chargers voltages, the correspondence between the vehicle and the chargers, quick charging frameworks and particular charging, safety efforts for vehicle wellbeing and wellbeing of people against electric stun, on-board electric vitality stockpiling for drive and vehicle vitality execution and vitality estimations. At an overall level, normalization is primarily under the ability of two foundations: the International Electro-mechanical Commission (IEC), established on 1906, and the International Organization for Standardization (ISO), established on 1946.

7 Financial Examination of Remote Charging and Social Ramifications, Maintainability and Wellbeing Issues

EVs starting theory incorporates the electrical transportation outlay and charging establishment, separating electrical transportation outlay into battery, vehicle parts, power train and charging sections. For the assessment of remote charging systems, just battery and charging parts are thought of. Here, we consider an assistance course of movement for EVs in order to look at remote charging plans. Ordinarily EVs have three sort that will talked about above static, semi and dynamic contact less charging incorporates limit outlay and charging unit foundation outlay to set up a base enthusiasm for gratifying the charging need for EVs. It is doubtlessly clear the charging way for DWC requires the most imperative number of establishment yet extraordinarily little battery is adequate. For SWC generous gigantic size battery is obligatory to absolute the whole outing since there is no charging in the center. In case of QWC each transport stops are having charging structures presented more diminutive battery size is required. In this fragment making an allowance for these factors a succinct examination is showed up.

Remote force move is generally pertinent for the remote charging of electric vehicles. Today, the serious issue for nature is gas and diesel motor vehicles, and diesel worked immense machines. In EVs scenario major benefit machine which are use in EVs never pollute our environment. Be that as it may, electrical apparatus is having the issue of taking care of intensity. Subsequently, remote charging plan will drive that apparatus remotely or, if the battery is used for driving the machine, using remote charging plan that can be charged with no issue. Remote charging plan diminishes the use of battery by driving the apparatus remotely, along these lines, reducing the largeness of the structure. Later on, as researchers have planned, the force transmission up to meters will improve the supportable portability with a decrease in the utilization of links just as batteries. It has product life cycle in this manner (Fig. 10).

Outlay of electrical vehicle is not a major issue because in terms of wired and wireless, wireless is little bit outlay because it has magnetic coupler and that equipment take a outlay of approx $400 USD. If we compare safeguarding outlay of wired and wireless so we found wireless has low outlay. But in wireless charging major outlay have battery but if we have sufficient amount of charging infrastructure the on board battery requirement will be less. Safety is major question arises in this type of vehicle and many person asked is safe for human or not. So many standards are over come into a nature that define a vital role in safety standards like IEEE, IEC, SAE, UL which was discussed in earlier.

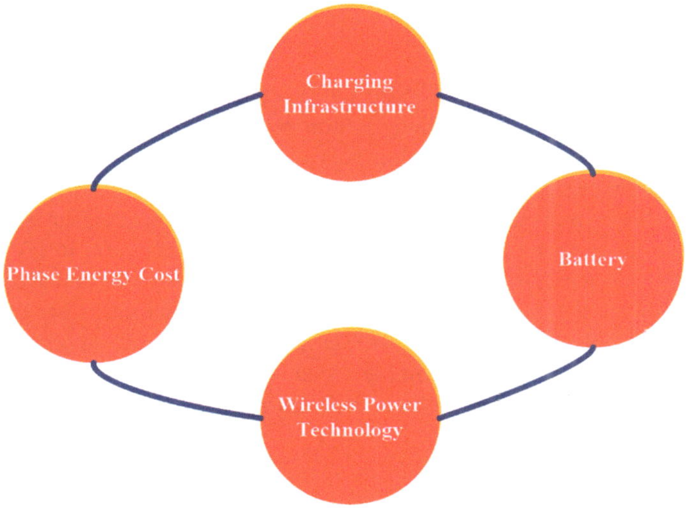

Fig. 10 Product life cycle of wireless power transfer for EVs

8 Future of EVs in Power Transmission to Grid

To the development of EVs has brought about the requirement for quick and productive charging and force move strategies. With the expanding number of EVs, the force prerequisites from dissemination systems has risen quickly and made a negative effect on it. So as to make up for the extra force necessities, sustainable power sources or renewable energy source have been acquainted with the AC micro grid however they have less to maintain. Vehicle to Grid (V2G) concept come into a scenario, in which power is transfer to grid when vehicle is on standby condition and being charged from AC wall socket. Process of power flow from source to load and load to source in this manner, first AC supply is converted into DC and fed into contactless or isolated DDC (DC–DC Converter) (in step-down mode battery charges, and boosts when its libration) for afford supplementary safety to the user. That DC power is now send to battery through battery management system (BMS). In this process only one major disadvantage is physical contact, which makes risk in terms of Distress. In order to overcome this major problem researchers makes contactless power transfer V2G system introduced. In this system bi directional converter are used. Receiver coil pad install bottom the vehicle and residual part is mounted into a vehicle. Those circuits have a additional isolation circuit between transmitter and receiver pad. When battery has surplus energy then that power is transfer to grid without circuit connection.

9 Conclusion

In this paper, it contains various variety of Electric vehicles technology, level of charging infrastructure, circuit configuration of power transfer, challenges, safety and future scope and sustainability has been discussed. We discussed on board and off board charging scheme and study of Static, Dynamic and Quasi-Dynamic wireless charging discussed and concludes that contactless charging conduction is more effective as compare to wired technology. In sustainable development in EVs charging infrastructure has a major role. In which researchers develop new charging topology circuit, for safety feature new standards have to be introduced. It has also offered V2G and G2V technology in which we can flow power in both direction with bidirectional converter. So in next upcoming decades older transportation facility system mostly replaced by newly one by Electrical Vehicle.

References

1. W.R. Cawthorne, P.Famouri, J. Chen, N.N. Clark, T.I. McDaniel, R.J. Atkinson, S. Nandkumar, C.M. Atkinson, S. Petreanu, Development of a linear alternator–engine for hybrid electric vehicle applications. IEEE Trans. Veh. Technol. **48**(6) (1999)
2. P. Brooking, J.R. Bumby, An integrated engine-generator set with power electronic interface for hybrid electric vehicle applications, in *Power Electronics, Machines and Drives Conference*, vol. 487, 16–18 Apr 2022
3. J.Merkisz, J.Pielecha, Emissions and fuel consumption during road test from diesel and hybrid buses under real road conditions, in *IEEE Conference on Power and Propulsion Conference (VPPC)*, pp. 1–5 (2010)
4. L. Parsa, H.A. Toliyat, A self reconfigurable electric motor controller for hybrid electric vehicle applications, in *IEEE Conference on Industrial Electronics Society*, pp. 919–924 (2003)
5. L. Shi, A. Meintz, M. Ferdowsi, Single-Phase bidirectional AC-DC converters for plug-in hybrid electric vehicle applications, in *IEEE Vehicle Power and Propulsion Conference (VPPC)*, Harbin, China, 3–5 Sept (2008)
6. N. Priyadarshi, A. Anand, A.K. Sharma, F. Azam, V.K. Singh, R.K. Sinha, An experimental implementation and testing of GA based maximum power point tracking for PV system under varying ambient conditions using dSPACE DS 1104 controller. Int J Renew Energy Res **7**(1), 255–265 (2017)
7. N. Priyadarshi, A.K. Sharma, F. Azam, A hybrid firefly-asymmetrical fuzzy logic controller based MPPT for pv-wind-fuel grid integration. Int. J. Renew. Energy Res. **7**(4) (2017)
8. N. Priyadarshi, S. Padmanaban, L. Mihet-Popa, F. Blaabjerg, F. Azam, Maximum power point tracking for brushless DC motor-driven photovoltaic pumping systems using a hybrid anfis-flower pollination optimization algorithm. MDPI Energies **11**(1), 1–16 (2018)
9. N. Priyadarshi, F. Azam, A.K. Bhoi, S. Alam, An artificial fuzzy logic intelligent controller based MPPT for PV grid utility. Lecture Notes in Networks and Systems, vol. 46. https://doi.org/10.1007/978-981-13-1217-5_88
10. S. Padmanaban, N. Priyadarshi, J.B. Holm-Nielsen, M.S. Bhaskar, F. Azam, A.K. Sharma, A novel modified sine-cosine optimized MPPT algorithm for grid integrated PV system under real operating conditions. IEEE Access **7**, 10467–10477 (2019). https://doi.org/10.1109/ACCESS.2018.2890533
11. S. Padmanaban, N. Priyadarshi, J.B. Holm-Nielsen, M.S. Bhaskar, E. Hossain, F. Azam, A hybrid photovoltaic-fuel cell for grid integration with jaya-based maximum power point

tracking: experimental performance evaluation. IEEE Access **7**, 82978–82990 (2019). https://doi.org/10.1109/access.2019.2924264

12. N. Priyadarshi, N. Padmanaban, J.B. Holm-Nielsen, F. Blaabjerg, M.S. Bhaskar, An experimental estimation of hybrid ANFIS–PSO-based MPPT for PV grid integration under fluctuating sun irradiance. IEEE Syst. J. **14**(1), 1218–1229 (2020). https://doi.org/10.1109/jsyst.2019.2949083
13. N. Priyadarshi, N. Padmanaban, M.S. Bhaskar, F. Blaabjerg, J.B. Holm-Nielsen, F. Azam, A.K. Sharma, A hybrid photovoltaic-fuel cell-based single-stage grid integration with lyapunov control scheme. IEEE Syst. J. https://doi.org/10.1109/jsyst.2019.2948899
14. N. Priyadarshi, M.S. Bhaskar, N. Padmanaban, F. Blaabjerg, F. Azam, New CUK–SEPIC converter based photovoltaic power system with hybrid GSA–PSO algorithm employing MPPT for water pumping applications. IET Power Electron. 1–0 (2020). https://doi.org/10.1049/iet-pel.2019.1154
15. N. Priyadarshi, N. Padmanaban, J.B. Holm-Nielsen, M.S. Bhaskar, F. Azam, Internet of things augmented a novel PSO-employed modified zeta converter-based photovoltaic maximum power tracking system: hardware realisation. IET Power Electron. 1–0 (2020). https://doi.org/10.1049/iet-pel.2019.1121